# Universitext

**Series Editors**

Nathanaël Berestycki, Universität Wien, Vienna, Austria

Carles Casacuberta, Universitat de Barcelona, Barcelona, Spain

John Greenlees, University of Warwick, Coventry, UK

Angus MacIntyre, Queen Mary University of London, London, UK

Claude Sabbah, École Polytechnique, CNRS, Université Paris-Saclay, Palaiseau, France

Endre Süli, University of Oxford, Oxford, UK

*Universitext* is a series of textbooks that presents material from a wide variety of mathematical disciplines at master's level and beyond. The books, often well class-tested by their author, may have an informal, personal, or even experimental approach to their subject matter. Some of the most successful and established books in the series have evolved through several editions, always following the evolution of teaching curricula, into very polished texts.

Thus as research topics trickle down into graduate-level teaching, first textbooks written for new, cutting-edge courses may find their way into *Universitext*.

Geraldo Botelho • Daniel Pellegrino •
Eduardo Teixeira

# Introduction to Functional Analysis

Geraldo Botelho
Universidade Federal de Uberlândia
Uberlândia, Minas Gerais, Brazil

Daniel Pellegrino
Universidade Federal da Paraiba
Joao Pessoa, Paraíba, Brazil

Eduardo Teixeira
University of Central Florida
Orlando, FL, USA

ISSN 0172-5939 ISSN 2191-6675 (electronic)
Universitext
ISBN 978-3-031-81790-8 ISBN 978-3-031-81791-5 (eBook)
https://doi.org/10.1007/978-3-031-81791-5

Mathematics Subject Classification: 46-01

English translation of the 3rd original Brazilian Portuguese edition published by Sociedade Brasileira de Matemática, 2023
The original submitted manuscript has been translated into English. The translation was done using artificial intelligence. A subsequent revision was performed by the author(s) to further refine the work and to ensure that the translation is appropriate concerning content and scientific correctness. It may, however, read stylistically different from a conventional translation.
Translation from the Portuguese language edition: "Fundamentos de Análise Funcional" by Geraldo Botelho et al., © Sociedade Brasileira de Matemática 2023. Published by Sociedade Brasileira de Matemática. All Rights Reserved.

© Sociedade Brasileira de Matemática 2025

This work is subject to copyright. All rights are solely and exclusively licensed by the Publisher, whether the whole or part of the material is concerned, specifically the rights of reprinting, reuse of illustrations, recitation, broadcasting, reproduction on microfilms or in any other physical way, and transmission or information storage and retrieval, electronic adaptation, computer software, or by similar or dissimilar methodology now known or hereafter developed.
The use of general descriptive names, registered names, trademarks, service marks, etc. in this publication does not imply, even in the absence of a specific statement, that such names are exempt from the relevant protective laws and regulations and therefore free for general use.
The publisher, the authors and the editors are safe to assume that the advice and information in this book are believed to be true and accurate at the date of publication. Neither the publisher nor the authors or the editors give a warranty, expressed or implied, with respect to the material contained herein or for any errors or omissions that may have been made. The publisher remains neutral with regard to jurisdictional claims in published maps and institutional affiliations.

This Springer imprint is published by the registered company Springer Nature Switzerland AG
The registered company address is: Gewerbestrasse 11, 6330 Cham, Switzerland

If disposing of this product, please recycle the paper.

# Preface

The title of this book carries two equally significant pieces of information: firstly, it indicates that the book pertains to Functional Analysis; secondly, it conveys that this is an introductory text, specifically customized for students who are new to the field and seek to understand its fundamental principles. While we hope the book also becomes a reference for experienced readers, its primary aim is to guide students new to the subject and provide a roadmap for both introductory and advanced courses tailored to sciences and engineering students.

In simple terms, Linear Algebra concerns about vector spaces and the linear transformations operating between them. Similarly, Functional Analysis examines normed vector spaces, particularly complete spaces, and the study of continuous linear operators between them. These definitions highlight the parallels and distinctions between these mathematical domains. While both disciplines explore vector spaces and linear transformations, Functional Analysis distinguishes itself by emphasizing the necessity of completeness of the vector space and continuity of the linear transformations. Here arises a pivotal question that piques the curiosity of all students: Why this distinction? Why is it that in Linear Algebra, continuity is not a primary concern, unlike in Functional Analysis where it is essential? Similarly, why is completeness of the vector space never required in Linear Algebra?

The answers await within these pages, but here is a glimpse: Linear Algebra primarily focuses on finite-dimensional vector spaces, which are automatically complete, and continuity of linear transformations in such spaces comes for free. To put it differently, in finite dimensions, continuity is a consequence of linearity. However, in infinite-dimensional spaces, the story is very different. Not every infinite-dimensional space is complete, and there exists a plethora of linear operators defined in such spaces that fail to be continuous; various examples of such phenomena are explored throughout this text. Infused with a touch of poetic freedom, one could suggest that Linear Algebra symbolizes Functional Analysis in finite dimensions, while Functional Analysis encapsulates Linear Algebra in infinite dimensions.

The necessity to study linear operators between infinite-dimensional spaces arises from the expansive development of mathematics between the seventeenth and nineteenth centuries. During this period, mathematics began to tackle a diverse array of problems that had previously eluded satisfactory mathematical treatment. Many of these problems involve functions (rather than numbers of geometric figures) as solutions, such as functional, ordinary, or partial differential equations. Seeking solutions to these problems naturally leads us to sets of functions, which are inherently infinite-dimensional spaces.

Let us delve into a classic historical example to illustrate this point. Among the myriad mathematical models that emerged after the advent of Calculus, one of the most renowned was the problem of the vibrating string. Let us denote by $u(x, t)$ the vertical displacement in relation to the rest position experienced by a point at abscissa $x$ at time $t$ (positive displacement upwards and negative downwards). In 1747, the eminent French mathematician Jean-le-Rond d'Alembert made a profound observation regarding solutions to the vibrating string problem. He asserted that every solution $u$ of this problem must satisfy the partial differential equation:

$$\frac{\partial^2 u}{\partial t^2} - \frac{\partial^2 u}{\partial x^2} = 0.$$

Furthermore, he proposed that functions of the form:

$$u(x, t) = f(x + t) + g(x - t),$$

where $f$ and $g$ are arbitrary real functions of a real variable, serve as solutions to the differential equation and, thus, as potential solutions to the original problem. The term "arbitrary real functions" is imprecise, reflecting the understanding of d'Alembert's time when the concepts of continuity and differentiability had yet to be rigorously introduced. Over time, mathematicians realized that the functions $f$ and $g$ should be sought in the set of functions that have at least two continuous derivatives. This set, like other useful sets of functions, forms vector spaces of infinite dimension. The collection of polynomial functions already generates an infinite-dimensional space, simply because the set $\{1, t, t^2, \cdots\}$, as per the Fundamental Theorem of Algebra, is linearly independent and contains infinitely many elements.

Here is another example: just as a solution of a linear system with $n$ unknowns with real coefficients is an $n$-tuple $(x_1, \ldots, x_n)$ of $\mathbb{R}^n$, a solution of a linear system with infinite unknowns with real coefficients is a sequence $(x_n)_{n=1}^{\infty}$ of real numbers. Sets of sequences also constitute vector spaces of infinite dimension, which may or may not be complete depending upon the norm introduced in such spaces.

The expanding horizons of mathematics over the last two centuries have brought spaces of infinite dimension into the spotlight, as they house solutions to a plethora of significant problems spanning various domains. Since continuity, an essential concept, is not inherently guaranteed by linearity in infinite dimensions, the importance of studying continuous linear operators between these spaces becomes evident in the pursuit of a comprehensive mathematical theory.

# Preface

Selecting the topics for a book is always a challenging endeavor. In this book, we faced similar dilemmas, and regrettably, some fascinating topics had to be left out. Undoubtedly, some readers may find this omission disappointing; indeed, each reader likely has at least one topic they wish we had included. Nevertheless, let us delve into the topics we did include.

Our selection aimed to cover the essential subjects of an introductory Functional Analysis course while also incorporating some standard topics and outlining recent advancements. The mandatory subjects are covered in the first seven chapters. Additionally, within these chapters, we present several topics that are not mandatory and can be omitted based on the instructor's and/or student's objectives. These include Corollary 2.3.5, Sect. 3.2, Proposition 3.3.3, Theorem 5.4.4, Sect. 5.6, Theorem 6.2.12, Sect. 6.5, Sect. 6.6, Theorem 7.2.7, and Proposition 7.2.8.

Once equipped with the fundamental concepts of the discipline, students are faced with three potential paths: (i) exploring a broader theory, (ii) applying the acquired knowledge, and (iii) deepening their understanding of the subject matter already covered. In the forthcoming chapters, we present options for each of these directions, providing the reader with avenues for further exploration and application of the material.

In Chap. 8, we introduce topological vector spaces, which represent a natural extension of the study of normed spaces and Banach spaces. Throughout this chapter, readers will recognize that many concepts previously explored remain applicable in this broader context. Additionally, they will observe that certain earlier constructions emerge as special cases within these more encompassing frameworks.

Chapter 9 serves as an introduction to Non-Linear Analysis techniques, providing readers with an opportunity to apply their newly acquired knowledge of Functional Analysis. In Applied Mathematics, particularly in Mathematical Modeling, linearity is often absent. However, within this chapter, readers will discover that even in the absence of linearity, Functional Analysis concepts and techniques remain highly applicable and fruitful.

In Chap. 10, we explore further into the study of Banach spaces by introducing some significant theorems not typically covered in introductory Functional Analysis courses. The topics highlighted in this chapter exemplify the heightened sophistication attained by Banach space theory throughout the twentieth century. This evolution culminated in the recent theory of hereditarily indecomposable spaces, notably advanced by W. T. Gowers and B. Maurey. Gowers' contributions earned him a Fields Medal in 1994. Initially devised to address core issues within Banach space theory, this theory has revitalized and fortified the field of Functional Analysis. Its impact extends beyond the discipline, fostering novel techniques and forging new connections with diverse areas of mathematics.

It is widely recognized that if a mathematical research area becomes disconnected from others, it risks becoming "art for art's sake," as von Neumann aptly noted [65]. The aim of the final three chapters of this book is to dispel any notion that Functional Analysis has fallen into this category. Beyond providing a foundation for broader theories, Functional Analysis finds diverse applications and lends support to various fields, including Partial Differential Equations, Calculus of Variations,

Approximation Theory, and Harmonic Analysis. Today, Functional Analysis is not only esteemed for its intrinsic beauty but also for its strengthened connections with other domains, nurturing powerful techniques of universal applicability.

Regarding the book's structure, each chapter concludes with two sections: one featuring comments and historical notes, and the other comprising exercises. The comments and historical notes serve multiple purposes: they contextualize and interconnect the chapter's content with other subjects, supplement the presented material with additional results and examples, and shed light on the historical aspects of the developed theory. The exercise section is an integral and pivotal component of the book. It is through solving these exercises that students can gauge their comprehension of the studied theory. Furthermore, tackling the exercises familiarizes students with important results and examples that couldn't be incorporated into the main text; for instance, the theory of quotient spaces is explored in the exercises across various chapters. Exercises marked with an asterisk (*) are particularly challenging and/or laborious. On occasion (albeit infrequent), we may utilize results from these exercises. At the end of the book there is a section — Appendix D — offering hints for selected exercises. We emphasize the importance of consulting exercise hints only after making several attempts to solve them independently; otherwise, the benefits gained from solving the exercises may be compromised.

Before suggesting course pathways, it is important to clarify our approach in writing this book. We aimed to distance ourselves from the typical "compiled lecture notes" model. While drawing from our experiences teaching Functional Analysis courses, the primary focus remains on the student. We trust that instructors will adapt the text to their classes, understanding that students may not be ready for a more condensed and dry presentation typical of lecture notes.

For a first course in Functional Analysis, we recommend covering the first seven chapters, possibly excluding non-mandatory sections listed above. Chapters 8–10 can form the core of a second course in Functional Analysis and also serve as a foundation for student projects.

Let us discuss the prerequisites needed for studying this book. It is assumed that the reader possesses a solid understanding of real analysis at the level of [57] and basic concepts of Linear Algebra. In Appendix A, readers will find what they need regarding Zorn's Lemma. In terms of Topology, familiarity with Metric Spaces is essential for tackling the initial chapters, while rudimentary knowledge of General Topology from Chap. 6 onward is recommended. More advanced familiarity with Topology becomes necessary for a thorough grasp of Chaps. 8–10. Appendix B serves as a concise summary of the essential Topology concepts.

Understanding important examples and theorems may require knowledge of Measure Theory and Integration, akin to what is outlined in Appendix C.

This book evolved from a previous work by the same three authors, originally written in Portuguese and published by the Brazilian Mathematical Society (SBM). Now in its third reprint, the book received a warm reception from colleagues in Brazil, inspiring us to expand it into this new English version. We express our deep appreciation to SBM for publishing the original book and to the numerous

Brazilian colleagues who have used it in their institutions and provided invaluable feedback. Given the extensive list of contributors, we extend our heartfelt gratitude to all who have supported the book, avoiding the potential embarrassment of omitting anyone by name. A special thank you to Anselmo Raposo Jr. for his meticulous proofreading, which helped identify errors, refine theorem statements, and enhance the clarity of several proofs. His expertise with LaTeX was invaluable in streamlining the typesetting process, and we are especially grateful for the figures he created, which improve the presentation of key concepts. His contributions have been instrumental in elevating the overall quality of this book.

| | |
|---|---:|
| Uberlândia, Brazil | Geraldo Botelho |
| João Pessoa, Brazil | Daniel Pellegrino |
| Orlando, FL, USA | Eduardo Teixeira |
| February 2025 | |

# Contents

1. **Normed Vector Spaces** .................................................... 1
   1.1 Definitions and First Examples .................................... 1
   1.2 The Spaces $L_p(X, \Sigma, \mu)$ ................................. 6
   1.3 The Space $L_\infty(X, \Sigma, \mu)$ ............................. 10
   1.4 Sequence Spaces ................................................... 12
   1.5 Compact Sets in Normed Vector Spaces .............................. 13
   1.6 Separable Normed Spaces ........................................... 16
   1.7 Comments and Historical Notes ..................................... 19
   1.8 Exercises ......................................................... 21

2. **Continuous Linear Operators** ............................................ 27
   2.1 Characterizations of Continuous Linear Operators .................. 28
   2.2 Examples .......................................................... 30
   2.3 The Banach–Steinhaus Theorem ...................................... 31
   2.4 The Open Mapping Theorem .......................................... 35
   2.5 The Closed Graph Theorem .......................................... 38
   2.6 Comments and Historical Notes ..................................... 39
   2.7 Exercises ......................................................... 41

3. **Hahn–Banach Theorems** .................................................. 47
   3.1 The Hahn–Banach Extension Theorem ................................. 47
   3.2 Continuous Extensions of Linear Operators ......................... 51
   3.3 Applications of the Hahn–Banach Theorem for Separable Spaces ...... 54
   3.4 Geometric Forms of the Hahn–Banach Theorem (Real Case) ............ 56
   3.5 Comments and Historical Notes ..................................... 64
   3.6 Exercises ......................................................... 65

4. **Duality and Reflexive Spaces** ........................................... 69
   4.1 The Dual of the $L_p$ Spaces ...................................... 69

xi

|     |     |                                                                      |     |
| --- | --- | -------------------------------------------------------------------- | --- |
|     | 4.2 | The Duals of the Spaces $\ell_p$ and $c_0$                           | 72  |
|     | 4.3 | Bidual and Reflexive Spaces                                          | 74  |
|     | 4.4 | Comments and Historical Notes                                        | 78  |
|     | 4.5 | Exercises                                                            | 79  |
| **5** | **Hilbert Spaces**                                                       |     | 85  |
|     | 5.1 | Spaces with Inner Product                                            | 86  |
|     | 5.2 | Orthogonality                                                        | 90  |
|     | 5.3 | Orthonormal Sets in Hilbert Spaces                                   | 94  |
|     | 5.4 | Orthogonalization Process and Its Consequences                       | 101 |
|     | 5.5 | Linear Functionals and the Riesz–Fréchet Theorem                     | 104 |
|     | 5.6 | The Lax-Milgram Theorem                                              | 107 |
|     | 5.7 | Comments and Historical Notes                                        | 110 |
|     | 5.8 | Exercises                                                            | 112 |
| **6** | **Weak Topologies**                                                      |     | 117 |
|     | 6.1 | The Topology Generated by a Family of Functions                      | 118 |
|     | 6.2 | The Weak Topology on a Normed Space                                  | 119 |
|     | 6.3 | The Weak-Star Topology                                               | 125 |
|     | 6.4 | Weak Compactness and Reflexivity                                     | 130 |
|     | 6.5 | Metrizability and Separability                                       | 134 |
|     | 6.6 | Uniformly Convex Spaces                                              | 137 |
|     | 6.7 | Comments and Historical Notes                                        | 144 |
|     | 6.8 | Exercises                                                            | 146 |
| **7** | **Spectral Theory of Compact Self-adjoint Operators**                    |     | 151 |
|     | 7.1 | Spectrum of a Continuous Linear Operator                             | 152 |
|     | 7.2 | Compact Operators                                                    | 154 |
|     | 7.3 | Spectral Theory of Compact Operators                                 | 159 |
|     | 7.4 | Self-adjoint Operators on Hilbert Spaces                             | 164 |
|     | 7.5 | Spectral Theory of Self-adjoint Operators                            | 169 |
|     | 7.6 | Comments and Historical Notes                                        | 173 |
|     | 7.7 | Exercises                                                            | 175 |
| **8** | **Topological Vector Spaces**                                            |     | 179 |
|     | 8.1 | Examples and First Properties                                        | 180 |
|     | 8.2 | Locally Convex Spaces                                                | 187 |
|     | 8.3 | Seminorms and Topologies                                             | 192 |
|     | 8.4 | Revisiting Hahn–Banach and Goldstine                                 | 198 |
|     | 8.5 | Comments and Historical Notes                                        | 203 |
|     | 8.6 | Exercises                                                            | 204 |
| **9** | **Introduction to Nonlinear Analysis**                                   |     | 209 |
|     | 9.1 | Continuity in the Norm Topology                                      | 209 |
|     | 9.2 | Continuity in the Weak Topology                                      | 212 |
|     | 9.3 | Minimization Problems                                                | 215 |
|     | 9.4 | Fixed Point Theorems                                                 | 217 |

|       | 9.5  | Compact Nonlinear Operators | 220 |
|-------|------|------------------------------|-----|
|       | 9.6  | Elements of Differential Calculus in Banach Spaces | 225 |
|       | 9.7  | A Quick Foray into Vector Integration | 232 |
|       | 9.8  | Comments and Historical Notes | 237 |
|       | 9.9  | Exercises | 240 |
| **10** | | **Elements of Banach Space Theory** | **245** |
|       | 10.1 | Series and the Dvoretzky-Rogers Theorem | 245 |
|       | 10.2 | Grothendieck's Inequality | 257 |
|       | 10.3 | Schauder Bases and Basic Sequences | 263 |
|       | 10.4 | The Bessaga–Pełczyński Selection Principle | 274 |
|       | 10.5 | Comments and Historical Notes | 282 |
|       | 10.6 | Exercises | 284 |
| **A** | | **Zorn's Lemma** | **289** |
| **B** | | **Concepts of General Topology** | **293** |
| **C** | | **Measure and Integration** | **301** |
| **D** | | **Answers/Hints for Selected Exercises** | **309** |
| **Bibliography** | | | **333** |
| **Index** | | | **337** |

# Chapter 1
# Normed Vector Spaces

In vector spaces, we know how to add their elements and multiply their elements by scalars, while in metric spaces we know how to calculate the distance between two of their elements. A normed space is the mathematical structure in which all this can be done in the same environment.

## 1.1 Definitions and First Examples

The symbol $\mathbb{K}$ will denote, indistinctly, the field $\mathbb{R}$ of real numbers or the field $\mathbb{C}$ of complex numbers. The elements of $\mathbb{K}$ are called scalars. The modulus function — or absolute value —

$$a \in \mathbb{K} \mapsto |a| \in \mathbb{R}$$

satisfies the following conditions:

- $|a| \geq 0$ for every $a$ and $|a| = 0 \iff a = 0$.
- $|a \cdot b| = |a| \cdot |b|$ for all $a$ and $b$.
- $|a + b| \leq |a| + |b|$ for all $a$ and $b$.

The number $|a - b|$ is interpreted as the distance between $a$ and $b$. Let $E$ be a vector space over $\mathbb{K}$. Any function that reproduces the conditions above will serve to induce a notion of distance in $E$. More precisely, a function

$$\|\cdot\|: E \longrightarrow \mathbb{R}$$

is a *norm* if the following properties are satisfied:

(N1) $\|x\| \geq 0$ for every $x \in E$ and $\|x\| = 0 \iff x = 0$.
(N2) $\|ax\| = |a| \cdot \|x\|$ for every scalar $a$ and every $x \in E$.
(N3) $\|x + y\| \leq \|x\| + \|y\|$ for any $x, y \in E$.

A vector space equipped with a norm will be called a *normed vector space*, or simply *normed space*. Just like in the case of the scalar field, a normed space is a metric space with the metric given by

$$d(x, y) = \|x - y\|. \tag{1.1}$$

In this case, we say that the metric $d$ is induced by the norm $\|\cdot\|$. Therefore, all the theory of metric spaces (see Appendix B) applies to normed spaces. In particular, a sequence $(x_n)_{n=1}^\infty$ in a normed space $E$ converges to the vector $x \in E$ if

$$\lim_{n \to \infty} \|x_n - x\| = 0,$$

and we say that $x$ is the limit of the sequence $(x_n)_{n=1}^\infty$ and we write $x = \lim_{n \to \infty} x_n = \lim_n x_n$ or $x_n \longrightarrow x$.

What makes the algebraic and topological structures of a normed space compatible is the fact that in a normed space the algebraic operations are continuous functions (see Exercise 1.8.13).

A normed space $E$ is called a *Banach space* when it is a complete metric space with the metric induced by the norm. From elementary courses in Analysis, we know that $(\mathbb{R}, |\cdot|)$ and $(\mathbb{C}, |\cdot|)$ are Banach spaces.

The proposition below highlights the importance of closed subspaces in a Banach space.

**Proposition 1.1.1** *Let $E$ be a Banach space and let $F$ be a vector subspace of $E$. Then $F$ is a Banach space, with the norm induced from $E$, if and only if $F$ is closed in $E$.*

**Proof** Suppose $F$ is a Banach space and consider $(x_n)_{n=1}^\infty$ a sequence in $F$ such that $x_n \longrightarrow x \in E$. Then $(x_n)_{n=1}^\infty$ is Cauchy in $F$, therefore convergent, because $F$ is complete by assumption; let $y \in F$ be such that $x_n \longrightarrow y$. From the uniqueness of the limit we have $x = y \in F$, proving that $F$ is closed in $E$.

Conversely, suppose $F$ is closed in $E$ and let $(x_n)_{n=1}^\infty$ be a Cauchy sequence in $F$. Therefore $(x_n)_{n=1}^\infty$ is Cauchy in $E$, and so there exists $x \in E$ such that $x_n \longrightarrow x$. As $F$ is closed it follows that $x \in F$, which proves that $F$ is complete. ∎

**Example 1.1.2** Let $X$ be a non-empty set. A function $f \colon X \longrightarrow \mathbb{K}$ is *bounded* if its range is a bounded subset of $\mathbb{K}$, that is, if there exists $M \geq 0$ such that $|f(x)| \leq M$ for every $x \in X$. The set $B(X)$ of all bounded functions $f \colon X \longrightarrow \mathbb{K}$, which is a vector space with the usual operations of functions, becomes a Banach space with the norm

## 1.1 Definitions and First Examples

$$\|f\|_\infty = \sup_{x \in X} |f(x)|.$$

In fact, it is easy to see that $\|\cdot\|_\infty$ is a norm. To show completeness, consider a Cauchy sequence $(f_n)_{n=1}^\infty$ in $B(X)$ and check that for every $x \in X$, the corresponding sequence $(f_n(x))_{n=1}^\infty$ is Cauchy in $\mathbb{K}$. The result follows easily from the completeness of $\mathbb{K}$.

For an interval $[a, b]$, for the sake of simplicity we shall write $B[a, b]$ instead of $B([a, b])$.

**Example 1.1.3** Since $[a, b]$ is compact in $\mathbb{R}$, the set $C[a, b]$ of all continuous functions from $[a, b]$ to $\mathbb{K}$ is a vector subspace of the Banach space $B[a, b]$, therefore it is a normed space with the norm

$$\|f\|_\infty = \sup\{|f(x)| : x \in [a, b]\} = \max\{|f(x)| : x \in [a, b]\}.$$

Moreover, $C[a, b]$ is a Banach space. In fact, by Proposition 1.1.1 it is enough to prove that $C[a, b]$ is a closed subspace of $B[a, b]$. To do so, observe that if $f_n \longrightarrow f$ in $B[a, b]$, then $(f_n)_{n=1}^\infty$ converges uniformly to $f$. It follows that $f$ is continuous as it is the uniform limit of continuous functions.

The reader probably knows that functions with certain differentiability properties are very useful in applications. One of the main objectives of Calculus is the study of the instantaneous rate of change of a real function $f$. This rate of change is called the derivative of $f$ and denoted by $f'$. We learned in Calculus that if $f$ and $g$ are differentiable, then $f + g$ is a differentiable function and $(f + g)' = f' + g'$ and if $\lambda \in \mathbb{R}$, $(\lambda f)' = \lambda f'$. From our perspective, this means that the set of all differentiable functions has a natural structure of vector space. We study in the next example the spaces of continuously differentiable functions.

**Example 1.1.4** The space of continuously differentiable functions is defined by

$$C^1[a, b] := \{f : [a, b] \longrightarrow \mathbb{R} : f \text{ is differentiable in } [a, b] \text{ and } f' \in C[a, b]\}.$$

As usual, at the endpoints of the interval, we consider the respective one-sided limits. As we saw above, $C^1[a, b]$ is a subspace of $C[a, b]$. However, as we will prove in Example 2.4.4, $C^1[a, b]$ is not closed in $C[a, b]$, therefore it is not complete with respect to the norm $\|\cdot\|_\infty$ by Proposition 1.1.1. It is somewhat routine to see that the norm

$$\|f\|_{C^1} = \|f\|_\infty + \|f'\|_\infty$$

makes $C^1[a, b]$ a Banach space. We leave to the reader the verification of the norm axioms. We prove that $(C^1[a, b], \|\cdot\|_{C^1})$ is complete. In fact, given a Cauchy sequence $(f_n)_{n=1}^\infty$ in $(C^1[a, b], \|\cdot\|_{C^1})$, as $\|f\|_\infty \leq \|f\|_{C^1}$ and $\|f'\|_\infty \leq \|f'\|_{C^1}$, it follows that the sequences $(f_n)_{n=1}^\infty$ and $(f_n')_{n=1}^\infty$ are Cauchy in $C[a, b]$. From the previous example we know that $C[a, b]$ is complete, so there exist $f, g \in C[a, b]$

such that $f_n \longrightarrow f$ and $f'_n \longrightarrow g$ uniformly. By the Fundamental Theorem of Calculus we can write

$$f_n(x) - f_n(a) = \int_a^x f'_n(t)dt$$

for all $x \in [a,b]$ and $n \in \mathbb{N}$. Letting $n \longrightarrow \infty$, we conclude that $f \in C^1[a,b]$ and that $f' = g$. It follows that $f_n \longrightarrow f$ in $C^1[a,b]$.

We also know from Calculus that the second derivative provides information about local concavity or convexity of the function. We can extend our treatment to functions that are twice differentiable within the framework of normed spaces. Accordingly, we define the space of functions that are twice continuously differentiable as follows:

$$C^2[a,b] := \left\{ f : [a,b] \longrightarrow \mathbb{R} : f \text{ is differentiable in } [a,b] \text{ and } f' \in C^1[a,b] \right\}.$$

Similarly, $C^2[a,b]$ becomes a Banach space with the norm

$$\|f\|_{C^2} = \|f\|_\infty + \|f'\|_\infty + \|f''\|_\infty.$$

Inductively, we define the space of $k$ times continuously differentiable functions, $k \in \mathbb{N}$, by

$$C^k[a,b] := \left\{ f : [a,b] \longrightarrow \mathbb{R} : f \text{ is differentiable in } [a,b] \text{ and } f' \in C^{k-1}[a,b] \right\}.$$

As the reader has already noticed, $C^k[a,b]$ becomes a Banach space with the norm

$$\|f\|_{C^k} = \|f\|_\infty + \|f'\|_\infty + \cdots + \|f^{(k)}\|_\infty,$$

where $f^{(j)}$ denotes the $j$-th derivative of $f$.

Considering finite-dimensional normed spaces, it is likely that the reader is familiar with the usual norms on the spaces $\mathbb{K}^n$, $n \in \mathbb{N}$:

$\|(a_1, \ldots, a_n)\|_1 = |a_1| + \cdots + |a_n|$,
$\|(a_1, \ldots, a_n)\|_2 = \left( |a_1|^2 + \cdots + |a_n|^2 \right)^{\frac{1}{2}}$, and
$\|(a_1, \ldots, a_n)\|_\infty = \max\{|a_1|, \ldots, |a_n|\}$.

Perhaps you also know that these spaces are complete. Our immediate goal now is to show that every finite-dimensional normed space is complete. To do so, we need the following lemma:

**Lemma 1.1.5** *Let $B = \{x_1, \ldots, x_n\}$ be a set of linearly independent vectors in a normed space $E$. Then there exists a constant $c > 0$, which depends on the set $B$, such that*

## 1.1 Definitions and First Examples

$$\|a_1 x_1 + \cdots + a_n x_n\| \geq c \left( |a_1| + \cdots + |a_n| \right),$$

*for any scalars $a_1, \ldots, a_n$.*

**Proof** As the reader will verify in Exercise 1.8.3, given two norms $\|\cdot\|_a$ and $\|\cdot\|_b$ in $\mathbb{R}^n$, there exists a constant $c > 0$, which depends on the norms, such that $\|x\|_a \leq c\|x\|_b$ for every $x \in \mathbb{R}^n$. The same, with a completely analogous proof, holds for $\mathbb{C}^n$. The reader will have no difficulty in verifying that the correspondence

$$(a_1, \ldots, a_n) \in \mathbb{K}^n \longrightarrow \left\| \sum_{j=1}^n a_j x_j \right\| \in \mathbb{R}$$

is a norm on $\mathbb{K}^n$. The result follows because, as we saw above, the expression $\|(a_1, \ldots, a_n)\|_1 = |a_1| + \cdots + |a_n|$ is also a norm on $\mathbb{K}^n$. ∎

**Theorem 1.1.6** *Every finite-dimensional normed space is a Banach space. Consequently, every finite-dimensional subspace of a normed space $E$ is closed in $E$.*

**Proof** Let $E$ be a normed space of dimension $n$ and let $\{\beta_1, \ldots, \beta_n\}$ be a normalized basis (that is, the vectors have norm 1) of $E$. Given a Cauchy sequence $(x_k)_{k=1}^\infty$ in $E$, for each $k \in \mathbb{N}$ there are unique scalars $a_1^k, \ldots, a_n^k$ such that $x_k = a_1^k \beta_1 + \cdots + a_n^k \beta_n$. Given $\varepsilon > 0$, choose $n_0 \in \mathbb{N}$ such that $\|x_k - x_m\| < c\varepsilon$ whenever $k, m \geq n_0$, where $c$ is the constant from Lemma 1.1.5 for the linearly independent set $\{\beta_1, \ldots, \beta_n\}$. It follows that

$$\sum_{j=1}^n |a_j^k - a_j^m| \leq \frac{1}{c} \left\| \sum_{j=1}^n (a_j^k - a_j^m) \beta_j \right\| = \frac{1}{c} \|x_k - x_m\| < \varepsilon,$$

whenever $k, m \geq n_0$. We conclude that, for each $j = 1, \ldots, n$, the scalar sequence $(a_j^k)_{k=1}^\infty$ is a Cauchy sequence, therefore convergent. Let $b_j = \lim_k a_j^k$, $j = 1, \ldots, n$. In this case we have $\lim_k \sum_{j=1}^n |a_j^k - b_j| = 0$. Defining $x = b_1 \beta_1 + \cdots + b_n \beta_n$ we have $x \in E$ and

$$\lim_k \|x_k - x\| = \lim_k \left\| \sum_{j=1}^n (a_j^k - b_j) \beta_j \right\| \leq \lim_k \sum_{j=1}^n |a_j^k - b_j| = 0,$$

proving that $x_k \longrightarrow x$ and completing the proof that $E$ is Banach. The second statement follows from the first and from Proposition 1.1.1. ∎

As expected, there are normed spaces, of infinite dimension, of course, that are not complete, as the following example illustrates:

**Example 1.1.7** By $c_0$ we denote the set of all scalar sequences that converge to zero, that is, for a fixed $\mathbb{K} = \mathbb{R}$ or $\mathbb{C}$,

$$c_0 = \{(a_k)_{k=1}^\infty : a_k \in \mathbb{K} \text{ for every } k \in \mathbb{N} \text{ and } a_k \longrightarrow 0\}.$$

It is clear that $c_0$ is a vector space with the usual operations. It is easy to prove that the expression

$$\|(a_k)_{k=1}^\infty\|_\infty = \sup\{|a_k| : k \in \mathbb{N}\}$$

makes $c_0$ a normed space. The reason for the subscript $\infty$ will become clear in Sect. 1.4. Let $(x_n)_{n=1}^\infty$ be a Cauchy sequence in $c_0$, say $x_n = (a_n^k)_{k=1}^\infty$ for each $n \in \mathbb{N}$. For each $j \in \mathbb{N}$, the inequality

$$|a_n^j - a_m^j| \le \sup\{|a_n^k - a_m^k| : k \in \mathbb{N}\} = \|x_n - x_m\|_\infty$$

makes it clear that the scalar sequence $(a_n^j)_{n=1}^\infty$ is Cauchy in $\mathbb{K}$, therefore convergent, say $a_n^j \xrightarrow{n \to \infty} a_j$ for each $j \in \mathbb{N}$. Calling $x = (a_j)_{j=1}^\infty$ it is not difficult to check that $x \in c_0$ and that $x_n \longrightarrow x$ in $c_0$. We conclude that $c_0$ is a Banach space.

Now call $c_{00}$ the subspace of $c_0$ comprising by the eventually null sequences, that is,

$$c_{00} = \{(a_k)_{k=1}^\infty \in c_0 : \text{there exists } k_0 \in \mathbb{N} \text{ such that } a_k = 0 \text{ for every } k \ge k_0\}.$$

Consider the following vectors of $c_{00}$: $x_1 = (1, 0, 0, 0, \dots,)$, $x_2 = \left(1, \frac{1}{2}, 0, 0, \dots\right)$, $\dots$, $x_n = \left(1, \frac{1}{2}, \dots, \frac{1}{n}, 0, 0 \dots,\right)$, $\dots$. It is clear that $(x_n)_{n=1}^\infty$ is a sequence in $c_{00}$. Taking $x = (\frac{1}{k})_{k=1}^\infty \in c_0$, from $\|x_n - x\|_\infty = \frac{1}{n+1} \longrightarrow 0$ we conclude that $x_n \longrightarrow x$ in $c_0$. As $x \notin c_{00}$, it follows that $c_{00}$ is a non-closed subspace of $c_0$. Therefore, from Proposition 1.1.1, $c_{00}$ is an incomplete normed space.

## 1.2 The Spaces $L_p(X, \Sigma, \mu)$

In this section and the next one we will use the concepts and terminology of Measure Theory, which the reader can find in Appendix C. The next examples involve integrable functions, and some preliminary results are necessary. We will establish the Hölder and Minkowski inequalities for integrals, which will be useful later.

Let $(X, \Sigma, \mu)$ be a measure space and $1 \le p < \infty$. The set of all measurable functions from $X$ to $\mathbb{K}$ such that

$$\|f\|_p := \left(\int_X |f|^p \, d\mu\right)^{\frac{1}{p}} < \infty$$

## 1.2 The Spaces $L_p(X, \Sigma, \mu)$

will be denoted by $\mathcal{L}_p(X, \Sigma, \mu)$.

**Theorem 1.2.1 (Hölder's Inequality for Integrals)** *Let $p, q > 1$ be such that $\frac{1}{p} + \frac{1}{q} = 1$ and let $(X, \Sigma, \mu)$ be a measure space. If $f \in \mathcal{L}_p(X, \Sigma, \mu)$ and $g \in \mathcal{L}_q(X, \Sigma, \mu)$, then $fg \in \mathcal{L}_1(X, \Sigma, \mu)$ and*

$$\|fg\|_1 \leq \|f\|_p \cdot \|g\|_q.$$

*Proof* The case $\|f\|_p = 0$ or $\|g\|_q = 0$ is simple. Suppose $\|f\|_p \neq 0 \neq \|g\|_q$. First it is convenient to show that for any $a, b > 0$, we have

$$a^{\frac{1}{p}} \cdot b^{\frac{1}{q}} \leq \frac{a}{p} + \frac{b}{q}. \tag{1.2}$$

To do so, consider, for each $0 < \alpha < 1$, the function $f = f_\alpha : (0, \infty) \longrightarrow \mathbb{R}$ given by $f(t) = t^\alpha - \alpha t$. Note that $f$ has a maximum at $t = 1$, hence $t^\alpha \leq \alpha t + (1 - \alpha)$ for every $t > 0$. Letting $t = \frac{a}{b}$ and $\alpha = \frac{1}{p}$ we get (1.2). It is clear that (1.2) is also valid if $a = 0$ or $b = 0$. Choosing

$$a = \frac{|f(x)|^p}{\|f\|_p^p} \quad \text{and} \quad b = \frac{|g(x)|^q}{\|g\|_q^q}$$

in (1.2), the result follows. ∎

**Theorem 1.2.2 (Minkowski's Inequality for Integrals)** *Let $1 \leq p < \infty$ be given and let $(X, \Sigma, \mu)$ be a measure space. If $f, g \in \mathcal{L}_p(X, \Sigma, \mu)$, then $f + g \in \mathcal{L}_p(X, \Sigma, \mu)$ and*

$$\|f + g\|_p \leq \|f\|_p + \|g\|_p. \tag{1.3}$$

*Proof* If $p = 1$ or $\|f + g\|_p = 0$, the result is clear. So, we can assume $\|f + g\|_p \neq 0$ and $p > 1$. Notice that for every $x \in X$, we have

$$|f(x) + g(x)|^p \leq (|f(x)| + |g(x)|)^p \leq (2\max\{|f(x)|, |g(x)|\})^p$$
$$= 2^p (\max\{|f(x)|, |g(x)|\})^p \leq 2^p \left(|f(x)|^p + |g(x)|^p\right).$$

It follows that $f + g \in \mathcal{L}_p(X, \Sigma, \mu)$. Now we will prove (1.3). Observe first that

$$|f(x) + g(x)|^p = |f(x) + g(x)| \cdot |f(x) + g(x)|^{p-1}$$
$$\leq |f(x) + g(x)|^{p-1} \cdot (|f(x)| + |g(x)|)$$
$$= |f(x)| \cdot |f(x) + g(x)|^{p-1} + |g(x)| \cdot |f(x) + g(x)|^{p-1} \tag{1.4}$$

for every $x \in X$. Choosing $q > 1$ such that $1/p + 1/q = 1$ we have $(p-1)q = p$, therefore $|f+g|^{p-1} = |f+g|^{\frac{p}{q}} \in \mathcal{L}_q(X, \Sigma, \mu)$. From Hölder's inequality, we have

$$\int_X |f| \cdot |f+g|^{p-1}\, d\mu \leq \left(\int_X |f|^p\, d\mu\right)^{\frac{1}{p}} \left(\int_X |f+g|^{(p-1)q}\, d\mu\right)^{\frac{1}{q}},$$

$$\int_X |g| \cdot |f+g|^{p-1}\, d\mu \leq \left(\int_X |g|^p\, d\mu\right)^{\frac{1}{p}} \left(\int_X |f+g|^{(p-1)q}\, d\mu\right)^{\frac{1}{q}}.$$

Adding the two inequalities above and combining with (1.4) we have

$$\int_X |f+g|^p\, d\mu \leq \left(\int_X |f+g|^p\, d\mu\right)^{\frac{1}{q}} \left[\left(\int_X |f|^p\, d\mu\right)^{\frac{1}{p}} + \left(\int_X |g|^p\, d\mu\right)^{\frac{1}{p}}\right],$$

which completes the proof. ∎

Note that $\|\cdot\|_p$ is not, in general, a norm on $\mathcal{L}_p(X, \Sigma, \mu)$, because it may happen that $\|f\|_p = 0$ for $f$ not identically zero. In general, if $(X, \Sigma, \mu)$ is a measure space, we introduce an equivalence relation saying that two functions $f, g\colon X \longrightarrow \mathbb{K}$ are equivalent if $f = g$ $\mu$-almost everywhere, that is, if there exists a set $A \in \Sigma$ such that $\mu(A) = 0$ and $f(x) = g(x)$ for every $x \notin A$. Denoting the equivalence class of a function $f$ by $[f]$, it is immediate that in the quotient set

$$L_p(X, \Sigma, \mu) := \{[f] : f \in \mathcal{L}_p(X, \Sigma, \mu)\},$$

the operations

$$[f] + [g] = [f+g] \text{ and } c[f] = [cf]$$

are well defined and make $L_p(X, \Sigma, \mu)$ a vector space. Moreover, defining

$$\|[f]\|_p := \|f\|_p,$$

we correct what was missing for $\|\cdot\|_p$ to be a norm. Thus $(L_p(X, \Sigma, \mu), \|\cdot\|_p)$ is a normed vector space. Next, we will show that $L_p(X, \Sigma, \mu)$ is a Banach space.

We should bear in mind that the vectors of $L_p(X, \Sigma, \mu)$ are equivalence classes, but in everyday life the equivalence classes are omitted and we write $f$ instead of $[f]$. We will, on purpose, make the proof below writing $[f]$ so that the reader realizes that this is an unnecessary care.

**Theorem 1.2.3** *If $1 \leq p < \infty$, then $L_p(X, \Sigma, \mu)$ is a Banach space with the norm*

$$\|[f]\|_p = \left(\int_X |f|^p\, d\mu\right)^{\frac{1}{p}}.$$

## 1.2 The Spaces $L_p(X, \Sigma, \mu)$

**Proof** We already know that $L_p(X, \Sigma, \mu)$ is a normed space. It remains to prove that it is a complete space. To do so, let $([f_n])_{n=1}^{\infty}$ be a Cauchy sequence in $L_p(X, \Sigma, \mu)$. Then, given $\varepsilon > 0$ there exists $M = M(\varepsilon) \in \mathbb{N}$ such that

$$\int_X |f_n - f_m|^p \, d\mu = \|f_n - f_m\|_p^p < \varepsilon^p \text{ whenever } m, n \geq M.$$

Let $(g_k)_{k=1}^{\infty}$ be a subsequence of $(f_n)_{n=1}^{\infty}$ such that $\|g_{k+1} - g_k\|_p < 2^{-k}$ for each $k \in \mathbb{N}$. Consider the function

$$g \colon X \longrightarrow \mathbb{R} \cup \{\infty\}, \quad g(x) = |g_1(x)| + \sum_{k=1}^{\infty} |g_{k+1}(x) - g_k(x)|. \tag{1.5}$$

We know that $g$ is measurable and non-negative. Furthermore,

$$|g(x)|^p = \lim_{n \to \infty} \left( |g_1(x)| + \sum_{k=1}^{n} |g_{k+1}(x) - g_k(x)| \right)^p.$$

By Fatou's Lemma (see Theorem C.17), it follows that

$$\int_X |g|^p \, d\mu \leq \liminf_{n \to \infty} \int_X \left( |g_1| + \sum_{k=1}^{n} |g_{k+1} - g_k| \right)^p d\mu.$$

Raising both sides to $\frac{1}{p}$ and using Minkowski's inequality, we obtain

$$\left( \int_X |g|^p \, d\mu \right)^{\frac{1}{p}} \leq \liminf_{n \to \infty} \left( \|g_1\|_p + \sum_{k=1}^{n} \|g_{k+1} - g_k\|_p \right) \leq \|g_1\|_p + 1. \tag{1.6}$$

Then, defining $A = \{x \in X : g(x) < \infty\}$, from (1.6) we conclude that $\mu(X - A) = 0$. Therefore, the series in (1.5) converges up to the null set $X - A$, that is, the series converges $\mu$-almost everywhere. It follows that $g \cdot \chi_A \in \mathcal{L}_p(X, \Sigma, \mu)$, where $\chi_A$ is the characteristic function of $A$ (see Appendix C). Define $f \colon X \longrightarrow \mathbb{K}$ by

$$f(x) = \begin{cases} g_1(x) + \displaystyle\sum_{k=1}^{\infty} (g_{k+1}(x) - g_k(x)), & \text{if } x \in A, \\ 0, & \text{if } x \notin A. \end{cases}$$

As $g_k = g_1 + (g_2 - g_1) + (g_3 - g_2) + \cdots + (g_k - g_{k-1})$, we have

$$|g_k(x)| \leq |g_1(x)| + \sum_{j=1}^{k-1} |g_{j+1}(x) - g_j(x)| \leq g(x)$$

for every $x$ and $g_k(x) \longrightarrow f(x)$ for every $x \in A$, that is, the sequence $(g_k)_{k=1}^\infty$ converges to $f$ $\mu$-almost everywhere. By the Dominated Convergence Theorem (see Theorem C.22), it follows that $f \in \mathcal{L}_p(X, \Sigma, \mu)$. As

$$|f - g_k|^p \leq (|f| + |g_k|)^p \leq (2g)^p \cdot \chi_A \ \mu\text{-almost everywhere and}$$

$$\lim_{k \to \infty} |f - g_k|^p = 0 \ \mu\text{-almost everywhere,}$$

again by the Dominated Convergence Theorem we have

$$\lim_{k \to \infty} \int_X |f - g_k|^p \, d\mu = \int_X 0 \, d\mu = 0.$$

We conclude that $g_k \longrightarrow f$ in $\mathcal{L}_p(X, \Sigma, \mu)$, therefore $[g_k] \longrightarrow [f]$ in $L_p(X, \Sigma, \mu)$. Then $([f_n])_{n=1}^\infty$ is a Cauchy sequence having a subsequence $([g_k])_{k=1}^\infty$ that converges to $[f]$. It immediately follows that $[f_n] \longrightarrow [f]$ in $L_p(X, \Sigma, \mu)$. ∎

## 1.3 The Space $L_\infty(X, \Sigma, \mu)$

Let $\mathcal{L}_\infty(X, \Sigma, \mu)$ be the set of all measurable functions that are bounded $\mu$-almost everywhere, that is, there exist a set $N \in \Sigma$ and a real number $K$ such that $\mu(N) = 0$ and $|f(x)| \leq K$ for every $x \notin N$. If $f \in \mathcal{L}_\infty(X, \Sigma, \mu)$ and $N \in \Sigma$ is a set of null measure, we define

$$S_f(N) = \sup\{|f(x)| : x \notin N\} \text{ and}$$

$$\|f\|_\infty = \inf\{S_f(N) : N \in \Sigma \text{ and } \mu(N) = 0\}. \tag{1.7}$$

Note that, again, it may happen that $\|f\|_\infty = 0$ with $f$ not identically zero. To overcome this problem, we also resort to equivalence classes. As in the previous section, we say that two functions are equivalent (belong to the same equivalence class) if they coincide $\mu$-almost everywhere. The set $L_\infty(X, \Sigma, \mu)$ is the set of all equivalence classes of measurable functions $f : X \longrightarrow \mathbb{K}$ that are bounded $\mu$-almost everywhere. It is clear that $L_\infty(X, \Sigma, \mu)$ is a vector space with the same operations as in the previous section. If $[f] \in L_\infty(X, \Sigma, \mu)$, we define

$$\|[f]\|_\infty = \|f\|_\infty. \tag{1.8}$$

In Exercise 1.8.23 the reader will prove that if the measure $\mu$ is finite and $f \in L_\infty(X, \Sigma, \mu)$, then $f \in L_p(X, \Sigma, \mu)$ for every $p \geq 1$ and $\lim_{p \to \infty} \|f\|_p = \|f\|_\infty$. This justifies the notation $L_\infty(X, \Sigma, \mu)$.

## 1.3 The Space $L_\infty(X, \Sigma, \mu)$

We leave it to the reader to prove that $\|\cdot\|_\infty$ is well defined and is a norm on $L_\infty(X, \Sigma, \mu)$. We will see that if $[f] \in L_\infty(X, \Sigma, \mu)$, then

$$|f(x)| \leq \|f\|_\infty \quad \mu\text{-almost everywhere.} \tag{1.9}$$

Indeed, by the definition of $\|f\|_\infty$ there is a sequence $(N_n)_{n=1}^\infty$ of sets of measure zero such that

$$\lim_{n \to \infty} S_f(N_n) = \|f\|_\infty \text{ and } |f(x)| \leq S_f(N_n) \text{ for every } x \notin N_n.$$

Therefore, defining $N = \bigcup_{n=1}^\infty N_n$, it follows that $N$ has measure zero and

$$|f(x)| \leq S_f(N_n) \text{ for every } x \notin N.$$

Taking the limit as $n \longrightarrow \infty$ we obtain (1.9).

**Theorem 1.3.1** $L_\infty(X, \Sigma, \mu)$ *is a Banach space.*

**Proof** It only remains to show that the space is complete. To do so, let $([f_n])_{n=1}^\infty$ be a Cauchy sequence in $L_\infty(X, \Sigma, \mu)$. By (1.9), for each pair $(n, m) \in \mathbb{N}^2$ there is $M_{n,m} \in \Sigma$ with $\mu(M_{n,m}) = 0$ and

$$|f_n(x) - f_m(x)| \leq \|f_n - f_m\|_\infty \text{ for every } x \notin M_{n,m}.$$

Let $M = \bigcup_{n,m=1}^\infty M_{n,m}$. Then $\mu(M) = 0$ and, for any $m, n \in \mathbb{N}$, we have

$$|f_n(x) - f_m(x)| \leq \|f_n - f_m\|_\infty \text{ for every } x \notin M. \tag{1.10}$$

Hence, for each $x \notin M$, the sequence $(f_n(x))_{n=1}^\infty$ is a Cauchy sequence in $\mathbb{K}$, therefore convergent. Define

$$f: X \longrightarrow \mathbb{K}, \quad f(x) = \begin{cases} \lim_{n \to \infty} f_n(x), & \text{if } x \notin M \\ 0, & \text{if } x \in M. \end{cases}$$

It follows that $f$ is measurable. By (1.10) and as $([f_n])_{n=1}^\infty$ is a Cauchy sequence, given $\varepsilon > 0$ there is $n_0 \in \mathbb{N}$ such that $\sup_{x \notin M} |f_n(x) - f_m(x)| \leq \varepsilon$ whenever $m, n \geq n_0$. Letting $m \longrightarrow \infty$ it follows that

$$\sup_{x \notin M} |f_n(x) - f(x)| \leq \varepsilon \tag{1.11}$$

whenever $n \geq n_0$. Thus, $(f_n)_{n=1}^\infty$ is uniformly convergent to $f$ in $X - M$. From (1.11) it follows that $f_n - f \in \mathcal{L}_\infty(X, \Sigma, \mu)$ for sufficiently large $n$. Hence, as

$f = f_n - (f_n - f)$, it follows that $[f] \in L_\infty(X, \Sigma, \mu)$. Rewriting (1.11) we conclude that

$$\|[f_n] - [f]\|_\infty = \|f_n - f\|_\infty \leq \sup_{x \notin M} |f_n(x) - f(x)| \leq \varepsilon$$

whenever $n \geq n_0$. Therefore $([f_n])_{n=1}^\infty$ converges to $[f]$ in $L_\infty(X, \Sigma, \mu)$. ∎

**Notation** Usually, we deal with $L_p(X, \Sigma, \mu)$ when $X = [a, b] \subseteq \mathbb{R}$, $\Sigma$ is the $\sigma$-algebra of Lebesgue-measurable sets (which contains the Borel sets) and $\mu$ is the Lebesgue measure. In this case we write $L_p[a, b]$.

## 1.4 Sequence Spaces

We have already studied the sequence spaces $c_0$ and $c_{00}$. In this section we will study other remarkable sequence spaces. For each real number $p \geq 1$, we define

$$\ell_p = \left\{ (a_j)_{j=1}^\infty : a_j \in \mathbb{K} \text{ for every } j \in \mathbb{N} \text{ and } \sum_{j=1}^\infty |a_j|^p < \infty \right\}.$$

Considering the set $\mathcal{P}(\mathbb{N})$ of parts of $\mathbb{N}$ and the counting measure $\mu_c$ on $\mathcal{P}(\mathbb{N})$, it is not difficult to check that $\ell_p$ is actually the space $L_p(\mathbb{N}, \mathcal{P}(\mathbb{N}), \mu_c)$. Moreover, in this case the usual operations of functions become the usual operations of sequences and the norm $\|\cdot\|_p$ becomes

$$\left\| (a_j)_{j=1}^\infty \right\|_p = \left( \sum_{j=1}^\infty |a_j|^p \right)^{\frac{1}{p}}.$$

In this way, it turns out that $\ell_p$ is a Banach space with the usual operations of sequences and with the norm $\|\cdot\|_p$. In particular, from Theorems 1.2.1 and 1.2.2 we have:

**Proposition 1.4.1 (Hölder's Inequality for Sequences)** *Let $n \in \mathbb{N}$ and let $p, q > 1$ be such that $\frac{1}{p} + \frac{1}{q} = 1$. Then*

$$\sum_{j=1}^n |a_j b_j| \leq \left( \sum_{j=1}^n |a_j|^p \right)^{\frac{1}{p}} \cdot \left( \sum_{j=1}^n |b_j|^q \right)^{\frac{1}{q}}$$

*for any scalars $a_1, \ldots, a_n, b_1, \ldots, b_n$.*

**Proposition 1.4.2 (Minkowski's Inequality for Sequences)** *For $p \geq 1$, we have*

$$\left(\sum_{j=1}^n |a_j+b_j|^p\right)^{\frac{1}{p}} \leq \left(\sum_{j=1}^n |a_j|^p\right)^{\frac{1}{p}} + \left(\sum_{j=1}^n |b_j|^p\right)^{\frac{1}{p}}$$

*for any $n \in \mathbb{N}$ and scalars $a_1, \ldots, a_n, b_1, \ldots, b_n$.*

For $p = \infty$, we define $\ell_\infty$ as the space of bounded scalar sequences, that is:

$$\ell_\infty = \left\{(a_j)_{j=1}^\infty : a_j \in \mathbb{K} \text{ for every } j \in \mathbb{N} \text{ and } \sup_{j \in \mathbb{N}} |a_j| < \infty \right\}.$$

We can conclude that $\ell_\infty$ is a Banach space with the usual operations of functions and with the norm

$$\|(a_j)_{j=1}^\infty\|_\infty = \sup\{|a_j| : j \in \mathbb{N}\}$$

in two ways: noting that $\ell_\infty = L_\infty(\mathbb{N}, \mathcal{P}(\mathbb{N}), \mu_c)$ or that $\ell_\infty = B(\mathbb{N})$ (Example 1.1.2). The fact that every convergent sequence is bounded implies that $c_0$ is a subspace of $\ell_\infty$. Since $c_0$ is a Banach space, it is a closed subspace of $\ell_\infty$ by Proposition 1.1.1.

## 1.5 Compact Sets in Normed Vector Spaces

Compactness, in addition to being one of the most important topological invariants, plays a fundamental role in Analysis. For example, a central result—and very useful—of elementary analysis says that every continuous function defined on a compact subset of $\mathbb{R}$ is bounded and assumes maximum and minimum.

In addition to its natural importance, in Functional Analysis compactness gains even more prominence because it represents a rupture between finite and infinite dimensions. We will see later that the *closed unit ball* $\{x \in E : \|x\| \leq 1\}$ of a normed space $E$, henceforth denoted by $B_E$, is compact whenever $E$ is finite-dimensional. In infinite dimension, it will be proven that the ball is never compact in the norm topology. To overcome the problem of non-compactness of the unit ball in infinite dimension, we will study in Chap. 6 the weak and weak star topologies, which, by having fewer open sets, facilitate the occurrence of compact sets.

By definition, a subset $K$ of a topological space is compact if every open cover of $K$ admits a finite subcover. In metric spaces, the following characterization holds: a subset $K$ of a metric space is compact if and only if it is sequentially compact, that is, every sequence formed by elements of $K$ admits a subsequence that converges to an element of $K$.

**Proposition 1.5.1** *If $E$ is a finite-dimensional normed vector space, then the compact sets in $E$ are precisely the bounded closed sets.*

**Proof** Compact sets in metric spaces are always closed and bounded. So, it is enough to prove that every bounded closed set $K \subseteq E$ is compact. Suppose that $\dim E = n$ and let $\{e_1, \ldots, e_n\}$ be a normalized basis of $E$. As we are in metric spaces, it is enough to show that every sequence in $K$ admits a convergent subsequence in $K$. Let $(x_m)_{m=1}^{\infty}$ be a sequence in $K$. For each $m$ there are scalars $a_1^{(m)}, \ldots, a_n^{(m)}$ such that $x_m = \sum_{j=1}^{n} a_j^{(m)} e_j$. Since $K$ is bounded, there exists $L > 0$ such that $\|x_m\| \leq L$ for every $m$. Consider in $\mathbb{K}^n$ the sum norm $\|\cdot\|_1$. By Lemma 1.1.5 there exists $c > 0$ such that

$$L \geq \|x_m\| = \left\|\sum_{j=1}^{n} a_j^{(m)} e_j\right\| \geq c \left(\sum_{j=1}^{n} |a_j^{(m)}|\right) = c \cdot \|(a_1^{(m)}, \ldots, a_n^{(m)})\|_1$$

for every $m \in \mathbb{N}$. Thus, the sequence $\left((a_1^{(m)}, \ldots, a_n^{(m)})\right)_{m=1}^{\infty}$ is bounded in $\mathbb{K}^n$. By the Bolzano–Weierstrass Theorem this sequence has a subsequence $\left((a_1^{(m_k)}, \ldots, a_n^{(m_k)})\right)_{k=1}^{\infty}$ that converges to a certain $b = (b_1, \ldots, b_n) \in \mathbb{K}^n$. From

$$\left\|\sum_{j=1}^{n} a_j^{(m_k)} e_j - \sum_{j=1}^{n} b_j e_j\right\| \leq \sum_{j=1}^{n} |a_j^{(m_k)} - b_j| = \left\|\left(a_1^{(m_k)}, \ldots, a_n^{(m_k)}\right) - b\right\|_1 \longrightarrow 0,$$

we conclude that $x_{m_k} \longrightarrow \sum_{j=1}^{n} b_j e_j$. Since $K$ is closed we know that $\sum_{j=1}^{n} b_j e_j \in K$. ∎

**Corollary 1.5.2** *The closed unit ball of a finite-dimensional normed space is compact.*

Next, we present a famous result, due to F. Riesz, which will be useful to show that the closed unit ball in an infinite-dimensional spaces is never compact.

**Lemma 1.5.3 (Riesz's Lemma)** *Let $M$ be a proper closed subspace of a normed space $E$ and let $\theta$ be a real number such that $0 < \theta < 1$. Then there exists $y \in E - M$ such that $\|y\| = 1$ and $\|y - x\| \geq \theta$ for every $x$ in $M$.*

**Proof** Let $y_0 \in E - M$ be given and consider the number

$$d = \mathrm{dist}(y_0, M) := \inf_{x \in M} \|y_0 - x\|.$$

Since $M$ is closed and $y_0 \notin M$, we have $d > 0$. In fact, assuming $d = 0$ there would be a sequence of elements of $M$ converging to $y_0$, and in this case we would have $y_0 \in M$ because $M$ is closed. As $\frac{d}{\theta} > d$, we can choose $x_0 \in M$ such that

## 1.5 Compact Sets in Normed Vector Spaces

$$\|y_0 - x_0\| \leq \frac{d}{\theta}.$$

We will show that choosing

$$y = \frac{y_0 - x_0}{\|y_0 - x_0\|},$$

the required conditions are satisfied. It is clear that $y$ has norm 1 and does not belong to $M$. Let $x \in M$. Since $M$ is a vector subspace, $(x_0 + \|y_0 - x_0\| x) \in M$, therefore $d \leq \|y_0 - (x_0 + \|y_0 - x_0\| x)\|$. Finally

$$\|y - x\| = \left\| \frac{y_0 - x_0}{\|y_0 - x_0\|} - x \right\| = \frac{\|y_0 - (x_0 + \|y_0 - x_0\| x)\|}{\|y_0 - x_0\|} \geq \frac{d}{\|y_0 - x_0\|} \geq \theta.$$

∎

Given a subset $A$ of a vector space $E$, by $[A]$ we denote the subspace of $E$ generated by $A$, that is, the set of all (finite) linear combinations of elements of $A$.

**Theorem 1.5.4** *A normed space is finite-dimensional if and only if its closed unit ball is compact.*

**Proof** From Corollary 1.5.2 it remains to prove that if the ball is compact, then the space is finite-dimensional. Suppose that $E$ is infinite-dimensional. Choose $x_1 \in E$ with norm 1. Since $\dim E = \infty$, the subspace $[x_1]$ generated by $x_1$ is a proper subspace of $E$. Being finite-dimensional, $[x_1]$ is a closed subspace of $E$ by Theorem 1.1.6. By Riesz's Lemma, with $\theta = \frac{1}{2}$, there exists $x_2 \in E - [x_1]$ of norm 1 such that

$$\|x_2 - x_1\| \geq \frac{1}{2}.$$

Applying the argument again to the subspace $[x_1, x_2]$ generated by $\{x_1, x_2\}$, which is a proper closed subspace for the same reasons, there exists $x_3 \in E - [x_1, x_2]$ of norm 1 such that

$$\|x_3 - x_j\| \geq \frac{1}{2} \text{ for } j = 1, 2.$$

We can continue this procedure indefinitely because at all steps we have a finite-dimensional subspace of $E$, hence closed and proper. Proceeding in this way, we construct a sequence $(x_n)_{n=1}^{\infty}$ in $B_E$ such that

$$\|x_m - x_n\| \geq \frac{1}{2}$$

whenever $m \neq n$. Then $(x_n)_{n=1}^{\infty}$ is a sequence in $B_E$ that does not have a convergent subsequence, which prevents $B_E$ from being compact. ∎

## 1.6 Separable Normed Spaces

Throughout your mathematical training, you have undoubtedly encountered several situations where the denseness of the set of rational numbers on the real line plays a crucial role. Similarly, the presence of a countable dense subset in a normed space is useful in several situations. It turns out, as we shall see soon, that not every normed space contains a countable dense subset. That is why we emphasize the class of spaces enjoying this property.

**Definition 1.6.1** A normed space containing a countable dense subset is said to be *separable*. More generally, a metric space is separable when it contains a countable dense subset.

**Example 1.6.2** Finite-dimensional normed spaces are separable. Let $\{x_1, \ldots, x_n\}$ be a basis of the normed space $E$ and, in the real case, consider the set

$$A = \left\{ \sum_{j=1}^{n} a_j x_j : n \in \mathbb{N} \text{ and } a_1, \ldots, a_n \in \mathbb{Q} \right\},$$

of linear combinations of the basic vectors with rational scalars. In the complex case replace $\mathbb{Q}$ by $\mathbb{Q} + i\mathbb{Q} = \{p + iq : p, q \in \mathbb{Q}\}$. The reader can easily check that $A$ is countable and dense in $E$.

The following lemma will be very useful in proving that many of the spaces we have been working with are separable.

**Lemma 1.6.3** *A normed space $E$ is separable if and only if there exists a countable subset $A \subseteq E$ such that $[A]$ is dense in $E$.*

*Proof* If $A$ is countable and dense in $E$, then $E = \overline{A} \subseteq \overline{[A]} \subseteq E$, therefore $[A]$ is dense in $E$. Conversely, suppose there exists a countable subset $A \subseteq E$ such that $\overline{[A]} = E$. Call $B$ the set spanned by all finite linear combinations of elements of $A$ with coefficients in $\mathbb{Q}_\mathbb{K}$, where $\mathbb{Q}_\mathbb{R} = \mathbb{Q}$ and $\mathbb{Q}_\mathbb{C} = \mathbb{Q} + i\mathbb{Q}$, that is:

$$B = \{a_1 x_1 + \cdots + a_n x_n : x_1, \ldots, x_n \in A, \ a_1, \ldots, a_n \in \mathbb{Q}_\mathbb{K} \text{ and } n \in \mathbb{N}\}.$$

It is easy to see that $B$ is countable: as $\mathbb{Q}_\mathbb{K}$ is countable, for each $n \in \mathbb{N}$, the set of linear combinations of $n$ elements of $A$ with coefficients in $\mathbb{Q}_\mathbb{K}$ is countable. It follows that $B$ is countable as it is the countable union of countable sets. We will now prove that $B$ is dense in $E$. To do so, let $x \in E$ and $\varepsilon > 0$ be given. As $\overline{[A]} = E$, there exists $y_0 \in [A]$ such that $\|x - y_0\| < \frac{\varepsilon}{2}$. We can write $y_0 = b_1 x_1 + \cdots + b_k x_k$, where $k \in \mathbb{N}$, $b_1, \ldots, b_k \in \mathbb{K}$ and $x_1, \ldots, x_k \in A$. As $\mathbb{Q}_\mathbb{K}$ is dense in $\mathbb{K}$, there exist $a_1, \ldots, a_k \in \mathbb{Q}_\mathbb{K}$ such that

## 1.6 Separable Normed Spaces

$$|a_j - b_j| < \frac{\varepsilon}{2\left(1 + \sum_{i=1}^{k} \|x_i\|\right)} \quad \text{for each } j = 1, \ldots, k.$$

Considering $y = a_1 x_1 + \cdots + a_k x_k$ we have $y \in B$ and

$$\|x - y\| = \|x - y_0 + y_0 - y\| \leq \|x - y_0\| + \|y_0 - y\|$$

$$< \frac{\varepsilon}{2} + \|(b_1 - a_1)x_1 + \cdots + (b_k - a_k)x_k\|$$

$$\leq \frac{\varepsilon}{2} + \max_{j=1,\ldots,k} |b_j - a_j| (\|x_1\| + \cdots + \|x_k\|)$$

$$< \frac{\varepsilon}{2} + \frac{\varepsilon}{2\left(1 + \sum_{i=1}^{k} \|x_i\|\right)} \cdot \sum_{i=1}^{k} \|x_i\| \leq \frac{\varepsilon}{2} + \frac{\varepsilon}{2} = \varepsilon,$$

which proves that $\overline{B} = E$. Thus $B \subseteq E$ is countable and dense in $E$, completing the proof that $E$ is separable. ∎

**Example 1.6.4** $c_0$ and $\ell_p$, $1 \leq p < \infty$, are separable. For each $n \in \mathbb{N}$, consider

$$e_n = (0, 0, \ldots, 0, 1, 0, 0, \ldots)$$

the sequence formed by 1 in the $n$-th coordinate and 0 in the other coordinates. The vectors $e_1, e_2, \ldots$ are called *canonical unit vectors of sequence spaces*. Given $x = (a_j)_{j=1}^{\infty} \in c_0$, we have

$$\lim_k \left\| x - \sum_{j=1}^{k} a_j e_j \right\|_\infty = \lim_k \|(0, 0, \ldots, 0, a_{k+1}, a_{k+2}, \ldots)\|_\infty = \lim_k \sup_{j > k} |a_j| = 0,$$

because $a_j \xrightarrow{j \to \infty} 0$. It follows that $\sum_{j=1}^{k} a_j e_j \xrightarrow{k \to \infty} x$ in $c_0$. As $\sum_{j=1}^{k} a_j e_j \in [e_1, e_2, \ldots]$ for every $k$, from Lemma 1.6.3 it follows that $c_0$ is separable. Given $x = (a_j)_{j=1}^{\infty} \in \ell_p$, as

$$\lim_k \left\| x - \sum_{j=1}^{k} a_j e_j \right\|_p^p$$

$$= \lim_k \|(0, 0, \ldots, 0, a_{k+1}, a_{k+2}, \ldots)\|_p^p = \lim_k \sum_{j=k+1}^{\infty} |a_j|^p = 0,$$

because the series $\sum_{j=1}^{\infty} |a_j|^p$ is convergent, the same argument shows that $\ell_p$ is separable.

According to the terminology to be introduced in Definition 5.3.7, we showed above that $x = \sum_{j=1}^{\infty} a_j e_j$ in $c_0$ and in $\ell_p$. In the case $p = 2$, this series is called the *Fourier series of $x$* (for further details, see Sect. 5.7).

**Example 1.6.5** $\ell_\infty$ is not separable. Suppose that $\ell_\infty$ contains a dense sequence $(x_n)_{n=1}^{\infty}$. For each $n \in \mathbb{N}$, consider $x_n = (a_j^{(n)})_{j=1}^{\infty}$. Let $y = (b_j)_{j=1}^{\infty}$ be the sequence defined by $b_j = 0$ if $|a_j^{(j)}| \geq 1$; and $b_j = a_j^{(j)} + 1$ if $|a_j^{(j)}| < 1$. Since $|b_j| < 2$ for every $j$, we have $y \in \ell_\infty$. From the denseness of the sequence $(x_n)_{n=1}^{\infty}$ there exists $n_0 \in \mathbb{N}$ such that $\|y - x_{n_0}\|_\infty < 1$. But

$$\|y - x_n\|_\infty = \sup\{|b_1 - a_1^{(n)}|, |b_2 - a_2^{(n)}|, \ldots, |b_n - a_n^{(n)}|, |b_{n+1} - a_{n+1}^{(n)}|, \ldots\}$$

$$\geq |b_n - a_n^{(n)}|$$

$$= \begin{cases} |0 - a_n^{(n)}| & \text{if } |a_n^{(n)}| \geq 1 \\ |a_n^{(n)} + 1 - a_n^{(n)}| & \text{if } |a_n^{(n)}| < 1 \end{cases} = \begin{cases} |a_n^{(n)}| & \text{if } |a_n^{(n)}| \geq 1 \\ 1 & \text{if } |a_n^{(n)}| < 1 \end{cases} \geq 1$$

for every $n \in \mathbb{N}$. This contradiction proves that $\ell_\infty$ is not separable.

To prove the separability of the space $C[a,b]$ we need a classic theorem whose real case is well known and proved in books of Real Analysis:

**Theorem 1.6.6 (Weierstrass Approximation Theorem)** *Let $f : [a,b] \longrightarrow \mathbb{K}$ be a continuous function. Then for every $\varepsilon > 0$ there is a polynomial $P : \mathbb{K} \longrightarrow \mathbb{K}$ such that $|P(x) - f(x)| < \varepsilon$ for every $x \in [a,b]$.*

**Example 1.6.7** $C[a,b]$ is separable. For each $n \in \mathbb{N}$ consider the function $t \in [a,b] \mapsto f_n(t) = t^n \in \mathbb{K}$. Taking $A = \{f_n : n \in \mathbb{N}\}$ we have $A \subseteq C[a,b]$ and $[A]$ is the set of all polynomials. It follows from Theorem 1.6.6 that $\overline{[A]} = C[a,b]$, therefore $C[a,b]$ is separable by Lemma 1.6.3.

**Example 1.6.8** $L_p[a,b]$, $1 \leq p < \infty$, is separable. Let $f \in L_p[a,b]$ and $\varepsilon > 0$ be given. As we will see in Theorem 1.7.1, the set of continuous functions is dense in $L_p[a,b]$, so there is $g \in C[a,b]$ such that $\|f - g\|_p < \frac{\varepsilon}{2}$. From Theorem 1.6.6 there is a polynomial $P$ such that $\|g - P\|_\infty < \frac{\varepsilon}{2(b-a)}$. Then

$$\|f - P\|_p \leq \|f - g\|_p + \|g - P\|_p < \frac{\varepsilon}{2} + (b-a)\|g - P\|_\infty < \varepsilon,$$

which proves that the set of polynomials is dense in $L_p[a,b]$. The separability of $L_p[a,b]$ now follows exactly as in Example 1.6.7.

From the theory of metric spaces it is known that every subset of a separable metric space is also separable; we thus conclude the following:

**Proposition 1.6.9** *Every subspace of a separable normed space is also separable.*

## 1.7 Comments and Historical Notes

The axioms of vector space, over the real numbers, were first established by the Italian mathematician G. Peano in 1888. The abstract notion of field appeared with H. Weber in 1893. The first book on vector spaces, written by the also Italian mathematician S. Pincherle in collaboration with his student U. Amaldi in 1901, already worked with some function spaces. As Calculus deals with functions defined on subsets of $\mathbb{K}$ and his book dealt with functions defined on sets of functions, Pincherle called his study *Functional Calculus*. In 1903, J. Hadamard began to call a linear transformation taking values in the scalar field a *functional*, and the name of our discipline, *Functional Analysis*, appeared for the first time in a book by P. Lévy published in 1922.

The norm axioms appeared in the years 1920–1922 in works of N. Wiener, E. Helly, S. Banach, and H. Hahn. Wiener's axioms were confusing, Helly did not require the norm to be defined on all elements of the space, and Hahn included many additional requirements, including completeness. In a remarkable work of abstraction inspired by concrete problems studied by Volterra and Hadamard, Banach presented the axioms as we know them today in his thesis [5], defended in 1920 and published in 1922. Banach and his Polish colleagues referred to complete normed spaces as *type B spaces*. The term *Banach space* was coined by the French mathematician M. Fréchet in 1928. Details can be found in [18] and [12].

Hölder's inequality, for sequences, was actually first proven by L. C. Rogers in 1888. Hölder's proof appeared in 1889. The inequality (1.2) is called *Young's Inequality*.

As continuous functions on $[a, b]$ are Riemann-integrable and Riemann-integrable functions are Lebesgue-integrable, it is clear that the set of continuous functions is a subspace of $L_p[a, b]$. Moreover:

**Theorem 1.7.1** *For $1 \leq p < \infty$, $C[a, b]$ is dense in $L_p[a, b]$.*

For the proof see [56, Theorem 3.14] or [54, Theorem 7.4.12].

For $0 < p < 1$, the definitions of $L_p(X, \Sigma, \mu)$ and $\|\cdot\|_p$ are the same as for $p \geq 1$. In this case, $L_p(X, \Sigma, \mu)$ is still a vector space, the difference is that $\|\cdot\|_p$ is no longer a norm, as the triangle inequality does not hold. Interestingly, the triangle inequality not only does not hold, but it holds in reverse for non-negative functions, which gives what is known as *reverse Minkowski inequality*:

**Theorem 1.7.2** *Let $0 < p < 1$ and let $(X, \Sigma, \mu)$ be a measure space. If $f, g \in L_p(X, \Sigma, \mu)$ and $f, g \geq 0$, then*

$$\|f\|_p + \|g\|_p \le \|f + g\|_p.$$

For the proof see [11, p. 76] or [26, Proposition 5.3.1].

The spaces of differentiable functions, introduced in Example 1.1.4, as well as their analogous versions for functions of several variables $f \colon \Omega \subseteq \mathbb{R}^n \longrightarrow \mathbb{R}$, provide an appropriate framework to the study of partial differential equations.

On the other hand, in many cases we have to deal with functions that, despite not being differentiable, enjoy an appropriate modulus of continuity, as is the case with *Hölder continuous functions*:

**Definition 1.7.3** Given a real number $0 < \alpha < 1$, define

$$C^{0,\alpha}[a,b] := \{f \in C[a,b] : \exists\, C > 0 \text{ such that } |f(x) - f(y)| \le C|x - y|^\alpha$$

for all $x, y \in [a,b]\}$.

For $f \in C^{0,\alpha}[a,b]$, define

$$\|f\|_{C^{0,\alpha}} = \sup_{t \in [a,b]} |f(t)| + \sup_{x,y \in [a,b], x \ne y} \frac{|f(x) - f(y)|}{|x - y|^\alpha}.$$

It is not difficult to check that $\|\cdot\|_{C^{0,\alpha}}$ defines a norm on $C^{0,\alpha}[a,b]$ and, with some additional work, it can be proved that $(C^{0,\alpha}[a,b], \|\cdot\|_{C^{0,\alpha}})$ is a Banach space.

The limit case $\alpha = 1$ represents the space of *Lipschitz continuous functions* and is denoted by $\mathrm{Lip}[a,b]$. The case $\alpha > 1$ is uninteresting because, in this case, $C^{0,\alpha}[a,b]$ is 1-dimensional.

Many practical situations lead to the study of the spaces $C^{k,\alpha} := \{f \in C^k : D^k f \in C^{0,\alpha}\}$. As the following theorem makes clear, the spaces of Hölder continuous functions are particularly important in the theory of elliptic partial differential equations:

**Theorem 1.7.4 (Schauder Regularity Theorem)** *Let $B_1$ and $B_{1/2}$ be the open balls in $\mathbb{R}^n$ with radii $1$ and $1/2$, respectively, centered at the origin. Let $u \colon B_1 \longrightarrow \mathbb{R}$ be a function satisfying*

$$\Delta u := \sum_{i=1}^n \frac{\partial^2 u}{\partial x_i^2} = f \text{ in } B_1.$$

*If $f \in C^{0,\alpha}(B_1)$, then $u \in C^{2,\alpha}(B_{1/2})$ and $\|u\|_{C^{2,\alpha}(B_{1/2})} \le C\left(\|u\|_{L^\infty(B_1)} + \|f\|_{C^{0,\alpha}(B_1)}\right)$, where $C$ depends only on the dimension $n$.*

The theorem above, whose proof can be found in [27, Theorem 4.6], is fundamental in the study of partial differential equations and is sharp in several aspects. For example, if $f$ is merely continuous, we cannot guarantee that $u \in C^2(B_{1/2})$, which shows that the space of Hölder continuous functions is the appropriate environment for the regularity theory in second order elliptic problems.

Lemma 1.5.3 was proven by F. Riesz in 1918. In Exercise 1.8.33 the reader will check that the lemma extends to $\theta = 1$ in the case of finite-dimensional normed spaces. We will see in Exercise 1.8.34 that the same does not occur, in general, in infinite-dimensional spaces.

The term *separable space* was coined in the context of topological spaces by Fréchet in 1906, in the same article in which he introduced metric spaces. We were unable to find out the reason that led him to choose exactly this term.

## 1.8 Exercises

**Exercise 1.8.1** Complete the details of Example 1.1.2.

**Exercise 1.8.2** Show that the set

$$\{f \in C[a, b] : f(x) > 0 \text{ for every } x \in [a, b]\}$$

is open in $C[a, b]$.

**Exercise 1.8.3** Two norms $\|\cdot\|_1$ and $\|\cdot\|_2$ in a vector space $E$ are said to be *equivalent* if there exist positive constants $c_1$ and $c_2$ such that

$$c_1 \|x\|_1 \leq \|x\|_2 \leq c_2 \|x\|_1 \text{ for every } x \in E.$$

Prove that if $E$ is a finite-dimensional normed space, then any two norms on $E$ are equivalent.

**Exercise 1.8.4** Prove that the correspondence $f \in C[0, 1] \mapsto \int_0^1 |f(t)|\, dt \in \mathbb{R}$ is a norm on $C[0, 1]$ that is not equivalent to the norm $\|\cdot\|_\infty$.

**Exercise 1.8.5** Show that in any vector space $E$ over $\mathbb{K}$, a metric $d$ can be defined that is not associated with any norm on $E$ by equality (1.1).

**Exercise 1.8.6** Show that if $0 < \alpha_1 < \alpha_2 \leq 1$, then $C^{0,\alpha_2}[a, b] \subsetneq C^{0,\alpha_1}[a, b] \subsetneq C[a, b]$.

**Exercise 1.8.7** Show that it is possible to define a norm on any vector space.

**Exercise 1.8.8** If $A$ and $B$ are subsets of a normed space $E$, show that $a\overline{A} = \overline{aA}$ for every scalar $a$ and $\overline{A} + \overline{B} \subseteq \overline{A + B}$.

**Exercise 1.8.9** Let $E$ be a normed space. A subset $A \subseteq E$ is said to be *bounded* if there exists $M > 0$ such that $\|x\| \leq M$ for every $x \in A$. If $A$ is bounded in a normed space $E$, show that $\overline{A}$ is also bounded.

**Exercise 1.8.10** If $E$ is a normed vector space and $G$ is a subspace of $E$, show that the closure of $G$ in $E$ is also a subspace of $E$.

**Exercise 1.8.11** Prove that proper subspaces of normed spaces have empty interior.

**Exercise 1.8.12 (Product Space)** Let $E_1, \ldots, E_n$ be normed spaces.

(a) Prove that the expressions
$$\|(x_1, \ldots, x_n)\|_1 = \|x_1\| + \cdots + \|x_n\|,$$
$$\|(x_1, \ldots, x_n)\|_2 = \left(\|x_1\|^2 + \cdots + \|x_n\|^2\right)^{\frac{1}{2}}, \text{ and}$$
$$\|(x_1, \ldots, x_n)\|_\infty = \max\{\|x_1\|, \ldots, \|x_n\|\},$$
define equivalent norms on the Cartesian product $E_1 \times \cdots \times E_n$.

(b) Let $(x_j^1)_{j=1}^\infty, \ldots, (x_j^n)_{j=1}^\infty$ be sequences in $E_1, \ldots, E_n$, respectively. Prove that $x_j^1 \longrightarrow x_1 \in E_1, \ldots, x_j^n \longrightarrow x_n \in E_n$ if and only if $\left((x_j^1, \ldots, x_j^n)\right)_{j=1}^\infty$ converges to $(x_1, \ldots, x_n)$ in $E_1 \times \cdots \times E_n$ equipped with any (therefore all) of the norms above.

(c) Prove that $E_1 \times \cdots \times E_n$ equipped with any (therefore all) of the norms above is Banach if and only if $E_1, \ldots, E_n$ are Banach.

**Exercise 1.8.13** Let $E$ be a normed space. In the Cartesian product consider any of the norms from Exercise 1.8.12.

(a) Prove that the algebraic operations of $E$:

$$(x, y) \in E \times E \mapsto x + y \in E \text{ and } (x, a) \in E \times \mathbb{K} \mapsto ax \in E,$$

are continuous functions.

(b) Prove that the norm $x \in E \mapsto \|x\| \in \mathbb{R}$ is a continuous function.

**Exercise 1.8.14\* (Quotient Space)** Let $E$ be a normed space and let $M$ be a subspace of $E$. Given $x, y \in E$, we say that $x \sim y$ if $x - y \in M$. For each $x \in E$, we define $[x] = \{y \in E : x \sim y\}$. We also define $E/M = \{[x] : x \in E\}$. Prove that

(a) $\sim$ is an equivalence relation.
(b) The operations $[x] + [y] = [x + y]$ and $a[x] = [ax]$ are well defined and make $E/M$ a vector space.
(c) If $M$ is closed in $E$, then the expression

$$\|[x]\| = \text{dist}(x, M) = \inf\{\|x - y\| : y \in M\}$$

defines a norm on $E/M$.
(d) If $E$ is a Banach space and $M$ is closed in $E$, then $E/M$ is also a Banach space.

**Exercise 1.8.15** Let $E$ be a vector space. A subset $A \subseteq E$ is said to be *convex* if whenever $x, y \in A$, the "closed segment"

## 1.8 Exercises

$$\{ax + (1-a)y : 0 \leq a \leq 1\}$$

is entirely contained in $A$. Show that the set $\{x \in E : \|x\| \leq r\}$, called the closed ball centered at the origin with radius $r > 0$, is convex.

**Exercise 1.8.16** Show that a subset $C$ of a vector space is convex if and only if $\sum_{i=1}^{n} a_i x_i \in C$ whenever $x_1, \ldots, x_n \in C$ and $a_1, \ldots, a_n \geq 0$ satisfy $\sum_{i=1}^{n} a_i = 1$.

**Exercise 1.8.17** Let $A$ be a subset of a vector space $E$. The set conv($A$), called the *convex hull of* $A$, is defined as the intersection of all convex subsets of $E$ that contain $A$. Show that

(a) The convex hull of any subset of $E$ is a convex set.
(b) For any $A \subseteq E$, conv($A$) is precisely the set formed by the vectors of the form $\sum_{i=1}^{n} \lambda_i x_i$, where $n \in \mathbb{N}$, $\sum_{i=1}^{n} \lambda_i = 1$, with $\lambda_i \geq 0$ and $x_i \in A$ for each $i = 1, \ldots, n$.

**Exercise 1.8.18** Let $A$ and $B$ be convex compact subsets of a normed space. Prove that conv($A \cup B$) is compact.

**Exercise 1.8.19** Prove that the closure of a convex subset of a normed space is convex.

**Exercise 1.8.20** Let $1 \leq p < \infty$.

(a) Prove that the expression

$$\|f\|_p = \left( \int_a^b |f(x)|^p \, dx \right)^{\frac{1}{p}}$$

is a norm on the vector space of continuous functions $f : [a, b] \longrightarrow \mathbb{K}$.
(b) Show that the normed space from item (a) is not complete.

**Exercise 1.8.21** Show that $L_\infty(X, \Sigma, \mu) \subseteq L_1(X, \Sigma, \mu)$ if and only if $\mu(X) < \infty$.

**Exercise 1.8.22** If $(X, \Sigma, \mu)$ is a finite measure space and $1 \leq r \leq p$, then $L_p(X, \Sigma, \mu) \subseteq L_r(X, \Sigma, \mu)$ and $\|f\|_r \leq \|f\|_p \, \mu(X)^s$ where $s = \frac{1}{r} - \frac{1}{p}$.

**Exercise 1.8.23** Let $(X, \Sigma, \mu)$ be a finite measure space and let $f \in L_\infty(X, \Sigma, \mu)$ be given. Show that $f \in L_p(X, \Sigma, \mu)$ for each $p \geq 1$ and that $\lim_{p \to \infty} \|f\|_p = \|f\|_\infty$.

**Exercise 1.8.24 (Generalized Hölder's Inequality)** Let $n \in \mathbb{N}$ and let $p, q, s > 0$ be such that $\frac{1}{p} + \frac{1}{q} = \frac{1}{s}$. Prove that

$$\left( \sum_{j=1}^{n} |a_j b_j|^s \right)^{\frac{1}{s}} \leq \left( \sum_{j=1}^{n} |a_j|^p \right)^{\frac{1}{p}} \cdot \left( \sum_{j=1}^{n} |b_j|^q \right)^{\frac{1}{q}}$$

for any scalars $a_1, \ldots, a_n, b_1, \ldots, b_n$.

**Exercise 1.8.25** Show, without passing through the space $L_\infty(X, \Sigma, \mu)$, that $\ell_\infty$ is a Banach space.

**Exercise 1.8.26** An element $(a_j)_{j=1}^\infty \in \ell_\infty$ is called a simple function if $\{a_j : j \in \mathbb{N}\}$ is finite. Show that the set of simple functions is a dense subset of $\ell_\infty$.

**Exercise 1.8.27** Consider

$$c = \left\{ (a_j)_{j=1}^\infty : a_j \in \mathbb{K} \text{ for every } j \in \mathbb{N} \text{ and } \lim_{j \to \infty} a_j \text{ exists in } \mathbb{K} \right\},$$

the set of all convergent sequences consisting of elements of $\mathbb{K}$. Show that $c$ is a closed subspace of $\ell_\infty$.

**Exercise 1.8.28*** Let $K$ be a relatively compact subset, that is $\overline{K}$ is compact, of $\ell_1$. Prove that $K$ is bounded and that for every $\varepsilon > 0$ there exists $n_\varepsilon \in \mathbb{N}$ such that

$$\sup_{(a_n)_{n=1}^\infty \in K} \left( \sum_{n \geq n_\varepsilon} |a_n| \right) \leq \varepsilon.$$

**Exercise 1.8.29** Determine $\overline{c_{00}}$ in $c_0$.

**Exercise 1.8.30** Determine $\overline{c_{00}}$ in $\ell_p$.

**Exercise 1.8.31** Show that $\{e_1, e_2, \ldots\}$ is not an algebraic basis (or Hamel basis) of $\ell_p$, $1 \leq p \leq \infty$.

**Exercise 1.8.32** Prove if true or give a counterexample if false: if $K$ is a closed subset of a normed space, then the subspace $[K]$ generated by $K$ is also closed.

**Exercise 1.8.33** Let $E$ be a finite-dimensional normed vector space and let $M$ be a proper subspace of $E$. Show that there exists $y_0 \in E$ with $\|y_0\| = 1$ and $\|y_0 - x\| \geq 1$ for every $x \in M$.

**Exercise 1.8.34 (Limitation of Riesz's Lemma)** Let $E = \{f \in C[0,1] : f(0) = 0\}$ and $F = \left\{ f \in E : \int_0^1 f(t)dt = 0 \right\}$.

(a) Prove that $E$ is a closed subspace of $C[0, 1]$.
(b) Prove that $F$ is a closed subspace of $E$.
(c) Show that there does not exist $g \in E$ such that $\|g\| = 1$ and $\|g - f\| \geq 1$ for every $f \in F$.

**Exercise 1.8.35* (The Hilbert Cube)** Call *Hilbert cube* the subset of $\ell_2$ consisting of the sequences $(a_j)_{j=1}^\infty$ such that $|a_j| \leq \frac{1}{j}$ for every $j \in \mathbb{N}$. Prove that the Hilbert cube is compact in $\ell_2$.

**Exercise 1.8.36** Show that the following are equivalent for a Banach space $E$:

(a) $E$ is separable.
(b) The closed unit ball $B_E = \{x \in E : \|x\| \leq 1\}$ is separable.
(c) The unit sphere $S_E = \{x \in E : \|x\| = 1\}$ is separable.

**Exercise 1.8.37** Prove that $L_\infty[a, b]$ is not separable.

**Exercise 1.8.38** Let $E$ be a separable Banach space and let $F$ be a closed subspace of $E$. Prove that $E/F$ is separable.

**Exercise 1.8.39*** (**Reciprocal of Hölder's Inequality**) Let $(X, \Sigma, \mu)$ be a $\sigma$-finite measure space, and let $p \geq 1$ and $q > 1$ be such that $\frac{1}{p} + \frac{1}{q} = 1$. Let $S$ be the set of all simple measurable functions $f$ that are null outside a set of finite measure, that is, there exists a set $A \in \Sigma$ such that $\mu(A) < \infty$ and $f(x) = 0$ for every $x \in X - A$. Let $g : X \longrightarrow \mathbb{R}$ be a measurable function such that:

(i) $fg \in L_1(X, \Sigma, \mu)$ for every $f \in S$,
(ii) $\sup\{|\int_X fg| : f \in S \text{ and } \|f\|_p = 1\} < \infty$.

Prove that $g \in L_q(X, \Sigma, \mu)$ and $\|g\|_q = \sup\{|\int_X fg| : f \in S \text{ and } \|f\|_p = 1\}$.

# Chapter 2
# Continuous Linear Operators

Normed spaces have an algebraic structure of a vector space to which linear transformations are associated, and a topological structure of a metric space to which continuous functions are associated. Thus, the morphisms between normed spaces are functions that are simultaneously linear and continuous. Such functions are usually called continuous linear operators. Therefore, a *continuous linear operator* from the normed space $E$ to the normed space $F$, both over the same field $\mathbb{K}$, is a function $T \colon E \longrightarrow F$, which is linear, that is

- $T(x + y) = T(x) + T(y)$ for any $x, y \in E$ and
- $T(ax) = aT(x)$ for all $a \in \mathbb{K}$ and $x$ in $E$;

and continuous, that is, for all $x_0 \in E$ and $\varepsilon > 0$ there exists $\delta > 0$ such that $\|T(x) - T(x_0)\| < \varepsilon$ whenever $x \in E$ and $\|x - x_0\| < \delta$.

The set of all continuous linear operators from $E$ to $F$ will be denoted by $\mathcal{L}(E, F)$. It is clear that $\mathcal{L}(E, F)$ is a vector space over $\mathbb{K}$ with the usual operations of functions. When $F$ is the scalar field, we write $E'$ instead of $\mathcal{L}(E, \mathbb{K})$, we call this space the *topological dual of* $E$, or simply the *dual of* $E$, and we say that its elements are *continuous linear functionals*.

Following the line that the morphisms between normed spaces are the continuous linear operators, we say that two normed spaces $E$ and $F$ are *topologically isomorphic*, or simply *isomorphic*, if there exists a bijective continuous linear operator $T \colon E \longrightarrow F$ whose inverse operator $T^{-1} \colon F \longrightarrow E$—which is always linear—is also continuous. Such operator $T$ is called a *topological isomorphism*, or simply *isomorphism*.

A function $f \colon E \longrightarrow F$—not necessarily linear—such that $\|f(x)\| = \|x\|$ for every $x \in E$ is called an *isometry*. A linear operator $T \colon E \longrightarrow F$ that is an isometry is called a *linear isometry*. Note that every linear isometry is injective and continuous. An isomorphism that is also an isometry is called an *isometric isomorphism*, and in this case we say that the spaces are *isometrically isomorphic*.

## 2.1 Characterizations of Continuous Linear Operators

We start off by recalling two important classes of functions in the context of metric spaces: a function $f \colon M \longrightarrow N$ between metric spaces is

- *Lipschitz* if there exists a constant $L > 0$ such that $d(f(x), f(y)) \leq L \cdot d(x, y)$ for all $x, y \in M$;
- *uniformly continuous* if for every $\varepsilon > 0$ there exists $\delta > 0$ such that $d(f(x), f(y)) < \varepsilon$ whenever $x, y \in M$ and $d(x, y) < \delta$.

It is known that for functions between metric spaces, the implications
Lipschitz $\Longrightarrow$ uniformly continuous $\Longrightarrow$ continuous $\Longrightarrow$ continuous at a point
are true and that, in general, all the reverse implications are false. The next result, which dates back to the works of F. Riesz, shows that all these concepts are equivalent in the context of linear operators between normed spaces. That is, linearity, or ultimately the algebraic structure, simplifies the topological behavior. In addition, we will prove other equivalences that will be very useful throughout the text.

**Theorem 2.1.1** *Let $E$ and $F$ be normed spaces over $\mathbb{K}$ and let $T \colon E \longrightarrow F$ be linear. The following conditions are equivalent:*

(a) *$T$ is Lipschitz.*
(b) *$T$ is uniformly continuous.*
(c) *$T$ is continuous.*
(d) *$T$ is continuous at some point of $E$.*
(e) *$T$ is continuous at the origin.*
(f) $\sup\{\|T(x)\| : x \in E \text{ and } \|x\| \leq 1\} < \infty.$
(g) *There exists a constant $C \geq 0$ such that $\|T(x)\| \leq C\|x\|$ for every $x \in E$.*

**Proof** The implications (a) $\Longrightarrow$ (b) $\Longrightarrow$ (c) $\Longrightarrow$ (d) hold in the context of metric spaces, that is, they do not depend on the linearity of $T$.

(d) $\Longrightarrow$ (e) Suppose $T$ is continuous at $x_0 \in E$. Let $\varepsilon > 0$ be given. Then there exists $\delta > 0$ such that $\|T(x) - T(x_0)\| < \varepsilon$ whenever $\|x - x_0\| < \delta$. Let $x \in E$ be such that $\|x - 0\| = \|x\| < \delta$. Then $\|(x + x_0) - x_0\| = \|x\| < \delta$. Therefore

$$\|T(x) - T(0)\| = \|T(x) - 0\| = \|T(x)\| = \|T(x) + T(x_0) - T(x_0)\|$$
$$= \|T(x + x_0) - T(x_0)\| < \varepsilon,$$

proving that $T$ is continuous at the origin.

(e) $\Longrightarrow$ (f) From the continuity of $T$ at the origin there exists $\delta > 0$ such that $\|T(x)\| < 1$ whenever $\|x\| < \delta$. If $\|x\| \leq 1$, then $\frac{\delta}{2}\|T(x)\| = \|T(\frac{\delta}{2}x)\| < 1$. This proves that $\sup\{\|T(x)\| : x \in E \text{ and } \|x\| \leq 1\} \leq \frac{2}{\delta} < \infty$.

(f) $\Longrightarrow$ (g) For $x \in E$, $x \neq 0$, we have

## 2.1 Characterizations of Continuous Linear Operators

$$\frac{\|T(x)\|}{\|x\|} = \left\|T\left(\frac{x}{\|x\|}\right)\right\| \leq \sup\{\|T(y)\| : \|y\| \leq 1\},$$

therefore $\|T(x)\| \leq (\sup\{\|T(y)\| : \|y\| \leq 1\})\|x\|$ for every $x \neq 0$. The result follows because this inequality is trivially checked for $x = 0$.

(g) $\Longrightarrow$(a) Given $x_1, x_2 \in E$,

$$\|T(x_1) - T(x_2)\| = \|T(x_1 - x_2)\| \leq C \|x_1 - x_2\|,$$

so $T$ is Lipschitz with constant $C$. ∎

**Corollary 2.1.2** *Let $T: E \longrightarrow F$ be a bijective linear operator between normed spaces. Then $T$ is an isomorphism if and only if there exist constants $C_1, C_2 > 0$ such that $C_1\|x\| \leq \|T(x)\| \leq C_2\|x\|$ for every $x \in E$.*

**Remark 2.1.3** From condition (f) of Theorem 2.1.1, it follows that continuous linear operators are exactly the ones that transform bounded subsets of the domain space into bounded subsets of the target space. Consequently, these operators are usually referred to as *bounded linear operators*. However, we refrain from adopting this terminology due to its inconsistency with the notion of a bounded function from Example 1.1.2. In that example, a function is deemed bounded if its range is bounded, whereas an operator is considered bounded if it transforms bounded sets into bounded sets. To avoid ambiguity, we will continue using the term "continuous linear operator" and keeping the notion of bounded function as one having a bounded range.

On the other hand, the same condition (f) teaches us how to define a norm on the space $\mathcal{L}(E, F)$ of continuous linear operators from $E$ to $F$:

**Proposition 2.1.4** *Let $E$ and $F$ be normed spaces.*

(a) *The expression*

$$\|T\| = \sup\{\|T(x)\| : x \in E \text{ and } \|x\| \leq 1\}$$

*defines a norm on the space $\mathcal{L}(E, F)$.*
(b) $\|T(x)\| \leq \|T\| \cdot \|x\|$ *for all $T \in \mathcal{L}(E, F)$ and $x \in E$.*
(c) *If $F$ is Banach, then $\mathcal{L}(E, F)$ is also Banach.*

**Proof** We leave (a) and (b) to the reader. We prove (c). Let $(T_n)_{n=1}^{\infty}$ be a Cauchy sequence in $\mathcal{L}(E, F)$. Given $\varepsilon > 0$ there exists $n_0 \in \mathbb{N}$ such that $\|T_n - T_m\| \leq \varepsilon$ whenever $n, m \geq n_0$. Hence

$$\|T_n(x) - T_m(x)\| = \|(T_n - T_m)(x)\| \leq \|T_n - T_m\| \cdot \|x\| \leq \varepsilon \|x\| \quad (2.1)$$

for all $x \in E$ and $n, m \geq n_0$. It follows that for each $x \in E$, the sequence $(T_n(x))_{n=1}^{\infty}$ is a Cauchy sequence in $F$, hence convergent because $F$ is Banach. Define the linear operator

$$T : E \longrightarrow F, \ T(x) = \lim_{n \to \infty} T_n(x).$$

Allowing $m \longrightarrow \infty$ in (2.1) we get

$$\|(T_n - T)(x)\| = \|T_n(x) - T(x)\| \leq \varepsilon \|x\| \qquad (2.2)$$

for all $x \in E$ and $n \geq n_0$. In particular,

$$\|(T_{n_0} - T)(x)\| = \|T_{n_0}(x) - T(x)\| \leq \varepsilon \|x\|$$

for every $x \in E$, which assures that $(T - T_{n_0}) \in \mathcal{L}(E, F)$. Therefore $T = (T - T_{n_0}) + T_{n_0} \in \mathcal{L}(E, F)$. From (2.2) it also follows that $\|T_n - T\| \leq \varepsilon$ for every $n \geq n_0$, which proves that $T_n \longrightarrow T$ in $\mathcal{L}(E, F)$. ∎

In Exercise 2.7.8 the reader will prove several alternative expressions for $\|T\|$. One that will be especially useful is the following:

$$\|T\| = \inf\{C : \|T(x)\| \leq C \|x\| \text{ for every } x \in E\}.$$

From Proposition 2.1.4(c) we have

**Corollary 2.1.5** *The dual $E'$ of any normed space $E$ is a Banach space.*

## 2.2 Examples

It is clear that in any normed space $E$, the identity operator on $E$ is a norm 1 continuous linear operator, and that between any two normed spaces, the null operator is a norm 0 continuous linear operator. It is also very simple to check that if $\varphi \in E'$ and $y \in F$, then

$$\varphi \otimes y \colon E \longrightarrow F, \ \varphi \otimes y(x) = \varphi(x) y,$$

is a continuous linear operator and its norm is $\|\varphi\| \cdot \|y\|$. Next we present some examples in specific spaces:

**Example 2.2.1** Let $1 \leq p < \infty$ be given. Choose a sequence $(b_j)_{j=1}^\infty \in \ell_p$ and consider the operator

$$T \colon \ell_\infty \longrightarrow \ell_p, \ T((a_j)_{j=1}^\infty) = (a_j b_j)_{j=1}^\infty.$$

The linearity of $T$ is immediate. From

## 2.3 The Banach–Steinhaus Theorem

$$\left\|T((a_j)_{j=1}^\infty)\right\|_p = \left(\sum_{j=1}^\infty |a_j b_j|^p\right)^{1/p} \leq \left(\sum_{j=1}^\infty |b_j|^p\right)^{1/p} \cdot \sup_j |a_j|$$

$$= \left\|(b_j)_{j=1}^\infty\right\|_p \cdot \left\|(a_j)_{j=1}^\infty\right\|_\infty$$

for every sequence $(a_j)_{j=1}^\infty \in \ell_\infty$, we conclude that $T$ is continuous and that $\|T\| \leq \|(b_j)_{j=1}^\infty\|_p$. Taking $x = (1, 1, 1, \ldots,) \in \ell_\infty$ we have $\|x\|_\infty = 1$ and $\|T(x)\| = \|(b_j)_{j=1}^\infty\|_p$, therefore $\|T\| = \|(b_j)_{j=1}^\infty\|_p$. This operator $T$ is called a *diagonal operator* by the sequence $(b_j)_{j=1}^\infty$.

**Example 2.2.2** Transferring the example above to function spaces, consider $1 \leq p < \infty$ and choose a function $g \in L_p[0, 1]$. Just like before, the operator

$$T: C[0, 1] \longrightarrow L_p[0, 1], \ T(f) = fg,$$

is linear, continuous and $\|T\| = \|g\|_p$. In this case $T$ is called a *multiplication operator* by the function $g$.

We will see next, and even more in Exercise 2.7.3, that not all linear operators on infinite-dimensional spaces are continuous. This constitutes the fundamental distinction between Linear Algebra and Functional Analysis. If the reader suspects that for a linear operator to be discontinuous it must be something very complicated, the reader will certainly be surprised by the simplicity of the following example.

**Example 2.2.3** Consider the subset $\mathcal{P}[0, 1]$ of $C[0, 1]$ consisting of polynomial functions (or polynomials). It is clear that $\mathcal{P}[0, 1]$ is a subspace of $C[0, 1]$, therefore it is a normed space with the norm $\|\cdot\|_\infty$ inherited from $C[0, 1]$. It is also clear that the derivative operator

$$T: \mathcal{P}[0, 1] \longrightarrow \mathcal{P}[0, 1], \ T(f) = f' = \text{derivative of } f,$$

is linear. Suppose that $T$ is continuous. In this case there exists $C$ such that $\|T(f)\|_\infty \leq C\|f\|_\infty$ for every polynomial $f \in \mathcal{P}[0, 1]$. For each $n \in \mathbb{N}$ let $f_n \in \mathcal{P}[0, 1]$ be the function given by $f_n(t) = t^n$. Then

$$n = \|f_n'\|_\infty = \|T(f_n)\|_\infty \leq C\|f_n\|_\infty = C$$

for every $n \in \mathbb{N}$, and this sets up a contradiction; therefore $T$ is discontinuous.

## 2.3 The Banach–Steinhaus Theorem

One of the main reasons for the success of Functional Analysis is the existence of a set of structural theorems that, while making the theory flexible and allowing its interaction with several other areas, make it full of interesting examples. We will begin to describe and prove these theorems in this section and will only finish in the next chapter with the Hahn–Banach Theorem and its applications.

Many of the main theorems in Analysis are associated with some type of uniform control based on pointwise hypotheses. A classic example is the fact that continuous functions on compact sets are uniformly continuous. The main result of this section, the Banach–Steinhaus Theorem, ensures that a family of continuous linear operators is uniformly bounded whenever it is pointwisely bounded. The significant consequences of this theorem will unfold later in this chapter and in subsequent passages of the book. For its proof we need the following classic from the topology of metric spaces:

**Theorem 2.3.1 (Baire's Theorem)** *Let $(M, d)$ be a complete metric space and let $(F_n)_{n=1}^{\infty}$ be a sequence of closed subsets of $M$ such that $M = \bigcup_{n=1}^{\infty} F_n$. Then there exists $n_0 \in \mathbb{N}$ such that $F_{n_0}$ has non-empty interior.*

**Proof** Denote by $\text{int}(A)$ the interior of a subset $A$ of $M$. Assume, for the sake of contradiction, that $\text{int}(F_n) = \emptyset$ for every $n$. Calling $A_n = (F_n)^c = (M - F_n)$, each $A_n$ is open and

$$\overline{A_n} = \overline{(F_n)^c} = (\text{int}(F_n))^c = \emptyset^c = M$$

for every $n$. In particular, each $A_n$ is non-empty. Choose $x_1 \in A_1$ and use the fact that $A_1$ is open to guarantee the existence of $0 < \delta_1 < 1$ such that the closed ball of center $x_1$ and radius $\delta_1$, denoted by $B[x_1; \delta_1]$, is contained in $A_1$. From $\overline{A_2} = M$ we have $A_2 \cap B(x_1; \delta_1) \neq \emptyset$, where $B(x_1; \delta_1)$ denotes the open ball of center $x_1$ and radius $\delta_1$. And as $A_2 \cap B(x_1; \delta_1)$ is open, there exist $x_2 \in A_2$ and $0 < \delta_2 < \frac{1}{2}$ such that $B[x_2; \delta_2] \subseteq A_2 \cap B(x_1; \delta_1) \subseteq A_2 \cap B[x_1; \delta_1]$. Continuing the process, we construct a sequence $(x_n)_{n=1}^{\infty}$ in $M$ and a sequence $(\delta_n)_{n=1}^{\infty}$ of real numbers such that

$$0 < \delta_n < \frac{1}{n} \text{ and } B[x_{n+1}; \delta_{n+1}] \subseteq A_{n+1} \cap B[x_n; \delta_n]$$

for every $n$. Given $\varepsilon > 0$ choose $n_0 \in \mathbb{N}$ such that $n_0 > \frac{2}{\varepsilon}$. If $m, n \geq n_0$, as $B[x_n; \delta_n] \cap B[x_m; \delta_m] \subseteq B[x_{n_0}; \delta_{n_0}]$, we have

$$d(x_m, x_n) \leq d(x_m, x_{n_0}) + d(x_n, x_{n_0}) \leq 2\delta_{n_0} < \frac{2}{n_0} < \varepsilon.$$

This shows that $(x_n)_{n=1}^{\infty}$ is Cauchy in $M$, therefore convergent. Call $x = \lim_{k \to \infty} x_k$ and fix $n \in \mathbb{N}$. Since $x_m \in B[x_m; \delta_m] \subseteq B[x_n; \delta_n]$ for all $m \geq n$ and $B[x_n; \delta_n]$ is a closed set, it follows that $x \in B[x_n; \delta_n] \subseteq A_n$. Therefore

$$x \in \bigcap_{n=1}^{\infty} A_n = \bigcap_{n=1}^{\infty} (F_n)^c = \left(\bigcup_{n=1}^{\infty} F_n\right)^c = M^c = \emptyset,$$

## 2.3 The Banach–Steinhaus Theorem

a contradiction that completes the proof. ∎

**Theorem 2.3.2 (Banach–Steinhaus Theorem)** *Let $E$ be a Banach space, let $F$ be a normed space and let $(T_i)_{i \in I}$ be a family of operators in $\mathcal{L}(E; F)$ satisfying the condition that for each $x \in E$ there exists $C_x < \infty$ such that*

$$\sup_{i \in I} \|T_i(x)\| < C_x. \tag{2.3}$$

*Then $\sup_{i \in I} \|T_i\| < \infty$.*

**Proof** From the continuity of $T_i$ it follows that the set $\{x \in E : \|T_i(x)\| \leq n\} = (\|\cdot\| \circ T_i)^{-1}([0, n])$ is closed for each $n \in \mathbb{N}$ and each $i \in I$. Then the set

$$A_n := \left\{ x \in E : \sup_{i \in I} \|T_i(x)\| \leq n \right\} = \bigcap_{i \in I} \{x \in E : \|T_i(x)\| \leq n\}$$

is closed as it is an intersection of closed sets. From (2.3) it follows that $E = \bigcup_{n=1}^{\infty} A_n$ and, by Theorem 2.3.1, some $A_n$ has non-empty interior. Let $n_0$ be a natural number such that $A_{n_0}$ has non-empty interior, and let $a \in \mathrm{int}(A_{n_0})$ and $r > 0$ be such that $\{x \in E : \|x - a\| \leq r\} \subseteq \mathrm{int}(A_{n_0})$. Let $y \in E$ be so that $\|y\| \leq 1$. If $x = a + ry$, then $\|x - a\| = \|ry\| \leq r$, therefore $x \in A_{n_0}$. It follows that

$$\|T_i(x - a)\| \leq \|T_i(x)\| + \|T_i(a)\| \leq n_0 + n_0$$

for each $i \in I$. Therefore $\|T_i(ry)\| = \|T_i(x - a)\| \leq 2n_0$ and $\|T_i(y)\| \leq \frac{2n_0}{r}$ for each $i$ in $I$. Hence $\sup_{i \in I} \|T_i\| \leq \frac{2n_0}{r}$. ∎

From the course on Real Analysis we know that the uniform limit of continuous functions is continuous, while the pointwise limit of continuous functions may not be continuous. We will see below that, in the presence of linearity, pointwise convergence is enough.

**Corollary 2.3.3** *Let $E$ be a Banach space, let $F$ be a normed space and let $(T_n)_{n=1}^{\infty}$ be a sequence in $\mathcal{L}(E; F)$ such that $(T_n(x))_{n=1}^{\infty}$ is convergent in $F$ for every $x$ in $E$. If we define*

$$T : E \longrightarrow F, \quad T(x) = \lim_{n \to \infty} T_n(x),$$

*then $T$ is a continuous linear operator.*

**Proof** The linearity of $T$ follows from the arithmetic properties of limits. By hypothesis, for each $x \in E$ the sequence $(T_n(x))_{n=1}^{\infty}$ is convergent, therefore bounded. Thus $\sup_{n \in \mathbb{N}} \|T_n(x)\| < \infty$ for every $x \in E$. By Theorem 2.3.2 there exists $c > 0$ such that $\sup_{n \in \mathbb{N}} \|T_n\| \leq c$. It follows that

$$\|T_n(x)\| \le \|T_n\| \cdot \|x\| \le c\,\|x\|$$

for all $x \in E$ and $n \in \mathbb{N}$. Letting $n \longrightarrow \infty$ we get $\|T(x)\| \le c\,\|x\|$ for every $x \in E$, which completes the proof that $T$ is continuous. ∎

To appreciate the next application of the Banach–Steinhaus Theorem, we recall a classic example from Calculus: the function

$$f: \mathbb{R}^2 \longrightarrow \mathbb{R}, \ f(x,y) = \frac{xy}{x^2+y^2}, \ f(0,0) = 0,$$

is separately continuous, that is, $f(x,\cdot): \mathbb{R} \longrightarrow \mathbb{R}$ and $f(\cdot,y): \mathbb{R} \longrightarrow \mathbb{R}$ are continuous functions for any fixed $x, y$. However, $f$ is not continuous. In fact, the limit $\lim_{(x,y)\to(0,0)} f(x,y)$ does not exist, as for $y = tx$, we have $f(x,tx) = \frac{t}{1+t^2}$ whenever $x \ne 0$. We will check next that for *bilinear* operators, the phenomenon above does not occur, even in infinite dimension.

**Definition 2.3.4** Let $E_1$, $E_2$ and $F$ be vector spaces. A map $B: E_1 \times E_2 \longrightarrow F$ is said to be a *bilinear operator* if $B(x_1,\cdot): E_2 \longrightarrow F$ and $B(\cdot,x_2): E_1 \longrightarrow F$ are linear operators for any fixed $x_1 \in E_1$ and $x_2 \in E_2$.

**Corollary 2.3.5** *Let $E_1$, $E_2$ and $F$ be normed vector spaces with $E_2$ complete and let $B: E_1 \times E_2 \to F$ be a bilinear operator. Suppose that $B$ is separately continuous, meaning that $B(x,\cdot): E_2 \longrightarrow F$ and $B(\cdot,y): E_1 \longrightarrow F$ are continuous linear operators for any fixed $x \in E_1$ and $y \in E_2$. Then $B: E_1 \times E_2 \longrightarrow F$ is a continuous bilinear operator.*

**Proof** Consider $\mathcal{F} = \{B(x,\cdot) : x \in E_1, \|x\| \le 1\} \subseteq \mathcal{L}(E_2, F)$. By assumption, $B(\cdot, y)$ is linear and continuous for each $y \in E_2$, therefore, for every $x \in E_1$, $\|x\| \le 1$, we have

$$\|B(x,y)\| \le \|B(\cdot,y)\|_{\mathcal{L}(E_1,F)} =: C_y.$$

Thus, the family $\mathcal{F}$ is pointwisely bounded and, by the Banach–Steinhaus Theorem, there exists a constant $C$ such that

$$\sup_{\|x\|\le 1} \sup_{\|y\|\le 1} \|B(x,y)\| < C.$$

It follows from the bilinearity of $B$ that $\|B(x,y)\| \le C\|x\| \cdot \|y\|$ for all $x \in E_1$ and $y \in E_2$. The continuity of $B$ now follows from Exercise 2.7.25. ∎

We finish this section by checking the necessity of the completeness of the domain space in the statement of the Banach–Steinhaus Theorem.

**Example 2.3.6** For each $n \in \mathbb{N}$, consider the continuous linear functional

$$\varphi_n: c_{00} \longrightarrow \mathbb{K}, \ \varphi_n((a_j)_{j=1}^\infty) = na_n.$$

It is clear that $(\varphi_n)_{n=1}^\infty \subseteq (c_{00})'$ and $\|\varphi_n\| = n$ for every $n$. Given a sequence $(a_j)_{j=1}^\infty \in c_{00}$, taking $j_0$ such that $a_j = 0$ for each $j \geq j_0$, it follows that $\varphi_n((a_j)_{j=1}^\infty) = 0$ for every $n \geq j_0$. Thus, $\sup_n |\varphi_n(x)| < \infty$ for every $x \in c_{00}$ but $\sup_n \|\varphi_n\| = \infty$.

## 2.4 The Open Mapping Theorem

The Open Mapping Theorem, another famous result from Functional Analysis, guarantees that if $E$ and $F$ are Banach spaces, then every continuous surjective linear operator $T \colon E \longrightarrow F$ is an *open mapping*, that is, $T(A)$ is open in $F$ whenever $A$ is open in $E$. To prove it, we need the following lemma. By $B_E(x_0; r)$ we denote the open ball in $E$ centered at $x_0$ with radius $r$, that is, $B_E(x_0; r) = \{x \in E : \|x - x_0\| < r\}$.

**Lemma 2.4.1** *Let $E$ be a Banach space, let $F$ be a normed space and let $T \colon E \longrightarrow F$ be a continuous linear operator. If there exist $R, r > 0$ such that $\overline{T(B_E(0; R))} \supseteq B_F(0; r)$, then $T(B_E(0; R)) \supseteq B_F\left(0; \frac{r}{2}\right)$.*

**Proof** As for every subset $M \subseteq E$ and every scalar $a \in \mathbb{K}$ we have $\overline{aM} = a\overline{M}$ (Exercise 1.8.8), it follows from the hypothesis that

$$\overline{T(B_E(0; aR))} \supseteq B_F(0; ar) \tag{2.4}$$

for every positive $a \in \mathbb{R}$. Let $y \in B_F\left(0; \frac{r}{2}\right)$. By (2.4) there exists $x_1 \in B_E\left(0; \frac{R}{2}\right)$ such that $\|y - T(x_1)\| < \frac{r}{4}$, that is, $y - T(x_1) \in B_F\left(0; \frac{r}{4}\right)$. Again by (2.4), there exists $x_2 \in B_E\left(0; \frac{R}{4}\right)$ such that

$$\|(y - T(x_1)) - T(x_2)\| < \frac{r}{8}.$$

As (2.4) holds for every $a > 0$, we can continue this procedure indefinitely to construct a sequence $(x_n)_{n=1}^\infty$ in $E$ such that $x_n \in B_E\left(0; \frac{R}{2^n}\right)$ and

$$\|y - T(x_1) - \cdots - T(x_n)\| < \frac{r}{2^{n+1}} \tag{2.5}$$

for every $n \in \mathbb{N}$. The series $\sum_{n=1}^\infty \|x_n\|$ is convergent because $\|x_n\| < \frac{R}{2^n}$ for every $n$, hence

$$\left\|\sum_{j=n}^m x_j\right\| \leq \sum_{j=n}^m \|x_j\| \stackrel{m > n \to \infty}{\longrightarrow} 0,$$

therefore $\left(\sum_{j=1}^{n} x_j\right)_{n=1}^{\infty}$ is a Cauchy sequence in $E$. As $E$ is Banach, there exists $x \in E$ for which this sequence converges. It follows that

$$\|x\| = \left\|\lim_{n\to\infty} \sum_{j=1}^{n} x_j\right\| \le \lim_{n\to\infty} \sum_{j=1}^{n} \|x_j\| = \|x_1\| + \sum_{n=2}^{\infty} \|x_n\|$$

$$< \frac{R}{2} + \sum_{n=2}^{\infty} \|x_n\| \le \frac{R}{2} + \sum_{n=2}^{\infty} \frac{R}{2^n} = \sum_{n=1}^{\infty} \frac{R}{2^n} = R,$$

therefore $x \in B_E(0; R)$. Letting $n \longrightarrow \infty$ in (2.5) we get

$$\|y - T(x)\| = \left\| y - T\left(\lim_{n\to\infty} \sum_{j=1}^{n} x_j\right) \right\|$$

$$= \lim_{n\to\infty} \|y - T(x_1) - \cdots - T(x_n)\| \le \lim_{n\to\infty} \frac{r}{2^{n+1}} = 0,$$

which gives $y = T(x)$. Hence $y \in T(B_E(0; R))$ and the result is proven. ∎

**Theorem 2.4.2 (Open Mapping Theorem)** *Let $E$ and $F$ be Banach spaces and let $T: E \longrightarrow F$ be a surjective continuous linear operator. Then $T$ is an open mapping. In particular, every continuous bijective linear operator between Banach spaces is an isomorphism.*

**Proof** From the equality $E = \bigcup_{n=1}^{\infty} B_E(0; n)$ and the surjectivity of $T$ it follows that

$$F = T(E) = \bigcup_{n=1}^{\infty} T(B_E(0; n)) = \bigcup_{n=1}^{\infty} \overline{T(B_E(0; n))}.$$

By Theorem 2.3.1 there exists $n_0 \in \mathbb{N}$ such that $\overline{T(B_E(0; n_0))}$ has non-empty interior. Then there exist $b \in F$ and $r > 0$ such that $B_F(b; r) \subseteq \overline{T(B_E(0; n_0))}$. As

$$\overline{T(B_E(0; n_0))} = -\overline{T(B_E(0; n_0))},$$

it follows that

$$B_F(-b; r) = -B_F(b; r) \subseteq \overline{T(B_E(0; n_0))}.$$

As $x = \frac{1}{2}(b + x) + \frac{1}{2}(-b + x)$,

$$B_F(0; r) \subseteq \frac{1}{2} B_F(b; r) + \frac{1}{2} B_F(-b; r)$$

## 2.4 The Open Mapping Theorem

$$\subseteq \frac{1}{2}\overline{T(B_E(0; n_0))} + \frac{1}{2}\overline{T(B_E(0; n_0))} = \overline{T(B_E(0; n_0))},$$

where the last equality follows from the convexity of the set $\overline{T(B_E(0; n_0))}$. By Lemma 2.4.1 we know that $T(B_E(0; n_0)) \supseteq B_F(0; \rho)$ for $\rho = \frac{r}{2}$, therefore $T(B_E(0; cn_0)) \supseteq B_F(0; c\rho)$ for every $c > 0$. Note that

$$T(B_E(x; cn_0)) \supseteq B_F(T(x); c\rho)$$

for every $x \in E$ and every $c > 0$. In fact, $B_E(x; cn_0) = x + B_E(0, cn_0)$, so

$$T(B_E(x; cn_0)) = T(x) + T(B_E(0; cn_0)) \supseteq T(x) + B_F(0; c\rho) = B_F(T(x); c\rho).$$

Now we can prove that $T(U)$ is open in $F$ for each $U$ open in $E$. Let $x \in U$ and $c > 0$ be such that $B_E(x, cn_0) \subseteq U$. Then $T(U) \supseteq T(B_E(x; cn_0)) \supseteq B_F(T(x); c\rho)$, proving that $T(U)$ is open in $F$.

The second statement is an immediate consequence of the first one. ∎

The surjectivity of the operator is an essential hypothesis in the Open Mapping Theorem because every open linear map $T \colon E \longrightarrow F$ between normed spaces is necessarily surjective. Indeed, given an open ball centered at the origin of $E$, its image by $T$, being open, contains a ball centered at the origin of $F$. The surjectivity follows easily from the linearity of $T$.

The next example shows that the completeness of the underlying spaces is also essential.

**Example 2.4.3** Let $T \colon c_{00} \longrightarrow c_{00}$ be the linear operator given by $T((a_n)_{n=1}^\infty) = (a_1, \frac{a_2}{2}, \frac{a_3}{3}, \ldots)$. In Exercise 2.7.27 the reader will check that $T$ is linear, bijective and continuous, but that the inverse operator $T^{-1}$ is not continuous.

**Example 2.4.4 (Closed Subspaces of $C[0, 1]$)** We will show that every infinite-dimensional closed subspace $F$ of $C[0, 1]$ contains a function that does not belong to $C^1[0, 1]$. This proves, in particular, that $C^1[0, 1]$ is not closed in $C[0, 1]$. Suppose, for the purpose of contradiction, that $F \subseteq C^1[0, 1]$. As $F$ is closed in $C[0, 1]$, by Proposition 1.1.1 we know that the space $E_1 := (F, \|\cdot\|_\infty)$ is complete. Now we will prove that the space $E_2 := (F, \|\cdot\|_{C^1})$ is also complete: given a sequence $(f_n)_{n=1}^\infty \subseteq F$ and a function $f \in C^1[0, 1]$ such that $f_n \longrightarrow f$ in $C^1[0, 1]$, from the inequality $\|\cdot\|_\infty \leq \|\cdot\|_{C^1}$ it follows that $f_n \longrightarrow f$ in $C[0, 1]$. The fact that $F$ is closed in $C[0, 1]$ ensures that $f \in F$. This proves that $F$ is closed in $C^1[0, 1]$, so $E_2$ is complete by Proposition 1.1.1. The identity operator $i \colon E_2 \longrightarrow E_1$ is clearly bijective and linear. It is also continuous because, for every $g \in F$,

$$\|i(g)\|_{E_1} = \|g\|_\infty \leq \|g\|_\infty + \|g'\|_\infty = \|g\|_{E_2}.$$

The Open Mapping Theorem guarantees the existence of a constant $C > 0$ such that

$$\|g'\|_\infty \leq C\|g\|_\infty$$

for every function $g \in F$. Let $(f_n)_{n=1}^\infty$ be a sequence in the closed unit ball of $E_1$, that is, $\max_{t\in[0,1]} |f_n(t)| \leq 1$ for every $n$. By the inequality above, $\max_{t\in[0,1]} |f'_n(t)| \leq C$ for every $n$. Then the set formed by the functions $f_n$, $n \in \mathbb{N}$, satisfies conditions (a) and (b) of the Ascoli Theorem (Theorem B.7 of Appendix B); so its closure is compact in $C[0, 1]$. We thus obtain a subsequence $(f_{n_j})_{j=1}^\infty$ and a function $f \in C[0, 1]$ such that $f_{n_j} \longrightarrow f$ in $C[0, 1]$. We know that $f \in F$ because $F$ is closed. We have proved that the closed unit ball of $E_1$ is compact. By the Riesz Theorem (Theorem 1.5.4), $F$ would necessarily be finite-dimensional, which contradicts our initial assumption.

## 2.5 The Closed Graph Theorem

Let $E$ and $F$ be normed spaces and let $T: E \longrightarrow F$ be a linear operator. The *graph* of $T$ is the set

$$G(T) = \{(x, y) : x \in E \text{ and } y = T(x)\} = \{(x, T(x)) : x \in E\} \subseteq E \times F.$$

Note that $G(T)$ is a subspace of $E \times F$ (which will be equipped with any of the equivalent norms from Exercise 1.8.12). In general, $G(T)$ can be a closed subset of $E \times F$ or not. The next result shows that, for linear operators between Banach spaces, the continuity of $T$ is equivalent to the fact that $G(T)$ is closed.

**Theorem 2.5.1 (Closed Graph Theorem)** *Let $E$ and $F$ be Banach spaces and let $T: E \longrightarrow F$ be a linear operator. Then $T$ is continuous if and only if $G(T)$ is closed in $E \times F$.*

***Proof*** Suppose $T$ is continuous. In this case, the function

$$f: E \times F \longrightarrow \mathbb{R}, \ f(x, y) = \|T(x) - y\|$$

is continuous, so $G(T) = f^{-1}(\{0\})$ is closed because it is the inverse image of the closed set $\{0\}$ by the continuous function $f$.

Conversely, suppose $G(T)$ is closed. From Exercise 1.8.12 we know that $E \times F$ is a Banach space with the norm $\|\cdot\|_1$. The function

$$\pi: G(T) \longrightarrow E, \ \pi(x, T(x)) = x,$$

is clearly linear and bijective. In addition, $\pi$ is continuous because

$$\|\pi(x, T(x))\| = \|x\| \leq \|x\| + \|T(x)\| = \|(x, T(x))\|_1.$$

From the Open Mapping Theorem it follows that $\pi^{-1}$ is continuous, so there exists $C > 0$ such that $\|(x, T(x))\|_1 \leq C \|x\|$ for every $x \in E$. Hence

$$\|T(x)\| \leq \|T(x)\| + \|x\| = \|(x, T(x))\|_1 \leq C \|x\|$$

for every $x \in E$. This proves that $T$ is continuous. ∎

In Theorem 2.6.1 and Exercise 2.7.35 the reader will check that one and only one of the implications of the Closed Graph Theorem remains valid for operators between non-complete normed spaces. The usefulness of the Closed Graph Theorem lies in the following remark: to prove the continuity of an operator $T \colon E \longrightarrow F$ (linear or not) via definition, we must prove that for every convergent sequence $x_n \longrightarrow x$ in $E$, it is true that:

(a) $(T(x_n))_{n=1}^{\infty}$ is convergent in $F$, that is, there exists $y \in F$ such that $T(x_n) \longrightarrow y$;
(b) $y = T(x)$.

In general, proving (a), that is, proving that $(T(x_n))_{n=1}^{\infty}$ converges, is a difficult task, especially in infinite dimension. The Closed Graph Theorem provides, in the case where $T$ is linear and acts between Banach spaces, item (a) for free, that is, we can assume the convergence of $(T(x_n))_{n=1}^{\infty}$, only needing to check (b). The following example illustrates what we have just said:

**Example 2.5.2** Let $E$ be a Banach space and let $T \colon E \longrightarrow E'$ be a *symmetric* linear operator, that is, $T$ is linear and

$$T(x)(y) = T(y)(x) \text{ for all } x, y \in E.$$

In order to prove that $T$ is continuous, by the Closed Graph Theorem, it is enough to check that $G(T)$ is closed. To do so, let $(x_n)_{n=1}^{\infty}$ be a sequence converging to $x$ in $E$ and suppose that $T(x_n) \longrightarrow \varphi$ in $E'$. For a fixed $y \in E$, taking the limit when $n \longrightarrow \infty$,

$$\varphi(y) \longleftarrow T(x_n)(y) = T(y)(x_n) \longrightarrow T(y)(x) = T(x)(y),$$

from which we conclude that $T(x)(y) = \varphi(y)$. As $y \in E$ was chosen arbitrarily, it follows that $\varphi = T(x)$, therefore $G(T)$ is closed.

## 2.6 Comments and Historical Notes

Let $T \colon E \longrightarrow F$ be a continuous linear operator. From Proposition 2.1.4(b) we know that in the expression

$$\|T\| = \inf\{C : \|T(x)\| \leq C\|x\| \text{ for every } x \in E\},$$

the infimum is attained. It is natural then to imagine that in the expression

$$\|T\| = \sup\{\|T(x)\| : x \in E \text{ and } \|x\| = 1\},$$

the supremum is also attained. However, as the reader will check in Exercise 2.7.11, this does not always happen and the subject is much more delicate than it may seem at first glance.

A continuous linear operator $T \colon E \longrightarrow F$ is said to *attain the norm* if there exists $x \in E$ with norm 1 such that $\|T(x)\| = \|T\|$. A deep result, known as the Bishop–Phelps Theorem, guarantees that, if $E$ is a real normed space and $F = \mathbb{R}$, then the set of continuous linear *functionals* that attain the norm is dense in $E'$. For other versions of the Bishop–Phelps Theorem and relationships with variational methods and applications to elliptic partial differential equations we suggest the book [24].

The Baire Theorem has several equivalent formulations. We state in Theorem 2.3.1 the one that is most convenient for our purposes.

Theorem 2.3.2 was proven by Banach and Steinhaus in a joint article published in 1927. Corollary 2.3.3 is also known as the Banach–Steinhaus Theorem, and both are also known as the Uniform Boundedness Principle. There are known proofs of the Banach–Steinhaus Theorem that do not use the Baire Theorem—see, for example, [13, 14.1] and [32].

The Open Mapping Theorem and the Closed Graph Theorem were published by Banach in 1932 in the book [6], which is usually considered the milestone of the foundation of Functional Analysis as a mathematical discipline. The first proof of the Open Mapping Theorem is due to J. Schauder in 1930.

In the Closed Graph Theorem, the implication '$T$ continuous $\Longrightarrow$ graph of $T$ closed' holds in a much more general context, which will be proved and used in Chap. 6 (see Lemma 6.2.8):

**Theorem 2.6.1** *Let $X$ and $Y$ be topological spaces with $Y$ Hausdorff. If the function $f \colon X \longrightarrow Y$ is continuous, then the graph of $f$ is closed in $X \times Y$.*

One of the central lessons of this chapter is the fact that algebraic properties significantly affect the topological behavior of functions between normed spaces. There is a notable example, due to S. Mazur and S. Ulam (1932), of how topological properties can also affect algebraic behavior:

**Theorem 2.6.2** *Let $E$ and $F$ be real normed spaces and let $f \colon E \longrightarrow F$ be a surjective function such that $f(0) = 0$ and*

$$\|f(x) - f(y)\| = \|x - y\|$$

*for any $x, y$ in $E$. Then $f$ is linear.*

A proof can be found in [61, Chapitre X, Lecture 1].

## 2.7 Exercises

The more general version of Lemma 2.4.1 proposed in Exercise 2.7.28 was kindly communicated to the authors by Diogo Diniz.

## 2.7 Exercises

**Exercise 2.7.1**

(a) Prove that every normed space of finite dimension $n$ over $\mathbb{K}$ is isomorphic to the Euclidean space $(\mathbb{K}^n, \|\cdot\|_2)$.
(b) Prove that if $E$ and $F$ are normed spaces, over the same field, of the same finite dimension, then $E$ and $F$ are isomorphic.

**Exercise 2.7.2** Prove that every linear operator whose domain is a finite-dimensional normed space is continuous.

**Exercise 2.7.3** Show that for every infinite-dimensional normed space $E$ and every normed space $F \neq \{0\}$, there exists a discontinuous linear operator $T: E \longrightarrow F$.

**Exercise 2.7.4** Let $E$ be a normed space over $\mathbb{C}$ and let $\varphi$ be a discontinuous linear functional on $E$. Show that $\{\varphi(x) : x \in E \text{ and } \|x\| \leq 1\} = \mathbb{C}$.

**Exercise 2.7.5** Let $E$ be a normed space, let $F$ be a Banach space and let $G$ be a subspace of $E$. Show that, if $T: G \longrightarrow F$ is a continuous linear operator, then there exists a unique continuous linear extension of $T$ to the closure of $G$. Also show that the norm of the extension coincides with the norm of $T$ and that if $T$ is a linear isometry, then the same happens with its extension.

**Exercise 2.7.6** Let $E$ be a Banach space, let $F$ a normed space and let $T \in \mathcal{L}(E, F)$ be a linear isometry. Show that $T(E)$ is closed in $F$.

**Exercise 2.7.7**

(a) Prove that if a normed space $E$ is isomorphic to a Banach space, then $E$ is a Banach space.
(b) Show that item (a) does not hold in metric spaces, that is, there exist homeomorphic metric spaces $M$ and $N$ with $M$ complete and $N$ incomplete.
(c) How do you explain the discrepancy between items (a) and (b)?

**Exercise 2.7.8**

(a) Let $E$ and $F$ be normed spaces and let $T: E \longrightarrow F$ be a continuous linear operator. Show that

$$\|T\| = \sup_{x \neq 0} \frac{\|T(x)\|}{\|x\|} = \sup_{\|x\|=1} \|T(x)\| = \sup_{\|x\|<1} \|T(x)\|$$
$$= \inf\{C : \|T(x)\| \leq C\|x\| \text{ for every } x \in E\}.$$

(b) Prove that $T \mapsto \|T\|$ is a norm on $\mathcal{L}(E, F)$.

**Exercise 2.7.9** Let $T \in \mathcal{L}(E, F)$ and $R \in \mathcal{L}(F, G)$ be given. Prove that $R \circ T \in \mathcal{L}(E, G)$ and $\|R \circ T\| \leq \|R\| \cdot \|T\|$. Find an example where $\|R \circ T\| < \|R\| \cdot \|T\|$.

**Exercise 2.7.10** Let $E$ and $F$ be normed spaces over the same field $\mathbb{K}$. Show that $E' \times F'$ and $(E \times F)'$ are isometrically isomorphic considering the norm $\| \cdot \|_\infty$ on $E \times F$ and the norm $\| \cdot \|_1$ on $E' \times F'$ (see the definitions of these norms in Exercise 1.8.12).

**Exercise 2.7.11 (A Continuous Linear Functional That Does Not Attain the Norm)** Consider the map

$$\varphi: c_0 \longrightarrow \mathbb{K}, \quad \varphi((a_j)_{j=1}^\infty) = \sum_{j=1}^\infty \frac{a_j}{2^j}.$$

Prove that:

(a) $\varphi \in (c_0)'$ (do not forget to prove, first, that $\varphi$ is well defined).
(b) $\|\varphi\| = 1$.
(c) There is no $x \in c_0$ such that $\|x\| \leq 1$ and $\|\varphi\| = |\varphi(x)|$.

**Exercise 2.7.12** Let $E$ and $F$ be normed spaces and let $T: E \longrightarrow F$ be a continuous linear operator.

(a) Prove that the *kernel of* $T$, defined by $\ker(T) = \{x \in E : T(x) = 0\}$, is a closed subspace of $E$.
(b) Prove that the *range of* $T$, defined by $R(T) = \{T(x) : x \in E\}$, is a subspace of $F$. Is it closed?

**Exercise 2.7.13*** Consider $X = \mathbb{R}^2$ equipped with the norm $\|(x, y)\| := (x^4 + y^4)^{1/4}$. Calculate explicitly the norm of each functional $\varphi \in X'$.

**Exercise 2.7.14* (Quotient Map)** Let $M$ be a closed subspace of the normed space $E$ and let $\pi: E \longrightarrow E/M$ be given by $\pi(x) = [x]$ (see Exercise 1.8.14). Prove that:

(a) $\|\pi(x)\| \leq \|x\|$ for every $x \in E$.
(b) Given $x \in E$ and $\varepsilon > 0$, there exists $y \in E$ such that $\pi(x) = \pi(y)$ and $\|y\| \leq \|\pi(x)\| + \varepsilon$.
(c) The image of the open unit ball of $E$ by $\pi$ is the open unit ball of $E/M$.
(d) $\pi \in \mathcal{L}(E, E/M)$, $\ker(\pi) = M$ and $\pi$ is an open map.
(e) If $M \neq E$, then $\|\pi\| = 1$.
(f) For every normed space $F$ and every continuous linear operator $T: E \longrightarrow F$ there exists a unique continuous linear operator $\tilde{T}: E/\ker(T) \longrightarrow F$ such that $T = \tilde{T} \circ \pi$. Furthermore, $\|T\| = \|\tilde{T}\|$.

## 2.7 Exercises

**Exercise 2.7.15 (Isomorphism Theorem)** Let $E$ and $F$ be Banach spaces and let $T \in \mathcal{L}(E, F)$ be given. Prove that if $T(E)$ is closed in $F$ then $T(E)$ is isomorphic to $E/\ker(T)$.

**Exercise 2.7.16** Let $E$ and $F$ be normed spaces. If $T \in \mathcal{L}(E; F)$ and there exists $c > 0$ such that $\|T(x)\| \geq c\|x\|$ for every $x \in E$, show that the inverse operator $T^{-1} \colon T(E) \longrightarrow E$ exists and is continuous.

**Exercise 2.7.17** Let $E$ and $F$ be normed spaces. Show that a linear operator $T \colon E \longrightarrow F$ is continuous if and only if $T(A)$ is bounded in $F$ whenever $A$ is bounded in $E$ (see the definition of bounded set in Exercise 1.8.9).

**Exercise 2.7.18** Let $E$ and $F$ be normed spaces. Show that if $T \colon E \longrightarrow F$ is an isometric isomorphism, then $\|T\| = 1$. On the other hand, find a Banach space $E$ and a topological isomorphism $T \colon E \longrightarrow E$ with $\|T\| = 1$, which is not an isometric isomorphism.

**Exercise 2.7.19** Show that the spaces $(\mathbb{R}^2, \|\cdot\|_1)$ and $(\mathbb{R}^2, \|\cdot\|_\infty)$ are isometrically isomorphic and that both are not isometrically isomorphic to $(\mathbb{R}^2, \|\cdot\|_2)$.

**Exercise 2.7.20 (Finite Rank Operators)** A continuous linear operator $T \colon E \longrightarrow F$ is said to be of *finite rank* if the range of $T$ is a finite-dimensional subspace of $F$.

(a) Prove that $T$ is of finite rank if and only if there exist $n \in \mathbb{N}$, $\varphi_1, \ldots, \varphi_n \in E'$ and $y_1, \ldots, y_n \in F$ such that $T = \varphi_1 \otimes y_1 + \cdots + \varphi_n \otimes y_n$.
(b) Prove that the set of continuous linear operators of finite rank from $E$ to $F$ is a vector subspace of $\mathcal{L}(E, F)$. Is it closed?

**Exercise 2.7.21** A subspace $M$ of a vector space $E$ is said to have *finite deficiency* if there exist finitely many vectors $x_1, \ldots, x_n$ in $E - M$ such that $E = [M \cup \{x_1, \ldots, x_n\}]$. Show that the kernel of a non-zero linear functional $T \colon E \longrightarrow \mathbb{K}$ has (in $E$) finite deficiency with $n = 1$.

**Exercise 2.7.22** Let $1 \leq p \leq q \leq \infty$ be given. Show that $\ell_p \subseteq \ell_q$ and that the inclusion $\ell_p \hookrightarrow \ell_q$ is a continuous linear operator. Determine the norm of this inclusion.

**Exercise 2.7.23** Consider the set

$$cs = \left\{ (a_j)_{j=1}^\infty : a_j \in \mathbb{K} \text{ for every } j \in \mathbb{N} \text{ and } \sum_{j=1}^\infty a_j \text{ converges} \right\}.$$

(a) Show that $cs$ is a normed space with the usual operations of sequences and with the norm $\left\| (a_j)_{j=1}^\infty \right\| = \sup_{n \in \mathbb{N}} \left| \sum_{j=1}^n a_j \right|$.
(b) Prove that $cs$ is isometrically isomorphic to the space $c$ of convergent sequences defined in Exercise 1.8.27.

**Exercise 2.7.24** Consider the set

$$bs = \left\{ (a_j)_{j=1}^\infty : a_j \in \mathbb{K} \text{ for every } j \in \mathbb{N} \text{ and } \sup_{n \in \mathbb{N}} \left| \sum_{j=1}^n a_j \right| < \infty \right\}.$$

(a) Show that $bs$ is a normed space with the usual operations of sequences and with the norm $\left\| (a_j)_{j=1}^\infty \right\| = \sup_{n \in \mathbb{N}} \left| \sum_{j=1}^n a_j \right|.$

(b) Prove that $bs$ is isometrically isomorphic to $\ell_\infty$.

**Exercise 2.7.25** Let $E, F, G$ be normed spaces and let $A \colon E \times F \longrightarrow G$ be a bilinear operator. Consider in $E \times F$ any of the norms from Exercise 1.8.12. Prove that the following conditions are equivalent:

(a) $A$ is continuous.
(b) $A$ is continuous at the origin.
(c) $\sup\{\|A(x, y)\| : x \in B_E, y \in B_F\} < \infty$.
(d) There exists $C \geq 0$ such that $\|A(x, y)\| \leq C \|x\| \cdot \|y\|$ for all $x \in E$ and $y \in F$.

**Exercise 2.7.26** Let $E, F$ be Banach spaces and let $T, T_1, T_2, \ldots,$ be operators in $\mathcal{L}(E, F)$ such that $T_n(x) \longrightarrow T(x)$ for every $x \in E$. Show that, for every compact set $K \subseteq E$, $\sup_{x \in K} \|T_n(x) - T(x)\| \longrightarrow 0$.

**Exercise 2.7.27** Complete the details of Example 2.4.3.

**Exercise 2.7.28**\* Prove the following stronger version of Lemma 2.4.1: Let $E$ be a Banach space, let $F$ be a normed space and let $T \in \mathcal{L}(E; F)$ be given. If there exist $R, r > 0$ such that $\overline{T(B_E(0; R))} \supseteq B_F(0; r)$, then $T(B_E(0; R)) \supseteq B_F(0; r)$.

**Exercise 2.7.29** Let $\| \cdot \|_1$ and $\| \cdot \|_2$ be two norms on a vector space $E$ such that $(E, \| \cdot \|_1)$ and $(E, \| \cdot \|_2)$ are complete.

(a) Suppose that $\|x_n\|_1 \longrightarrow 0$ always implies that $\|x_n\|_2 \longrightarrow 0$. Prove that the two norms are equivalent (see definition in Exercise 1.8.3). In particular, the convergence of a sequence (not necessarily to zero) in one of the norms implies convergence (to the same limit) in the other norm.
(b) If there exists $c$ such that $\|x\|_1 \leq c \|x\|_2$ for every $x \in E$, then the two norms are equivalent.
(c) If the topology generated by the norm $\| \cdot \|_1$ is contained in the topology generated by $\| \cdot \|_2$, then the two norms are equivalent. In particular, the topologies generated by the two norms coincide.

**Exercise 2.7.30** Let $E, F$ be normed vector spaces and let $T \colon E \longrightarrow F$ be a linear operator with closed graph (such operators are called *closed operators*). Show that the kernel of $T$ is closed.

## 2.7 Exercises

**Exercise 2.7.31** Let $E$ and $F$ be normed spaces. If $T_1: E \longrightarrow F$ is a closed linear operator and $T_2: E \longrightarrow F$ is a continuous linear operator, then $T_1 + T_2$ is closed.

**Exercise 2.7.32** Let $E$ and $F$ be normed spaces and let $T \in \mathcal{L}(E; F)$ be given. For each subset $A$ of $E$, we write $\|T\|_A := \sup\{\|T(x)\| : x \in A\}$. Prove that

$$\|T\|_A = \|T\|_{\overline{A}} = \|T\|_{\text{conv}(A)} = \|T\|_{\overline{\text{conv}(A)}},$$

for every $A \subseteq E$. For the definition of the convex hull see Exercise 1.8.17.

**Exercise 2.7.33*** Let $E$ be a real Banach space and let $T: E \longrightarrow E'$ be a linear operator such that $T(x)(x) \geq 0$ for every $x \in E$. Prove that $T$ is continuous.

**Exercise 2.7.34** Show that, in the Closed Graph Theorem, the implication

$$T \text{ is continuous} \implies G(T) \text{ is closed in } E \times F$$

remains valid for linear operators between normed spaces not necessarily complete.

**Exercise 2.7.35** Prove that the discontinuous linear operator $T^{-1}$ from Example 2.4.3 has closed graph. Conclude that, in the Closed Graph Theorem, the implication

$$G(T) \text{ is closed in } E \times F \implies T \text{ is continuous}$$

does not remain valid for linear operators between normed spaces not necessarily complete.

**Exercise 2.7.36** Let $E$ be a Banach space, let $F$ a normed space and let $T: E \longrightarrow F$ be a linear operator with closed graph. Show that if the inverse operator $T^{-1}$ exists and is continuous, then the range $R(T)$ is a closed subspace of $F$.

**Exercise 2.7.37** Let $(x_j)_{j=1}^\infty$ be a sequence in the Banach space $E$ such that $\sum_{j=1}^\infty |\varphi(x_j)| < \infty$ for every $\varphi \in E'$. Show that $\sup_{\|\varphi\| \leq 1} \sum_{j=1}^\infty |\varphi(x_j)| < \infty$.

# Chapter 3
# Hahn–Banach Theorems

It is difficult to measure the relevance of the Hahn–Banach Theorem in Functional Analysis, given the numerous corollaries and applications it has. It also finds applications in other areas of mathematics; for example, Complex Analysis, Measure Theory, Control Theory, Convex Programming, and Game Theory (for precise references see [47, p. 194]). Recent applications to Combinatorial Analysis can be found in [29]. In fact, its impact goes beyond the boundaries of mathematics, see, for example, an application to Thermodynamics in [22]. In this chapter, we will prove several forms of the Hahn–Banach Theorem, starting with purely algebraic results, moving on to topological results: first in the real case, then in the complex case, and culminating in the geometric forms of the theorem, also called separation theorems. Some initial consequences of these several forms of the Hahn–Banach Theorem will also be explored.

## 3.1 The Hahn–Banach Extension Theorem

The essence of the Hahn–Banach Theorem, in its version for normed spaces, is that continuous linear functionals defined on a subspace $G$ of a normed space $E$ can be extended to the entire space $E$ preserving linearity, continuity, and even the norm. The fundamental piece of the proof of the Hahn–Banach Theorem is an algebraic argument that shows that such an extension is possible from $G$ to $G \oplus [v]$ with $v \notin G$. In parallel, Zorn's Lemma is applied as a technical tool of induction, which can be replaced by a canonical argument of finite induction when $E$ is separable.

To follow the proof of the theorem below, the reader unfamiliar with Zorn's Lemma should consult Appendix A.

**Theorem 3.1.1 (Hahn–Banach Theorem—Real Case)** *Let $E$ be a vector space over the field of real numbers and let $p \colon E \longrightarrow \mathbb{R}$ be a function that satisfies*

$$p(ax) = ap(x) \text{ for all } a > 0 \text{ and } x \in E, \text{ and} \tag{3.1}$$

$$p(x+y) \le p(x) + p(y) \text{ for any } x, y \in E.$$

*Also let G be a subspace of E and let* $\varphi \colon G \longrightarrow \mathbb{R}$ *be a linear functional such that* $\varphi(x) \le p(x)$ *for every* $x \in G$. *Then there exists a linear functional* $\widetilde{\varphi} \colon E \longrightarrow \mathbb{R}$ *that extends* $\varphi$, *that is* $\widetilde{\varphi}(x) = \varphi(x)$ *for every* $x \in G$, *and that satisfies* $\widetilde{\varphi}(x) \le p(x)$ *for every* $x \in E$.

***Proof*** Consider the following family $\mathcal{P}$ of linear functionals defined on subspaces of $E$ that contain $G$:

$$\mathcal{P} = \left\{ \begin{array}{l} \phi \colon D(\phi) \subseteq E \longrightarrow \mathbb{R} : D(\phi) \text{ is subspace of } E, \\ \phi \text{ is linear, } G \subseteq D(\phi), \phi(x) = \varphi(x) \text{ for every } x \in G \\ \text{and } \phi(x) \le p(x) \text{ for every } x \text{ in } D(\phi) \end{array} \right\}.$$

In $\mathcal{P}$, we define the partial order relation

$$\phi_1 \le \phi_2 \iff D(\phi_1) \subseteq D(\phi_2) \text{ and } \phi_2 \text{ extends } \phi_1.$$

Note that $\mathcal{P}$ is non-empty because $\varphi \in \mathcal{P}$. We will prove that every totally ordered subset of $\mathcal{P}$ admits an upper bound. Indeed, given $\mathcal{Q} \subseteq \mathcal{P}$ totally ordered, define $\phi \colon D(\phi) \longrightarrow \mathbb{R}$ by

$$D(\phi) = \bigcup_{\theta \in \mathcal{Q}} D(\theta) \text{ and } \phi(x) = \theta(x) \text{ if } x \in D(\theta).$$

Note that the function $\varphi$ is well defined due to the total ordering of $\mathcal{Q}$. It is immediate that $\phi \in \mathcal{P}$ and that $\phi$ is an upper bound for $\mathcal{Q}$. By Zorn's Lemma, $\mathcal{P}$ admits a maximal element, which will be denoted by $\widetilde{\varphi}$. Note that to obtain the result it is enough to show that $D(\widetilde{\varphi}) = E$. To do so, suppose that $D(\widetilde{\varphi}) \ne E$. In this case we can choose $x_0 \in E - D(\widetilde{\varphi})$ and define $\widetilde{\phi} \colon D(\widetilde{\phi}) \longrightarrow \mathbb{R}$ by

$$D(\widetilde{\phi}) = D(\widetilde{\varphi}) + [x_0] \text{ and } \widetilde{\phi}(x + tx_0) = \widetilde{\varphi}(x) + t\alpha,$$

where $\alpha$ is a constant that will be chosen later in order to ensure that $\widetilde{\phi} \in \mathcal{P}$. We want, for now, that $\alpha$ satisfies the following inequalities:

$$\widetilde{\varphi}(x) + \alpha = \widetilde{\phi}(x + x_0) \le p(x + x_0) \text{ for every } x \in D(\widetilde{\varphi}) \text{ and}$$

$$\widetilde{\varphi}(x) - \alpha = \widetilde{\phi}(x - x_0) \le p(x - x_0) \text{ for every } x \in D(\widetilde{\varphi}).$$

It is enough to choose $\alpha$ so that

$$\sup_{x \in D(\widetilde{\varphi})} \{\widetilde{\varphi}(x) - p(x - x_0)\} \le \alpha \le \inf_{x \in D(\widetilde{\varphi})} \{p(x + x_0) - \widetilde{\varphi}(x)\}.$$

3.1 The Hahn–Banach Extension Theorem

Fortunately, such a choice is possible because, for $x, y \in D(\widetilde{\varphi})$, we have

$$\widetilde{\varphi}(x) + \widetilde{\varphi}(y) = \widetilde{\varphi}(x+y) \leq p(x+y) = p(x+x_0+y-x_0) \leq p(x+x_0) + p(y-x_0),$$

hence

$$\widetilde{\varphi}(y) - p(y - x_0) \leq p(x + x_0) - \widetilde{\varphi}(x)$$

for any $x, y \in D(\widetilde{\varphi})$. It is established that it is possible to choose $\alpha$ meeting those two initial requirements. Having done that, we conclude that:

- For $t > 0$,

$$\widetilde{\phi}(x + tx_0) = \widetilde{\phi}\left(t\left(\frac{x}{t} + x_0\right)\right) = t\widetilde{\phi}\left(\frac{x}{t} + x_0\right) \leq tp\left(\frac{x}{t} + x_0\right) = p(x + tx_0).$$

- For $t < 0$,

$$\widetilde{\phi}(x + tx_0) = \widetilde{\phi}\left(-t\left(\frac{x}{-t} - x_0\right)\right) = -t\widetilde{\phi}\left(\frac{x}{-t} - x_0\right)$$

$$\leq -tp\left(\frac{-x}{t} - x_0\right) = p(x + tx_0).$$

- For $t = 0$,

$$\widetilde{\phi}(x + tx_0) = \widetilde{\phi}(x) = \widetilde{\varphi}(x) \leq p(x) = p(x + tx_0).$$

It follows that $\widetilde{\phi} \in \mathcal{P}$, $\widetilde{\varphi} \leq \widetilde{\phi}$ and $\widetilde{\varphi} \neq \widetilde{\phi}$. As this contradicts the maximality of $\widetilde{\varphi}$, we have $D(\widetilde{\varphi}) = E$ and the theorem is proven. ∎

Next, we will prove the version of the Hahn-Banach Theorem that is also valid for complex vector spaces:

**Theorem 3.1.2 (Hahn-Banach Theorem)** *Let $E$ be a vector space over the field $\mathbb{K} = \mathbb{R}$ or $\mathbb{C}$ and let $p \colon E \longrightarrow \mathbb{R}$ be a function that satisfies*

$$p(ax) = |a| p(x) \text{ for every } a \in \mathbb{K} \text{ and every } x \in E, \text{ and} \tag{3.2}$$

$$p(x + y) \leq p(x) + p(y) \text{ for any } x, y \in E. \tag{3.3}$$

*If $G \subseteq E$ is a vector subspace and $\varphi \colon G \longrightarrow \mathbb{K}$ is a linear functional such that $|\varphi(x)| \leq p(x)$ for every $x \in G$, then there exists a linear functional $\widetilde{\varphi} \colon E \longrightarrow \mathbb{K}$ that extends $\varphi$ to $E$ and that satisfies $|\widetilde{\varphi}(x)| \leq p(x)$ for every $x \in E$.*

**Proof** First of all, observe that from (3.2) and (3.3) it follows that $p(x) \geq 0$ for every $x \in E$, a fact that will be used in the final part of the proof. Indeed, from (3.3) it follows that $p(0) = p(0+0) \leq 2p(0)$, hence $p(0) \geq 0$. From (3.2) it follows that $p(x) = p(-x)$, so, for each $x \in E$,

$$2p(x) = p(x) + p(-x) \geq p(x + (-x)) = p(0) \geq 0.$$

We first consider $\mathbb{K} = \mathbb{R}$. In this case the assumption guarantees that $\varphi(x) \leq p(x)$ for every $x \in G$. By Theorem 3.1.1 there exists a linear functional $\widetilde{\varphi} \colon E \longrightarrow \mathbb{R}$ that extends $\varphi$ to $E$ and that satisfies $\widetilde{\varphi}(x) \leq p(x)$ for every $x \in E$. From this inequality and from (3.2) we have

$$-\widetilde{\varphi}(x) = \widetilde{\varphi}(-x) \leq p(-x) = |-1| p(x) = p(x),$$

for every $x \in E$. Therefore $|\widetilde{\varphi}(x)| \leq p(x)$ for every $x \in E$.

We now consider the complex case $\mathbb{K} = \mathbb{C}$. In this case $E$ is a complex vector space and $\varphi$ takes values in $\mathbb{C}$. Defining $\varphi_1, \varphi_2 \colon E \longrightarrow \mathbb{R}$ by $\varphi_1(x) = \mathrm{Re}(\varphi(x))$ and $\varphi_2(x) = \mathrm{Im}(\varphi(x))$, it is clear that $\varphi_1$ and $\varphi_2$ are linear real-valued maps and $\varphi = \varphi_1 + i\varphi_2$. It is convenient to call $E_\mathbb{R}$ and $G_\mathbb{R}$ the underlying real vector spaces (that is, as a set the respective space is the same, the addition operation is the same and scalar multiplication is done only with real scalars). Then $\varphi_1$ and $\varphi_2$ are linear functionals on $G_\mathbb{R}$. For $x \in G_\mathbb{R}$, we get

$$\varphi_1(x) \leq |\varphi_1(x)| \leq |\varphi(x)| \leq p(x).$$

By Theorem 3.1.1 there is a linear functional $\widetilde{\varphi}_1 \colon E_\mathbb{R} \longrightarrow \mathbb{R}$ that extends $\varphi_1$ to $E_\mathbb{R}$ and that satisfies $\widetilde{\varphi}_1(x) \leq p(x)$ for every $x \in E_\mathbb{R}$. Now we study the case of $\varphi_2$. Note that for $x \in G$,

$$i(\varphi_1(x) + i\varphi_2(x)) = i\varphi(x) = \varphi(ix) = \varphi_1(ix) + i\varphi_2(ix).$$

Therefore $\varphi_2(x) = -\varphi_1(ix)$ for every $x \in G$. Defining

$$\widetilde{\varphi} \colon E \longrightarrow \mathbb{C}, \ \widetilde{\varphi}(x) = \widetilde{\varphi}_1(x) - i\widetilde{\varphi}_1(ix),$$

it follows that $\widetilde{\varphi}(x) = \varphi(x)$ for every $x \in G$. We first note that $\widetilde{\varphi}$ is a linear functional on the complex space $E$. It is immediate that $\widetilde{\varphi}(x + y) = \widetilde{\varphi}(x) + \widetilde{\varphi}(y)$ for all $x, y \in E$. Given $(a + bi) \in \mathbb{C}$ and $x \in E$,

$$\widetilde{\varphi}((a + bi)x) = \widetilde{\varphi}_1(ax + ibx) - i\widetilde{\varphi}_1((i(a + bi)x))$$
$$= a\widetilde{\varphi}_1(x) + b\widetilde{\varphi}_1(ix) - i(a\widetilde{\varphi}_1(ix) - b\widetilde{\varphi}_1(x))$$
$$= (a + bi)(\widetilde{\varphi}_1(x) - i\widetilde{\varphi}_1(ix)) = (a + bi)\widetilde{\varphi}(x).$$

Finally, we show that $|\widetilde{\varphi}(x)| \leq p(x)$ for every $x \in E$. If $\widetilde{\varphi}(x) = 0$, then the inequality is obvious because $p(x) \geq 0$. Choose $x \in E$ such that $\widetilde{\varphi}(x) \neq 0$. Then

## 3.2 Continuous Extensions of Linear Operators

there exists $\theta$ such that $\widetilde{\varphi}(x) = |\widetilde{\varphi}(x)| e^{i\theta}$. It follows that $|\widetilde{\varphi}(x)| = e^{-i\theta}\widetilde{\varphi}(x) = \widetilde{\varphi}(e^{-i\theta}x)$. As $|\widetilde{\varphi}(x)|$ is real, by (3.2) we have

$$|\widetilde{\varphi}(x)| = \widetilde{\varphi}(e^{-i\theta}x) = \widetilde{\varphi}_1(e^{-i\theta}x) \leq p(e^{-i\theta}x) = \left|e^{-i\theta}\right| p(x) = p(x). \blacksquare$$

From now on, we return to the previous procedure in which a normed space means a normed space over $\mathbb{K} = \mathbb{R}$ or $\mathbb{C}$.

To be consistent with the terminology established in the literature, the following result will be also called Hahn-Banach Theorem.

**Corollary 3.1.3 (Hahn-Banach Theorem)** *Let $G$ be a subspace of a normed space $E$ and let $\varphi \colon G \longrightarrow \mathbb{K}$ be a continuous linear functional. Then there exists a continuous linear functional $\widetilde{\varphi} \colon E \longrightarrow \mathbb{K}$ whose restriction to $G$ coincides with $\varphi$ and $\|\widetilde{\varphi}\| = \|\varphi\|$.*

**Proof** It is enough to apply Theorem 3.1.2 with $p(x) = \|\varphi\| \cdot \|x\|$. $\blacksquare$

**Corollary 3.1.4** *Let $E$ be a normed space. For every $x_0 \in E$, $x_0 \neq 0$, there exists a linear functional $\varphi \in E'$ such that $\|\varphi\| = 1$ and $\varphi(x_0) = \|x_0\|$.*

**Proof** Apply Corollary 3.1.3 for $G = [x_0]$ and $\varphi(ax_0) = a\|x_0\|$. $\blacksquare$

The next application of the Hahn-Banach Theorem will be used several times throughout the text.

**Corollary 3.1.5** *For any vector $x$ in a normed space $E \neq \{0\}$, it holds*

$$\|x\| = \sup\{|\varphi(x)| : \varphi \in E' \text{ and } \|\varphi\| \leq 1\}$$
$$= \max\{|\varphi(x)| : \varphi \in E' \text{ and } \|\varphi\| = 1\}.$$

**Proof** For each functional $\varphi \in E'$ with $\|\varphi\| \leq 1$, it is immediate that $|\varphi(x)| \leq \|\varphi\| \cdot \|x\| \leq \|x\|$. This shows that $\sup\{|\varphi(x)| : \varphi \in E' \text{ and } \|\varphi\| \leq 1\} \leq \|x\|$. The fact that the supremum is attained in a norm 1 functional is guaranteed by Corollary 3.1.4. $\blacksquare$

## 3.2 Continuous Extensions of Linear Operators

Bearing the Hahn-Banach Theorem in mind, it is natural to speculate whether continuous linear operators can be extended to larger spaces. More precisely, if $G$ is a subspace of the normed space $E$ and $T \in \mathcal{L}(G, F)$, is there always an operator $\widetilde{T} \in \mathcal{L}(E, F)$ that coincides with $T$ on $G$? To address this issue we need the concepts of projection and complemented subspace.

**Definition 3.2.1** Let $E$ be a Banach space. A continuous linear operator $P: E \longrightarrow E$ is a *projection* if $P^2 := P \circ P = P$. It is clear that if $P \neq 0$ is a projection, then $\|P\| \geq 1$.

**Proposition 3.2.2** *Let $F$ be a subspace of the Banach space $E$. The following statements are equivalent:*

(a) *There exists a projection $P: E \longrightarrow E$ whose range coincides with $F$. In this case we say that $P$ is a projection of $E$ onto $F$.*
(b) *$F$ is closed and there exists a closed subspace $G$ of $E$ such that $E = F \oplus G$, that is, $E = F + G$ and $F \cap G = \{0\}$.*

*In this case $F = \{x \in E : P(x) = x\}$ and $G = \ker(P)$.*

**Proof** (a) $\Longrightarrow$ (b) We first prove the equality $F = \{x \in E : P(x) = x\}$. If $x \in F$, taking $y \in E$ such that $P(y) = x$ results in $x = P(y) = P(P(y)) = P(x)$. If $x = P(x)$, then clearly $x \in \text{Im}(P) = F$. Consequently, calling $\text{id}_E$ the identity operator on $E$, it is true that

$$F = \{x \in E : P(x) = x\} = \{x \in E : (P - \text{id}_E)(x) = 0\} = (P - \text{id}_E)^{-1}(\{0\}),$$

which is closed as the inverse image of the closed set $\{0\}$ by the continuous function $P - \text{id}_E$.

Now let $G = \ker(P)$. It is clear that $G$ is a closed subspace of $E$. For every $x \in E$ it holds that $(x - P(x)) \in G$, $P(x) \in F$ and $x = (x - P(x)) + P(x)$. If $x \in G \cap F$, then $x = P(x)$ since $x \in F$ and $P(x) = 0$ since $x \in G$. It follows that $x = 0$.

(b) $\Longrightarrow$ (a) For each $x \in E$, there are unique $x_1 \in F$ and $x_2 \in G$ such that $x = x_1 + x_2$. It is immediate that the operator $P: E \longrightarrow E$ given by $P(x) = x_1$ is well defined, is linear, $P^2 = P$, $\text{Im}(P) = F$, $\ker(P) = G$ and $F = \{x \in E : P(x) = x\}$. It remains to prove that $P$ is continuous. Let $(x_n)_{n=1}^\infty$ be a sequence in $E$ such that $x_n \longrightarrow x$ and $P(x_n) \longrightarrow y$. For each $n$, write $x_n = y_n + z_n$ with $y_n \in F$ and $z_n \in G$. Then $z_n = x_n - y_n = x_n - P(x_n) \longrightarrow x - y$. As $G$ is closed, $x - y \in G$, therefore $P(x) = P(y)$. On the other hand, $y_n = P(x_n) \longrightarrow y$. As $F$ is closed, $y \in F$. Thus $y = P(y) = P(x)$. By the Closed Graph Theorem, it follows that the linear operator $P$ is continuous. ∎

**Definition 3.2.3** A subspace $F$ of the Banach space $E$ is *complemented* if it satisfies the equivalent conditions of Proposition 3.2.2. We say that $F$ is $\lambda$-complemented, $\lambda \geq 1$, if $F$ is complemented by a projection of norm less than or equal to $\lambda$.

It follows from Proposition 3.2.2 that every complemented subspace of a Banach space is closed.

**Example 3.2.4**

(a) Every finite-dimensional subspace of a Banach space is complemented. Let $F$ be a subspace of dimension $n$ of the Banach space $E$ and let $\{e_1, \ldots, e_n\}$ be a

3.2 Continuous Extensions of Linear Operators

basis for $F$. For each $j = 1, \ldots, n$, consider the continuous linear functional $\varphi_j \in F'$ given by $\varphi_j \left( \sum_{k=1}^{n} a_k e_k \right) = a_j$. By the Hahn-Banach Theorem there exists $\widetilde{\varphi}_j \in E'$, extension of $\varphi_j$ to $E$, $j = 1, \ldots, n$. The reader will have no difficulty in checking that the operator $P := \sum_{j=1}^{n} \widetilde{\varphi}_j \otimes e_j$ is a projection from $E$ onto $F$.

(b) Let $E$ and $F$ be Banach spaces. By means of the projection $(x, y) \in E \times F \mapsto (x, 0) \in E \times F$, we see that $E = E \times \{0\}$ is 1-complemented in $E \times F$.

(c) We already know that a non-closed subspace of a Banach space is not complemented. For example, $c_{00}$ is not complemented in $c_0$. It is believed that the existence of non-complemented closed subspaces was first proven in 1937 by Murray [45], who proved that, for every $2 \neq p > 1$, $\ell_p$ contains a non-complemented closed subspace. In 1940, Phillips [50] proved that $c_0$ is not complemented in $\ell_\infty$.

It is easy to see that operators defined on complemented subspaces can be continuously extended to the whole space:

**Proposition 3.2.5** *Let $G$ be a normed vector space, let $F$ be a complemented subspace of the Banach space $E$ and let $T \in \mathcal{L}(F, G)$ be given. Then there exists an extension $\widetilde{T} \in \mathcal{L}(E, G)$ of $T$ to $E$.*

**Proof** Define $\widetilde{T} = T \circ P$, where $P$ is a projection from $E$ onto $F$. ∎

However, when the subspace is not complemented the situation can change radically. The next result shows that, unlike continuous linear functionals, continuous linear operators cannot always be extended continuously. That is, there is no vector-valued version of the Hahn–Banach Theorem:

**Proposition 3.2.6** *Let $E$ be a Banach space and let $F$ be a non-complemented subspace of $E$. Then there is no continuous linear operator $T: E \longrightarrow F$ such that $T(x) = x$ for every $x \in F$, that is, the identity operator on $F$ cannot be continuously extended to $E$.*

**Proof** Suppose there exists a continuous linear operator $T: E \longrightarrow F$ such that $T(x) = x$ for every $x \in F$. It is clear that the inclusion $i_F: F \longrightarrow E$ is a continuous linear operator, therefore $i_F \circ T$ is a continuous linear operator from $E$ to $E$. Since $T(x) \in F$ for every $x \in E$, it follows that $T^2(x) = T(T(x)) = T(x)$ for every $x \in E$. In this case, $i_F \circ T$ would be a projection onto $F$, which contradicts the fact that $F$ is not complemented in $E$. ∎

Even though there is no vector version of the Hahn–Banach Theorem, the possibility of continuous linear operators on certain spaces being continuously extendable is still open. For example, continuous linear operators taking values in $\ell_\infty$ can always be extended preserving the norm:

**Theorem 3.2.7 (Phillips' Theorem)** *Let $F$ be a subspace of the normed space $E$ and let $T \in \mathcal{L}(F, \ell_\infty)$ be given. Then there exists an extension $\widetilde{T} \in \mathcal{L}(E, \ell_\infty)$ of $T$ to $E$. Moreover, $\|\widetilde{T}\| = \|T\|$.*

*Proof* For each $n \in \mathbb{N}$, consider the norm 1 continuous linear functional given by

$$\varphi_n \colon \ell_\infty \longrightarrow \mathbb{K}, \quad \varphi_n\left((a_j)_{j=1}^\infty\right) = a_n.$$

Then $\varphi_n \circ T \in F'$ for every $n$ and $T(x) = ((\varphi_n \circ T)(x))_{n=1}^\infty$ for every $x \in F$. By the Hahn–Banach Theorem, for each $n$ there exists $\widetilde{\varphi}_n$, extension of $\varphi_n \circ T$ to $E$, with $\|\varphi_n \circ T\| = \|\widetilde{\varphi}_n\|$. It is easy to see that the operator

$$\widetilde{T} \colon E \longrightarrow \ell_\infty, \quad \widetilde{T}(x) = (\widetilde{\varphi}_n(x))_{n=1}^\infty,$$

is linear and extends $T$. Finally, we have

$$\|\widetilde{T}(x)\| = \sup_n |\widetilde{\varphi}_n(x)| \leq \sup_n \|\widetilde{\varphi}_n\| \cdot \|x\| = \sup_n \|\varphi_n \circ T\| \cdot \|x\|$$
$$\leq \sup_n \|\varphi_n\| \cdot \|T\| \cdot \|x\| = \|T\| \cdot \|x\|,$$

therefore $\widetilde{T}$ is continuous and $\|\widetilde{T}\| \leq \|T\|$. As the reverse inequality is immediate, the result follows. ∎

**Corollary 3.2.8** *If $\ell_\infty$ is a closed subspace of a Banach space $E$, then it is 1-complemented in $E$.*

*Proof* Apply Theorem 3.2.7 to the identity operator on $\ell_\infty$ to obtain an operator $T \colon E \longrightarrow \ell_\infty$ that extends the identity and $\|T\| = 1$. Calling $i$ the inclusion from $\ell_\infty$ in $E$ results in $i \circ T$ being a norm 1 projection from $E$ onto $\ell_\infty$. ∎

## 3.3 Applications of the Hahn–Banach Theorem for Separable Spaces

As we have already anticipated, separable spaces enjoy special properties that make the theory of separable spaces much richer than the general theory. We will see in this section the first signs of this phenomenon. Remember that if $A$ is a subset of a normed space $E$ and $x \in E$, then $\mathrm{dist}(x, A) = \inf\{\|x - y\| : y \in A\}$.

**Proposition 3.3.1** *Let $E$ be a normed space, let $M$ be a closed subspace of $E$, let $y_0 \in E - M$ be given and call $d = \mathrm{dist}(y_0, M)$. Then there exists a linear functional $\varphi \in E'$ such that $\|\varphi\| = 1$, $\varphi(y_0) = d$ and $\varphi(x) = 0$ for every $x \in M$.*

*Proof* Define $N = M + [y_0]$. For $z \in N$, there are unique $a \in \mathbb{K}$ and $x \in M$ such that $z = x + ay_0$. Define

## 3.3 Applications of the Hahn–Banach Theorem for Separable Spaces

$$\varphi_0 \colon N \longrightarrow \mathbb{K}, \quad \varphi_0(x + ay_0) = ad.$$

It is clear that $\varphi_0$ is linear, $\varphi_0(M) = \{0\}$ and that $\varphi_0(y_0) = d$. We prove that $\|\varphi_0\| = 1$. Let $z = x + ay_0 \in N$ be given. If $a \neq 0$, then

$$\|z\| = \|x + ay_0\| = |a| \cdot \left\| \frac{-x}{a} - y_0 \right\| \geq d\,|a| = |\varphi_0(z)|.$$

If $a = 0$, then the inequality $\|z\| \geq |\varphi_0(z)|$ is obvious. It follows that $\|\varphi_0\| \leq 1$. Given $\varepsilon > 0$, there exists $x_\varepsilon \in M$ such that $d \leq \|y_0 - x_\varepsilon\| \leq d + \varepsilon$. Putting $z_\varepsilon = \dfrac{y_0 - x_\varepsilon}{\|y_0 - x_\varepsilon\|}$, we have $z_\varepsilon \in N$, $\|z_\varepsilon\| = 1$ and

$$\varphi_0(z_\varepsilon) = \frac{d}{\|y_0 - x_\varepsilon\|} \geq \frac{d}{d + \varepsilon}.$$

As $\varepsilon > 0$ is arbitrary, it follows that $\|\varphi_0\| \geq 1$, hence $\|\varphi_0\| = 1$. By the Hahn-Banach Theorem, there exists $\varphi \in E'$ that extends $\varphi_0$ to $E$ such that $\|\varphi\| = \|\varphi_0\| = 1$. ∎

**Theorem 3.3.2** *If $E'$ is separable, then $E$ is also separable.*

**Proof** Let $S_{E'}$ be the unit sphere of $E'$, that is, $S_{E'} = \{\varphi \in E' : \|\varphi\| = 1\}$. As we have mentioned before, a subset of a separable metric space is also separable, so $S_{E'}$ is separable. Let $\{\varphi_n : n \in \mathbb{N}\}$ be a countable dense subset of $S_{E'}$. For each $n \in \mathbb{N}$, choose $x_n \in S_E$ such that $|\varphi_n(x_n)| \geq \frac{1}{2}$. We shall prove that $M := \overline{[x_1, x_2, \ldots]} = E$. To do so, suppose that $M \neq E$ and choose $y_0 \in E - M$. By Proposition 3.3.1 there exists a functional $\varphi \in E'$ with $\|\varphi\| = 1$ such that $\varphi(y_0) = d = \text{dist}(y_0, M)$ and $\varphi(x) = 0$ for every $x \in M$. Then

$$\|\varphi - \varphi_n\| = \sup_{x \in B_E} |(\varphi - \varphi_n)(x)| \geq |(\varphi - \varphi_n)(x_n)| = |\varphi_n(x_n)| \geq \frac{1}{2}.$$

This is a contradiction because $\{\varphi_n : n \in \mathbb{N}\}$ is dense in $S_{E'}$. Therefore $M = E$ and the separability of $E$ follows from Lemma 1.6.3. ∎

We will see in Remark 4.2.2 that the converse of Theorem 3.3.2 is not true.

It is surprising that there exists a Banach space within which we can find all separable spaces. Even more surprising is the fact that this super-space is one of our old acquaintances:

**Proposition 3.3.3** *Every separable normed space is isometrically isomorphic to a subspace of $\ell_\infty$.*

**Proof** Let $E$ be a separable normed space, of course we can assume that $E \neq \{0\}$, and let $\mathcal{D} = \{x_n : n \in \mathbb{N}\}$ be a countable dense subset of $E$. We can clearly assume that $0 \notin \mathcal{D}$. By Corollary 3.1.4, for each $n \in \mathbb{N}$ there exists a linear functional $\varphi_n \in E'$ such that $\|\varphi_n\| = 1$ and $\varphi_n(x_n) = \|x_n\|$. Consider the operator

$$T: E \longrightarrow \ell_\infty, \quad T(x) = (\varphi_n(x))_{n=1}^\infty.$$

It is clear that $|\varphi_n(x)| \leq \|\varphi_n\| \|x\| = \|x\|$ for all $n \in \mathbb{N}$ and $x \in E$, therefore $T$ is well defined in the sense that $(\varphi_n(x))_{n=1}^\infty \in \ell_\infty$ for every $x \in E$. In addition, $T$ is clearly linear and

$$\|T(x)\| = \sup\{|\varphi_n(x)| : n \in \mathbb{N}\} \leq \|x\|, \tag{3.4}$$

which proves, in particular, that $T$ is continuous. Note that

$$\|T(x_k)\| = \sup\{|\varphi_n(x_k)| : n \in \mathbb{N}\} \geq |\varphi_k(x_k)| = \|x_k\| \tag{3.5}$$

for every $k \in \mathbb{N}$. From (3.4) and (3.5) it follows that $\|T(x_k)\| = \|x_k\|$ for every $k \in \mathbb{N}$. From the denseness of the set $\{x_n : n \in \mathbb{N}\}$ and the continuity of the function $x \in E \mapsto \|T(x)\| \in \mathbb{R}$, it follows that $\|T(x)\| = \|x\|$ for every $x$ in $E$. Then $T$ is an isometric isomorphism between $E$ and $T(E) \subseteq \ell_\infty$. ∎

## 3.4 Geometric Forms of the Hahn–Banach Theorem (Real Case)

In this section we will prove two theorems, known as the first and second geometric forms of the Hahn–Banach theorem. These results say, essentially, that if $A$ and $B$ are subsets of the real normed space $E$ that are not too intertwined (in a sense that will become clear in due time), then there exists a continuous linear functional on $E$ such that

$$\varphi(x) < \varphi(y) \text{ for all } x \in A \text{ and } y \in B.$$

Under certain conditions on $A$ and $B$ one can go further and guarantee that

$$\sup\{\varphi(x) : x \in A\} < \inf\{\varphi(x) : x \in B\}.$$

We can interpret these results from the perspective that the functional $\varphi$ separates the sets $A$ and $B$. Therefore, these results are also called *separation theorems*. For simplicity, we will only deal with the real case, although there are similar results for the complex case (see Exercise 3.6.26). Thus, throughout this section all vector spaces are real, that is, $\mathbb{K} = \mathbb{R}$. In Chap. 8 the geometric forms of Hahn-Banach will be treated in a more general context.

## 3.4 Geometric Forms of the Hahn–Banach Theorem (Real Case)

**Definition 3.4.1** Let $V$ be a non-null vector space. A *hyperplane* of $V$ is a subspace $W \neq V$ such that if $W_1$ is a subspace of $V$ and $W \subseteq W_1$, then $W_1 = V$ or $W_1 = W$.

We start off by proving that hyperplanes are exactly the kernels of linear functionals:

**Proposition 3.4.2** *Let $V$ be a vector space, $V \neq \{0\}$, and let $W$ be a subspace of $V$. Then $W$ is a hyperplane of $V$ if and only if there is a non-null linear functional $\varphi: V \longrightarrow \mathbb{R}$ such that $\ker(\varphi) = W$.*

**Proof** Suppose that $W$ is a hyperplane of $V$. As $W \neq V$, we can choose $v_0 \in V - W$. Taking $W_1 = [W \cup \{v_0\}]$, as $W \subseteq W_1$ and $W \neq W_1$, we have $W_1 = V$. It follows easily that each $v \in V$ is uniquely written as $v = u + av_0$ with $u \in W$ and $a \in \mathbb{R}$. It is immediate that the functional

$$\varphi: V \longrightarrow \mathbb{R}, \quad \varphi(u + av_0) = a,$$

is non-null, linear and $\ker(\varphi) = W$.

Conversely, let $\varphi: V \longrightarrow \mathbb{R}$ be a non-null linear functional such that $\ker(\varphi) = W$. It is clear that $\ker(\varphi) \neq V$ because $\varphi \neq 0$. Let $W_1$ be a subspace of $V$ such that $\ker(\varphi) \subseteq W_1$. We just need to show that if $\ker(\varphi) \neq W_1$, then $W_1 = V$. Suppose $\ker(\varphi) \neq W_1$ and choose $v_0 \in W_1 - \ker(\varphi)$. Given $v \in V$, taking $u = v - \frac{\varphi(v)}{\varphi(v_0)} v_0$ we have $u \in \ker(\varphi) \subseteq W_1$. Now it follows that $v = u + \frac{\varphi(v)}{\varphi(v_0)} v_0 \in W_1$. ∎

If $H$ is a hyperplane of $V$ and $v_0 \in V$, the set

$$v_0 + H = \{v_0 + v : v \in H\}$$

is called an *affine hyperplane* of $V$. From Proposition 3.4.2 it follows that the affine hyperplanes of $V$ are precisely the sets of the form

$$\{v \in V : \varphi(v) = a\},$$

where $\varphi$ is a non-null linear functional on $V$ and $a$ is a real number. From now on, affine hyperplanes will be simply called hyperplanes.

**Proposition 3.4.3** *Let $H = \{x \in E : \varphi(x) = a\}$ be a hyperplane of a normed space $E$, where $\varphi$ is a linear functional on $E$ and $a \in \mathbb{R}$. Then the hyperplane $H$ is closed if and only if the linear functional $\varphi$ is continuous.*

**Proof** Assuming $\varphi$ is continuous, the hyperplane $H = \varphi^{-1}(\{a\})$ is closed as the inverse image of a closed set by a continuous function. Conversely, suppose that $H$ is closed. Then $E - H$ is open and non-empty, so there exist $x_0 \in E - H$ and $r > 0$ such that $B(x_0; r) \subseteq E - H$. As $\varphi(x_0) \neq a$, we can assume, without loss of generality, that $\varphi(x_0) < a$.

Claim: $\varphi(x) < a$ for every $x \in B(x_0; r)$.

Indeed, if there was $x_1 \in B(x_0; r)$ with $\varphi(x_1) > a$, choosing $t = \frac{\varphi(x_1)-a}{\varphi(x_1)-\varphi(x_0)}$ we would have $tx_0 + (1-t)x_1 \in B(x_0; r)$ and

$$\varphi(tx_0 + (1-t)x_1) = t\varphi(x_0) + (1-t)\varphi(x_1)$$
$$= \frac{\varphi(x_1) - a}{\varphi(x_1) - \varphi(x_0)}\varphi(x_0) + \left(1 - \frac{\varphi(x_1) - a}{\varphi(x_1) - \varphi(x_0)}\right)\varphi(x_1) = a.$$

Since this contradicts the fact that $B(x_0; r) \subseteq E - H$, the claim follows. Therefore, $\varphi(x_0) + r\varphi(z) = \varphi(x_0 + rz) < a$ for every $z \in E$ with $\|z\| < 1$. As $\|-z\| = \|z\|$, it follows that

$$\frac{\varphi(x_0) - a}{r} < \varphi(z) < \frac{a - \varphi(x_0)}{r},$$

for every $z \in E$ with $\|z\| < 1$. This implies that $\|\varphi\| \leq \frac{a-\varphi(x_0)}{r}$, which guarantees the continuity of $\varphi$. ∎

Next, we will introduce the Minkowski functional, which will be useful in obtaining the geometric forms of the Hahn-Banach Theorem. Besides, in Chap. 8 it will play a central role in the construction of the theory of Topological Vector Spaces.

**Definition 3.4.4** Let $C$ be an open convex subset of the normed space $E$, containing the origin. The map

$$p_C : E \longrightarrow \mathbb{R}, \quad p_C(x) = \inf\left\{a > 0 : \frac{x}{a} \in C\right\},$$

is called the *Minkowski functional of $C$*.

Next result shows that the Minkowski functional is *almost* a norm on the space. Under further conditions, it is indeed a norm equivalent (or even equal) to the original norm of the space, see Exercises 3.6.23 and 3.6.24.

**Proposition 3.4.5** *The Minkowski functional enjoys the following properties:*

(a) $p_C(bx) = bp_C(x)$ *for every $b > 0$ and every $x \in E$.*
(b) $C = \{x \in E : p_C(x) < 1\}$.
(c) *There exists $M > 0$ such that $0 \leq p_C(x) \leq M \|x\|$ for every $x$ in $E$.*
(d) $p_C(x + y) \leq p_C(x) + p_C(y)$ *for any $x, y \in E$.*

**Proof**

(a) $p_C(bx) = \inf\{a > 0 : \frac{bx}{a} \in C\} = b \cdot \inf\{a > 0 : \frac{x}{a} \in C\} = bp_C(x)$.
(b) Since $C$ is open, for every $x \in C$ there exists $\varepsilon > 0$ such that $(1+\varepsilon)x \in C$. Thus $\frac{x}{(1+\varepsilon)^{-1}} \in C$, therefore $p_C(x) \leq (1+\varepsilon)^{-1} < 1$. Conversely, if $p_C(x) < 1$, then from the definition of infimum there exists $0 < a < 1$ such that $\frac{x}{a} \in C$. Since $C$ is convex we have $x = a\left(\frac{x}{a}\right) + (1-a) \cdot 0 \in C$.

## 3.4 Geometric Forms of the Hahn–Banach Theorem (Real Case)

(c) Let $r > 0$ be such that $B(0; r) \subseteq C$. For $0 < s < r$ we have $s \frac{x}{\|x\|} \in C$ for every $x \in E$, $x \neq 0$. From (a) and (b) it follows that $p_C(x) \leq \frac{\|x\|}{s}$ for every non-null $x \in E$. It is clear that this last inequality also holds for $x = 0$, hence the result is proven with $M = 1/s$.

(d) Let $x, y \in E$ and $\varepsilon > 0$ be given. We first prove that $\frac{x}{p_C(x)+\varepsilon} \in C$. Indeed, by (a),

$$p_C\left(\frac{x}{p_C(x)+\varepsilon}\right) = \frac{1}{p_C(x)+\varepsilon} p_C(x) < 1,$$

and by (b) it follows that $\frac{x}{p_C(x)+\varepsilon} \in C$. Similarly, $\frac{y}{p_C(y)+\varepsilon} \in C$. As $C$ is convex, choosing $0 < t := \frac{p_C(x)+\varepsilon}{p_C(x)+p_C(y)+2\varepsilon} < 1$, we have

$$\frac{x+y}{p_C(x)+p_C(y)+2\varepsilon} = t \frac{x}{p_C(x)+\varepsilon} + (1-t)\frac{y}{p_C(y)+\varepsilon} \in C.$$

From (a) and (b) it follows that

$$\frac{1}{p_C(x)+p_C(y)+2\varepsilon} p_C(x+y) = p_C\left(\frac{x+y}{p_C(x)+p_C(y)+2\varepsilon}\right) < 1,$$

therefore $p_C(x+y) < p_C(x) + p_C(y) + 2\varepsilon$. Since $\varepsilon > 0$ is arbitrary, the proof is done. ∎

At this point the reader should have noticed that the idea is to apply the Hahn–Banach Theorem with the Minkowski functional playing the role of the functional $p$.

**Lemma 3.4.6** *Let $C$ be a convex, open, proper non-empty subset of the normed space $E$ and let $x_0 \in E - C$ be given. Then there exists a linear functional $\varphi \in E'$ such that $\varphi(x) < \varphi(x_0)$ for every $x \in C$.*

**Proof** Choose $z_0 \in C$, consider $D = \{x - z_0 : x \in C\}$ and call $y_0 = x_0 - z_0$. We have that $y_0 \notin D$, $0 \in D$ and $D$ is convex and open. So, we can consider the Minkowski functional $p_D$ of $D$. Defining $G = [y_0]$ and $g(ty_0) = t \cdot p_D(y_0)$ for every $t \in \mathbb{R}$, we have $g(x) \leq p_D(x)$ for every $x \in G$. In fact, for $t > 0$,

$$g(ty_0) = t \cdot p_D(y_0) = p_D(ty_0)$$

by Proposition 3.4.5(a). For $t \leq 0$,

$$g(ty_0) = t \cdot p_D(y_0) \leq 0 \leq p_D(ty_0),$$

because $p_D(x) \geq 0$ for every $x \in E$. By the Hahn–Banach Theorem there exists a linear functional $\varphi: E \longrightarrow \mathbb{R}$ such that $\varphi(x) = g(x)$ for every $x \in G$ and $\varphi(x) \leq p_D(x)$ for every $x \in E$. By Proposition 3.4.5(c), there exists $M > 0$ such

**Fig. 3.1** Geometric interpretation of Lemma 3.4.6

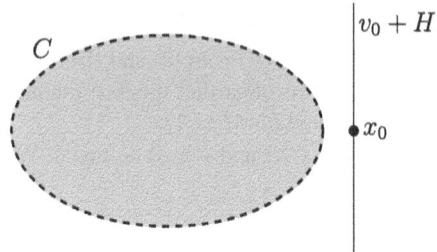

that $\varphi(x) \leq p_D(x) \leq M \|x\|$ for every $x \in E$, which guarantees the continuity of $\varphi$. From Proposition 3.4.5(b) it follows that $\varphi(y) \leq p_D(y) < 1$ for every $y \in D$. Since $p_D(y_0) \geq 1$ because $y_0 \notin D$,

$$\varphi(y) < 1 \leq p_D(y_0) = g(y_0) = \varphi(y_0) = \varphi(x_0 - z_0)$$

for every $y \in D$. From the previous inequality and the definition of $D$ it follows that

$$\varphi(x) = \varphi(x - z_0) + \varphi(z_0) < \varphi(x_0 - z_0) + \varphi(z_0) = \varphi(x_0)$$

for every $x$ in $C$ (Fig. 3.1). ∎

The last ingredient we need is an interesting property of continuous linear functionals on normed spaces:

**Lemma 3.4.7** *Let $E$ be a normed space, let $\varphi \in E'$ be a non-zero functional and let $A$ be a convex open non-empty subset of $E$. Then $\varphi(A)$ is an open interval (not necessarily bounded).*

**Proof** Since the range of a linear functional is a subspace of $\mathbb{R}$ and $\varphi \neq 0$, we have that $\varphi$ is surjective. From the linearity of $\varphi$ and the convexity of $A$ it easily follows that $\varphi(A)$ is a convex subset of $\mathbb{R}$, that is, an interval. It remains to prove that $\varphi(A)$ is open. We first prove that, if $a = \sup \varphi(A) < \infty$, then $a \notin \varphi(A)$. Suppose that $a \in \varphi(A)$. In this case, there exists $x \in A$ such that $\varphi(x) = a$ and $\varphi(y) \leq a$ for every $y \in A$. As $A$ is open, there exists $\varepsilon > 0$ such that the open ball $B(x; \varepsilon)$ is contained in $A$. Let $z \in E$ be a non-zero vector. From

$$\left\| x + \frac{\varepsilon}{2\|z\|} \cdot z - x \right\| = \frac{\varepsilon}{2\|z\|} \cdot \|z\| = \frac{\varepsilon}{2} < \varepsilon$$

it follows that $x + \frac{\varepsilon}{2\|z\|} \cdot z$ belongs to $A$. In this case,

$$a \geq \varphi\left(x + \frac{\varepsilon}{2\|z\|} \cdot z\right) = \varphi(x) + \frac{\varepsilon}{2\|z\|} \cdot \varphi(z) = a + \frac{\varepsilon}{2\|z\|} \cdot \varphi(z),$$

### 3.4 Geometric Forms of the Hahn–Banach Theorem (Real Case)

from which we conclude that $\varphi(z) \leq 0$. As $z$ is an arbitrary non-zero vector in $E$, this contradicts the surjectivity of $\varphi$, therefore $a \notin \varphi(A)$. Similarly, one can prove that if $b = \inf \varphi(A)$ is finite, then $b$ does not belong to $\varphi(A)$. The proof is complete. ∎

To simplify the notation, the hyperplane $H = \{x \in E : \varphi(x) = a\}$ will be denoted by the symbol $[\varphi = a]$.

**Theorem 3.4.8 (First Geometric Form of the Hahn–Banach Theorem)** *Let $A$ and $B$ be convex non-empty disjoint subsets of the normed space $E$. If $A$ is open, then there exist a functional $\varphi \in E'$ and $a \in \mathbb{R}$ such that*

$$\varphi(x) < a \leq \varphi(y) \text{ for all } x \in A \text{ and } y \in B.$$

*In this case, it is said that the closed hyperplane $[\varphi = a]$ separates $A$ and $B$.*

**Proof** Define $C = \{a - b : a \in A \text{ and } b \in B\}$ and note that:

(i) $C$ is open, because $C = \bigcup_{b \in B} (\{a - b : a \in A\})$.

(ii) $C$ is convex, because for $x_1, x_2 \in A$, $y_1, y_2 \in B$ and $0 < t < 1$,

$$t(x_1 - y_1) + (1 - t)(x_2 - y_2) = (tx_1 + (1 - t)x_2) - (ty_1 + (1 - t)y_2) \in C.$$

(iii) $\emptyset \neq C \neq E$, because $A$ and $B$ are disjoint and non-empty.

Since $0 \notin C$, by Lemma 3.4.6 there exists $\varphi \in E'$ such that $\varphi(z) < \varphi(0) = 0$ for every $z \in C$. Therefore,

$$\varphi(x) = \varphi(x - y) + \varphi(y) < \varphi(y) \tag{3.6}$$

for all $x \in A$ and $y \in B$. Since $B$ is non-empty, $a := \sup_{x \in A} \varphi(x)$ is a real number. Obviously it is true that $\varphi(A) \subseteq (-\infty, a]$, and from (3.6) it follows that $a \leq \varphi(y)$ for every $y \in B$. From Lemma 3.4.7 we know that $\varphi(A)$ is an open interval, therefore $\varphi(A) \subseteq (-\infty, a)$, that is, $\varphi(x) < a$ for every $x \in A$ (Fig. 3.2). ∎

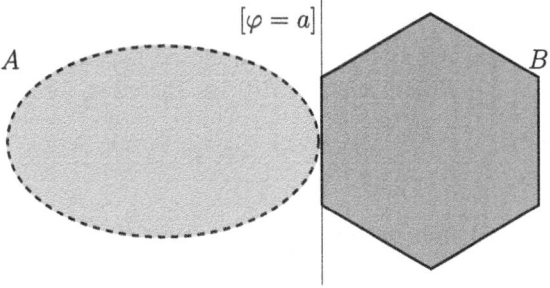

**Fig. 3.2** Geometric interpretation of the first geometric form of the Hahn–Banach Theorem

**Theorem 3.4.9 (Second Geometric Form of the Hahn–Banach Theorem)** *Let A and B be convex non-empty disjoint subsets of the normed space E. If A is closed and B is compact, then there exist a functional $\varphi \in E'$ and $a, b \in \mathbb{R}$ such that*

$$\varphi(x) \leq a < b \leq \varphi(y) \text{ for all } x \in A \text{ and } y \in B.$$

*For $c \in (a, b)$, it is said that the closed hyperplane $[\varphi = c]$ strictly separates A and B.*

**Proof** We start off by proving that it is possible to choose $\varepsilon > 0$ so that $A + B(0; \varepsilon)$ and $B + B(0; \varepsilon)$ are open, convex and disjoint:

(i) For any choice of $\varepsilon > 0$, the sets $A + B(0; \varepsilon)$ and $B + B(0; \varepsilon)$ are open. In fact,

$$A + B(0; \varepsilon) = \bigcup_{\alpha \in A} (\alpha + B(0; \varepsilon)) = \bigcup_{\alpha \in A} B(\alpha; \varepsilon),$$

and the same occurs for $B + B(0; \varepsilon)$.

(ii) For any choice of $\varepsilon > 0$, as $A$, $B$ and $B(0; \varepsilon)$ are convex, it is immediate that the sets $A + B(0; \varepsilon)$ and $B + B(0; \varepsilon)$ are convex.

(iii) It is enough to prove that it is possible to choose $\varepsilon$ so that $A + B(0; \varepsilon)$ and $B + B(0; \varepsilon)$ are disjoint. Suppose this is not possible. In this case, for each $n \in \mathbb{N}$, we have

$$\left(A + B\left(0; \tfrac{1}{n}\right)\right) \cap \left(B + B\left(0; \tfrac{1}{n}\right)\right) \neq \emptyset.$$

Therefore, there exist $(x_n)_{n=1}^\infty \subseteq A$, $(y_n)_{n=1}^\infty \subseteq B$ and $(z_n)_{n=1}^\infty, (w_n)_{n=1}^\infty \subseteq B\left(0; \tfrac{1}{n}\right)$ such that $x_n + z_n = y_n + w_n$ for every $n \in \mathbb{N}$. Then

$$\|x_n - y_n\| = \|z_n - w_n\| < \frac{2}{n} \text{ for every } n \in \mathbb{N}. \tag{3.7}$$

As $B$ is compact, the sequence $(y_n)_{n=1}^\infty$ has a convergent subsequence in $B$, say $y_{n_k} \longrightarrow \beta \in B$. From (3.7) it follows that $x_{n_k} \longrightarrow \beta$. As $A$ is closed, we know that $\beta \in A$, hence $\beta \in A \cap B$, which contradicts the fact that $A$ and $B$ are disjoint.

Fix $\varepsilon > 0$ such that $A + B(0; \varepsilon)$ and $B + B(0; \varepsilon)$ are disjoint. By Theorem 3.4.8 there is a closed hyperplane $[\varphi = c]$ in $E$ that separates $A + B(0; \varepsilon)$ and $B + B(0; \varepsilon)$. Therefore, for $x \in A$ and $y \in B$,

$$\varphi(x) + \sup_{\|z_1\| < \varepsilon} \varphi(z_1) \leq c \leq \varphi(y) + \inf_{\|z_2\| < \varepsilon} \varphi(z_2),$$

and, from the linearity of $\varphi$,

## 3.4 Geometric Forms of the Hahn–Banach Theorem (Real Case)

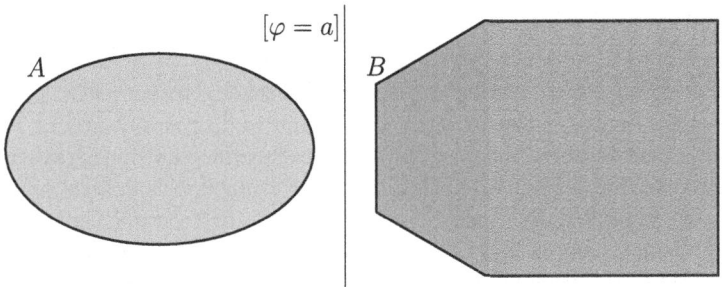

**Fig. 3.3** Geometric interpretation of the second geometric form of the Hahn–Banach Theorem

$$\varphi(x) + \varepsilon \cdot \sup_{\left\|\frac{z_1}{\varepsilon}\right\|<1} \varphi\left(\frac{z_1}{\varepsilon}\right) \leq c \leq \varphi(y) + \varepsilon \cdot \inf_{\left\|\frac{z_2}{\varepsilon}\right\|<1} \varphi\left(\frac{z_2}{\varepsilon}\right).$$

It follows that $\varphi(x) + \varepsilon \|\varphi\| \leq c \leq \varphi(y) - \varepsilon \|\varphi\|$ for all $x \in A$ and $y \in B$. It is enough to choose $a = c - \frac{\varepsilon \|\varphi\|}{2}$ and $b = c + \frac{\varepsilon \|\varphi\|}{2}$ (Fig. 3.3). ∎

The following variant of Proposition 3.3.1 follows very easily:

**Corollary 3.4.10** *Let $M$ be a closed subspace of the normed space $E$. Then for every $x_0 \in E - M$ there exists a functional $\varphi \in E'$ such that $\varphi(x_0) = 1$ and $\varphi(x) = 0$ for every $x \in M$.*

*Proof* Applying Theorem 3.4.9 for $A = M$ and $B = \{x_0\}$, there exist a functional $\Phi \in E'$ and $c \in \mathbb{R}$ such that $\Phi(x) < c < \Phi(x_0)$ for every $x \in M$. Since $\Phi(M)$ is a subspace of $\mathbb{R}$, we conclude that $\Phi(x) = 0$ for every $x \in M$. Thus $\Phi(x_0) > 0$ and it suffices to consider $\varphi = \frac{1}{\Phi(x_0)} \Phi$. ∎

**Corollary 3.4.11** *Let $M$ be a subspace of the normed space $E$. Then for every $x_0 \in E$, $x_0 \in \overline{M}$ if and only if $\varphi(x_0) = 0$ for every $\varphi \in E'$ such that $\varphi(x) = 0$ for each $x \in M$. That is,*

$$\overline{M} = \bigcap \{\ker(\varphi) : \varphi \in E' \text{ and } M \subseteq \ker(\varphi)\}.$$

*Proof* Suppose $x_0 \in \overline{M}$. For each $\varphi \in E'$ such that $M \subseteq \ker(\varphi)$, we have $\overline{M} \subseteq \overline{\ker(\varphi)} = \ker(\varphi)$ because $\ker(\varphi)$ is closed.

Conversely, suppose that $x_0 \notin \overline{M}$. As $\overline{M}$ is a closed subspace of $E$, by Corollary 3.4.10 there exists a functional $\varphi_0 \in E'$ such that $\varphi_0(x_0) = 1$ and $\varphi_0(x) = 0$ for every $x \in \overline{M}$. Hence $x_0 \notin \bigcap \{\ker(\varphi) : \varphi \in E' \text{ and } M \subseteq \ker(\varphi)\}$. ∎

## 3.5 Comments and Historical Notes

Simpler, or less general, versions of the Hahn–Banach Theorem were proven by Helly in 1912 and by Hahn in 1922. Hahn improved his result and in 1927 he published a version of the theorem for functionals defined on Banach spaces. The definitive form of the Hahn–Banach Theorem, for functionals defined on normed spaces, was published by Banach in 1929. These first proofs used transfinite induction; Zorn's Lemma appeared only in 1935. Two reliable historical sources, [48] and [51], emphasize that Helly had a much greater importance in the birth of Functional Analysis than is usually credited to him.

The complex case of the Hahn–Banach Theorem was first proved for the spaces $L_p$, $p > 1$, by F. Murray in 1936. The general case was proven independently in two works, one by G. A. Soukhomlinov and another by H. F. Bohnenblust and A. Sobczyk, both curiously published in the same year of 1938. In 1975, J. Holbrook proved a version that holds in any archimedean field. In the proof of the Hahn–Banach Theorem, we used the Axiom of Choice in the form of Zorn's Lemma. The reader may wonder if, like Zorn's Lemma itself, Zermelo's Theorem (Well-Ordering Principle), and several other known statements, the Hahn–Banach Theorem is equivalent to the Axiom of Choice. The answer is negative and the proof can be found in [48, 7.11].

The search for vector-valued versions of the Hahn–Banach Theorem, in the sense of Sect. 3.2, has become an important aspect of Banach space theory. As we saw in Theorem 3.2.7, operators with values in $\ell_\infty$ can always be extended, preserving the norm. The same applies to other spaces, for example, the space of bounded functions $B(X)$ from Example 1.1.2 (see Exercise 3.6.17) and the space $L_\infty(\mu)$, where $\mu$ is a probability measure (see [17, Theorem 4.14]). In this context, a Banach space $E$ is said to be *injective* if the thesis of the Hahn–Banach Theorem holds for operators with values in $E$, that is, if for all Banach spaces $F$ and $G$ with $F$ being a subspace of $G$, every continuous linear operator from $G$ to $E$ admits a continuous linear extension to $F$, preserving the norm. For now, we know that $\ell_\infty$, $B(X)$ and $L_\infty(\mu)$, with $\mu$ a probability measure, are injective. The same argument of the proof of Corollary 3.2.8 shows that every injective space is 1-complemented in any space that contains it as a subspace. A. Grothendieck proved in 1953 that every injective infinite-dimensional space is non-separable. Injectivity is a slippery and sometimes surprising property, for example, it is not hereditary: we know $\ell_\infty$ is injective but its closed subspace $c_0$ is not injective (because it is separable). The structure of injective spaces was disclosed by the work of several mathematicians during the period 1950–1958, including the eminent Brazilian mathematician Leopoldo Nachbin (1922–1993):

**Theorem 3.5.1 (Nachbin, Goodner, Kelley, Hasumi)** *The following statements are equivalent for an infinite-dimensional (real or complex) Banach space $E$:*

(a) *$E$ is injective.*
(b) *$E$ is 1-complemented in any Banach space that contains it as a subspace.*

(c) *E* has the binary intersection property, that is, every collection of closed balls in *E* with mutual intersection has a non-empty intersection.
(d) *E* is isometrically isomorphic to the space $C(K)$ of continuous scalar-valued functions defined on an extremely disconnected compact topological space $K$ with the supremum norm.

A topological space $K$ is extremely disconnected if the closure of any open subset of $K$ is open. L. Nachbin, in 1950, also definitively settled the question of real injective finite-dimensional spaces:

**Theorem 3.5.2 (Nachbin [46])** *A real n-dimensional Banach space $E$, $n \in \mathbb{N}$, is injective if and only if $E$ is isometrically isomorphic to $(\mathbb{R}^n, \|\cdot\|_\infty)$.*

The interest in the possibility of extending functions maintaining their properties dates back, at least, to Tietze's Theorem in topological spaces. For the case of extensions of Lipschitz functions, see [19, Theorem 6.1.1]. It is interesting to note that the Hahn-Banach Theorem can be obtained as a consequence of this theorem of extension of Lipschitz functions (see [19, Theorem 6.1.4]).

Exhibiting a non-complemented closed subspace of a Banach space, or even proving its existence, is never an easy task. Besides the examples mentioned in the text, it is also known that $\ell_1$ contains a non-complemented closed subspace (see [2, Corollary 2.3.3]), that $L_1[0, 1]$ contains a non-complemented subspace isomorphic to $\ell_2$ (see [17, p. 14]) and that for $1 < p < \infty$, $p \neq 2$, $L_p[0, 1]$ contains a non-complemented subspace isomorphic to $\ell_p$ (see [17, p. 26]).

The first version of Theorem 3.4.8 was proven by Mazur in 1933. To state it we need the following concept: an *affine subspace* (or *linear manifold*) of a vector space $V$ is the translation of a vector subspace, that is, it is a set of the form $x_0 + W$ where $x_0 \in V$ and $W$ is a vector subspace of $V$.

**Theorem 3.5.3 ([48, Theorem 7.7.3])** *Let $A$ be an open convex subset of a normed space $E$ and let $M$ be an affine subspace of $E$ such that $M \cap A = \emptyset$. Then there exists a closed hyperplane $H$ of $E$ containing $M$ such that $H \cap A = \emptyset$.*

This result due to Mazur was called the *geometric form of the Hahn-Banach Theorem* in the books of the Bourbaki collection. The reason for this denomination is that this result generalizes the fact that every line in $\mathbb{R}^3$ that does not pierce a given sphere is contained in a plane that does not intersect the interior of the sphere. Since then, results of this type are usually called geometric forms of the Hahn-Banach Theorem.

## 3.6 Exercises

**Exercise 3.6.1** Let $E$ be a normed space, $E \neq \{0\}$. Show that $E' \neq \{0\}$.

**Exercise 3.6.2** Let $E$ be a normed space. Show that $E'$ *separates points* of $E$, that is, if $x, y \in E$ and $x \neq y$, then there exists $\varphi \in E'$ such that $\varphi(x) \neq \varphi(y)$.

**Exercise 3.6.3** Let $x_1, \ldots, x_n$ be linearly independent vectors of the normed space $E$ and let $a_1, \ldots, a_n$ be given scalars. Show that there exists a functional $\varphi \in E'$ such that $\varphi(x_j) = a_j$ for each $j = 1, \ldots, n$.

**Exercise 3.6.4** Show that, in general, the Hahn–Banach extension is not unique.

**Exercise 3.6.5** In Theorem 3.1.2, if $E$ is a normed space and there exists $C > 0$ such that $p(x) \leq C\|x\|$ for every $x \in E$, prove that $\|\widetilde{\varphi}\| \leq C$.

**Exercise 3.6.6** Prove the Hahn–Banach Theorem, in the form of Corollary 3.1.3, in the case where $E$ is a separable space without using Zorn's Lemma, that is, without using Theorems 3.1.1 and 3.1.2.

**Exercise 3.6.7** Let $F$ be a normed space. Prove that if there exists a normed space $E \neq \{0\}$ such that $\mathcal{L}(E, F)$ is a Banach space, then $F$ is also a Banach space.

**Exercise 3.6.8** Let $E$ and $F$ be normed spaces and let $T \in \mathcal{L}(E, F)$ be given. Prove that $\|T\| = \sup\{|\varphi(T(x))| : \varphi \in B_{F'} \text{ and } x \in B_E\}$.

**Exercise 3.6.9** Let $F$ be a dense subspace of the normed space $E$. Prove that $F'$ and $E'$ are isometrically isomorphic.

**Exercise 3.6.10** Let $E, F$ be normed spaces, let $G$ be a subspace of $E$ and let $T \in \mathcal{L}(G, F)$ be a finite rank operator (see Exercise 2.7.20). Prove that there exists $\widetilde{T} \in \mathcal{L}(E, F)$ that extends $T$ and has the same range as $T$.

**Exercise 3.6.11** Let $E$ be a separable normed space. Prove that there exists a sequence $(\varphi_n)_{n=1}^\infty$ in $E'$ such that $\|\varphi_n\| = 1$ for every $n$ and, for every $x \in E$, $\|x\| = \sup_n |\varphi_n(x)|$ in the complex case and $\|x\| = \sup_n \varphi_n(x)$ in the real case.

**Exercise 3.6.12** Let $E$ be a normed space and let $F$ be a subspace of $E$. Show that there exists an injective map $T : F' \longrightarrow E'$ such that $\|T(\varphi)\| = \|\varphi\|$ for every functional $\varphi \in F'$. Does this mean that $E'$ contains a subspace isometrically isomorphic to $F'$?

**Exercise 3.6.13\*** (**Annihilator**) Let $E$ be a normed space and let $M$ be a non-empty subset of $E$. The *annihilator of $M$* is defined by

$$M^\perp = \{\varphi \in E' : \varphi(x) = 0 \text{ for every } x \in M\}.$$

Prove that:

(a) $M^\perp$ is a closed subspace of $E'$.
(b) If $M$ is a subspace of $E$, then $M'$ is isometrically isomorphic to $E'/M^\perp$.
(c) If $M$ is a closed subspace of $E$, then $M^\perp$ is isometrically isomorphic to $(E/M)'$.

**Exercise 3.6.14** Let $F$ and $G$ be distinct closed subspaces of the normed space $E$. Prove that $F^\perp \neq G^\perp$.

**Exercise 3.6.15** Let $F$ be a subspace of the Banach space $E$ whose dual $E'$ is separable. Prove that $F'$ is separable.

## 3.6 Exercises

**Exercise 3.6.16** Given a normed space $E$ and a vector $x_0 \in E$, prove that the set

$$\left\{\varphi \in E' : \|\varphi\| = \|x_0\| \text{ and } \varphi(x_0) = \|x_0\|^2\right\}$$

is non-empty, convex, and closed in $E'$.

**Exercise 3.6.17** Let $F$ be a subspace of the normed space $E$ and let $T \in \mathcal{L}(F, B(X))$ be given, where $B(X)$ is the space from Example 1.1.2. Prove that there exists an extension $\widetilde{T} \in \mathcal{L}(E, B(X))$ of $T$ to $E$ such that $\|\widetilde{T}\| = \|T\|$.

**Exercise 3.6.18** Let $E$ be a Banach space and let $F$ be a complemented subspace of $E$. Prove that if $E = F \oplus M$ and $E = F \oplus N$, then $M$ is isomorphic to $N$.

**Exercise 3.6.19** Sobczyk's Theorem states the following: Let $F$ be a closed subspace of the separable normed space $E$ and let $T \in \mathcal{L}(F, c_0)$ be given. Then there exists an extension $\widetilde{T} \in \mathcal{L}(E; c_0)$ of $T$ with $\|\widetilde{T}\| \leq 2\|T\|$. For a proof, see [2, Theorem 2.5.8].

Using Sobczyk's Theorem, show that if $c_0$ is a complemented subspace of a separable normed space $E$, then $c_0$ is 2-complemented in $E$.

**Exercise 3.6.20** Let $F$ be a closed subspace of the Banach space $E$ such that $E/F$ is finite-dimensional. Prove that $F$ is complemented in $E$.

**Exercise 3.6.21** Let $F$ be a subspace of the Banach space $E$, let $P: E \longrightarrow E$ be a projection onto $F$ and let $Q \in \mathcal{L}(E, E)$ be given. If $P \circ Q = Q$ and $Q \circ P = P$, then $Q$ is also a projection onto $F$.

**Exercise 3.6.22** Consider the following translation operator

$$T: \ell_\infty \longrightarrow \ell_\infty, \quad T((a_1, a_2, a_3, \ldots)) = (a_2, a_3, a_4, \ldots).$$

(a) Prove that $T \in \mathcal{L}(\ell_\infty, \ell_\infty)$ and calculate the norm of $T$.
(b) In the real case, prove that there exists a functional $\varphi \in (\ell_\infty)'$ such that

$$\varphi \circ T = \varphi \text{ and } \liminf_n a_n \leq \varphi((a_j)_{j=1}^\infty) \leq \limsup_n a_n \text{ for every } (a_j)_{j=1}^\infty \in \ell_\infty.$$

**Exercise 3.6.23** Let $C$ be a convex open subset containing the origin of the real normed space $E$. Prove that if $C$ is bounded and symmetric (that is, $C = -C$), then the Minkowski functional $p_C$ of $C$ is a norm equivalent to the original norm of $E$.

**Exercise 3.6.24** Let $(E, \|\cdot\|)$ be a real normed space. Prove that the Minkowski functional of the open ball $B(0; 1) = \{x \in E : \|x\| < 1\}$ coincides with $\|\cdot\|$.

**Exercise 3.6.25*** Let $A$ and $B$ be convex subsets of the real normed space $E$ such that

$$\delta := \mathrm{dist}(A, B) = \inf\{\|x - y\| : x \in A \text{ and } y \in B\} > 0.$$

Prove that there exists a functional $\varphi \in E'$ such that $\|\varphi\| = 1$ and $\inf \varphi(A) = \sup \varphi(B) + \delta$.

**Exercise 3.6.26 (Complex Case of the Geometric Forms of the Hahn–Banach Theorem)** Let $E$ be a complex normed space and let $A$ and $B$ be disjoint nonempty convex subsets of $E$.

(a) If $A$ is open, then there exist a functional $\varphi \in E'$ and a real number $a$ such that
$$\mathrm{Re}(\varphi(x)) \leq a \leq \mathrm{Re}(\varphi(y)) \text{ for all } x \in A \text{ and } y \in B.$$

(b) If $A$ is closed and $B$ is compact, then there exist a functional $\varphi \in E'$ and real numbers $a$ and $b$ such that
$$\mathrm{Re}(\varphi(x)) < a < b < \mathrm{Re}(\varphi(y)) \text{ for all } x \in A \text{ and } y \in B.$$

**Exercise 3.6.27** A *closed half-space* in a real normed space $E$ is a set of the form
$$\varphi^{-1}([a, \infty)) = \{x \in E : \varphi(x) \geq a\},$$
where $\varphi \in E'$ and $a \in \mathbb{R}$. Prove that every closed convex subset of a real normed space is the intersection of all closed half-spaces that contain it.

**Exercise 3.6.28**

(a) Prove that the operator
$$T : \ell_\infty \longrightarrow \mathcal{L}(\ell_2, \ell_2), \quad T((a_j)_{j=1}^\infty)((b_j)_{j=1}^\infty) = (a_j b_j)_{j=1}^\infty,$$
is a linear isometry.

(b) Conclude that $\mathcal{L}(\ell_2, \ell_2)$ is not separable.

**Exercise 3.6.29** Prove that a subspace $M$ of the normed space $E$ is dense in $E$ if and only if the only functional $\varphi \in E'$ that vanishes on $M$ is the zero functional.

**Exercise 3.6.30** Let $x$ and $(x_i)_{i \in I}$ be vectors of a normed space $E$. Prove that $x \in \overline{[x_i : i \in I]}$ if and only if $\varphi(x) = 0$ for every $\varphi \in E'$ such that $\varphi(x_i) = 0$ for every $i \in I$.

**Exercise 3.6.31** Let $K_1, \ldots, K_n$ be closed convex subsets of the real normed space $E$, each containing the origin, and let $c_1, \ldots, c_n$ be positive real numbers. Prove that if $x \in E$ and $x$ cannot be written in the form $c_1 x_1 + \cdots + c_n x_n$ with $x_i \in K_i$ for each $i = 1, \ldots, n$, then there exists $\varphi \in E'$ such that $\varphi(x) > 1$ and $\varphi(y) \leq \frac{1}{c_i}$ for each $i = 1, \ldots, n$ and every $y \in K_i$.

# Chapter 4
# Duality and Reflexive Spaces

We will begin this chapter by showing that it is possible to describe all continuous linear functionals on some of the classical spaces we have been working with. The relevant feature of this description of dual spaces is that the functionals are described as objects belonging to the same class as the elements of the space, namely, continuous linear functionals on function spaces shall be described as functions, and continuous linear functionals on sequence spaces shall be described as sequences.

Still in the context of duality, when we consider the dual of the dual of a normed space, we sometimes find ourselves back in the original space. Such spaces are called reflexive, the subject of Sect. 4.3. The study of reflexive spaces will provide an excellent opportunity to introduce and apply the important notion of adjoint of a continuous linear operator.

## 4.1 The Dual of the $L_p$ Spaces

From now on, given $1 < p < \infty$, by $p^*$ we will denote the number greater than 1 such that $\frac{1}{p} + \frac{1}{p^*} = 1$. For $p = 1$ we consider $p^* = \infty$. It is said that $p^*$ is the *conjugate* of $p$ and vice versa.

We first show how elements of $L_{p^*}$ generate linear functionals on $L_p$:

**Example 4.1.1** Let $1 \leq p < \infty$ and let $(X, \Sigma, \mu)$ be a measure space. Given a function $g \in L_{p^*}(X, \Sigma, \mu)$, consider the functional

$$\varphi_g \colon L_p(X, \Sigma, \mu) \longrightarrow \mathbb{K}, \ \varphi_g(f) = \int_X fg \, d\mu.$$

From Hölder's inequality, which is also valid for $p = 1$, it follows that

$$|\varphi_g(f)| = \left|\int_X fg\, d\mu\right| \leq \int_X |fg|\, d\mu \leq \|g\|_{p^*} \cdot \|f\|_p$$

for every $f \in L_p(X, \Sigma, \mu)$. This ensures that $\varphi_g$ is well defined and, as in this case linearity is clear, it also follows that $\varphi_g \in L_p(X, \Sigma, \mu)'$ and $\|\varphi_g\| \leq \|g\|_{p^*}$.

To prove the equality $\|\varphi_g\| = \|g\|_{p^*}$ we need to split the proof into two cases. If $g = 0$ in $L_{p^*}(X, \Sigma, \mu)$, there is nothing to do. We can assume $\|g\|_{p^*} \neq 0$. We begin by considering $p > 1$. In this case $p^* < \infty$, so we can define the function

$$f(x) = \frac{|g(x)|^{p^*-1} \cdot \overline{g(x)}}{\|g\|_{p^*}^{p^*-1} \cdot |g(x)|} \text{ if } g(x) \neq 0 \text{ and } f(x) = 0 \text{ if } g(x) = 0.$$

As $(p^* - 1)p = p^*$, it follows that $f \in L_p(X, \Sigma, \mu)$ and $\|f\|_p = 1$. A quick calculation gives $|\varphi_g(f)| = \|g\|_{p^*}$, which shows that $\|\varphi_g\| = \|g\|_{p^*}$.

For the case $p = 1$, suppose that the measure $\mu$ is $\sigma$-finite, that is, there are sets $(A_n)_{n=1}^\infty$ in $\Sigma$, all of them of finite measure, such that $X = \bigcup_{n=1}^\infty A_n$. Let $0 < c < \|g\|_\infty$ be given. Considering $A = \{x \in X : |g(x)| > c\}$, from the definition of $\|\cdot\|_\infty$ it follows that $\mu(A) > 0$. From $X = \bigcup_{n=1}^\infty A_n$ it turns out that $\mu(A \cap A_n) > 0$ for some $n$. Denoting $B = A \cap A_n$, we have $0 < \mu(B) < \infty$ and $|g(x)| > c$ for every $x \in B$. Defining

$$f(x) = \frac{\overline{g(x)}}{\mu(B)|g(x)|} \text{ if } x \in B \text{ and } f(x) = 0 \text{ if } x \notin B,$$

we have $\|f\|_1 = 1$, so $\|\varphi_g\| \geq |\varphi_g(f)| = \int_B \frac{|g|}{\mu(B)}\, d\mu \geq c$. As this holds for each $0 < c < \|g\|_\infty$, it follows that $\|\varphi_g\| \geq \|g\|_\infty$.

From the example above it immediately follows that the correspondence

$$g \in L_{p^*}(X, \Sigma, \mu) \mapsto \varphi_g \in L_p(X, \Sigma, \mu)', \ \varphi_g(f) = \int_X fg\, d\mu,$$

is a linear isometry (hence injective); bearing in mind that the measure must be $\sigma$-finite in the case $p = 1$. For the spaces $L_{p^*}(X, \Sigma, \mu)$ and $L_p(X, \Sigma, \mu)'$ to be isometrically isomorphic, it remains to check that this correspondence is surjective, that is, that every functional on $L_p(X, \Sigma, \mu)$ comes from a function of $L_{p^*}(X, \Sigma, \mu)$. This is exactly what the next theorem says, one of the several results known as *Riesz Representation Theorem*. In the case $\mathbb{K} = \mathbb{R}$ and $1 < p < \infty$, a proof that does not rely on elements of Measure Theory shall be presented in Theorem 6.6.13.

## 4.1 The Dual of the $L_p$ Spaces

**Theorem 4.1.2** *Let $1 \le p < \infty$ and let $(X, \Sigma, \mu)$ be a measure space. In the case $p = 1$ we suppose that $\mu$ is a $\sigma$-finite measure. Then the correspondence $g \mapsto \varphi_g$ establishes an isometric isomorphism between $L_{p^*}(X, \Sigma, \mu)$ and $L_p(X, \Sigma, \mu)'$ in which the duality relation is given by*

$$\varphi_g(f) = \int_X fg\, d\mu \text{ for every } f \in L_p(X, \Sigma, \mu).$$

**Proof** We will prove at this moment only the case in which the measure $\mu$ is $\sigma$-finite and $1 \le p < \infty$. This encompasses the most used measure spaces. Anyway, the case in which $p > 1$ and $\mu$ is an arbitrary measure will be proved in Appendix C.

From our previous discussion, it is enough to prove that given a functional $\varphi \in L_p(X, \Sigma, \mu)'$, there exists $g \in L_{p^*}(X, \Sigma, \mu)$ such that $\varphi_g = \varphi$. We first suppose that the measure $\mu$ is finite. It is not difficult to check that the correspondence $A \in \Sigma \mapsto \varphi(\mathcal{X}_A)$, where $\mathcal{X}_A$ is the characteristic function of $A$, is a measure (note that, since we are assuming $\mu$ finite, $\mathcal{X}_A \in L_p(X, \Sigma, \mu)$). As $\varphi(0) = 0$ we have $\varphi(\mathcal{X}_A) = 0$ for each $A \in \Sigma$ such that $\mu(A) = 0$. By the Radon–Nikodým Theorem (see Theorem C.18) there exists a non-negative function $g \in L_1(X, \Sigma, \mu)$, which in this case can be chosen to be real-valued, such that $\varphi(\mathcal{X}_A) = \int_A g\, d\mu$ for every $A \in \Sigma$. Therefore, $\varphi(f) = \int_X fg\, d\mu$ for every simple function $f$. Thus, for every simple function $f$ we have $fg \in L_1(X, \Sigma, \mu)$ and

$$\left| \int_X fg\, d\mu \right| = |\varphi(f)| \le \|f\|_p \cdot \|\varphi\|.$$

It follows from the reciprocal of Hölder's inequality (Exercise 1.8.39) that $g \in L_{p^*}(X, \Sigma, \mu)$. As the set of simple functions is dense in $L_p(X, \Sigma, \mu)$ [25, Proposition 6.7], from the linearity and continuity of $\varphi$ it follows that $\varphi(f) = \int_X fg\, d\mu$ for every $f \in L_p(X, \Sigma, \mu)$, that is, $\varphi_g = \varphi$.

We now consider the case in which $\mu$ is $\sigma$-finite. In this case we can consider an increasing sequence $(A_n)_{n=1}^\infty$ of measurable sets such that $0 < \mu(A_n) < \infty$ for every $n$ and $X = \bigcup_{n=1}^\infty A_n$. For each $n \in \mathbb{N}$, calling $\Sigma_n = \{A \cap A_n : A \in \Sigma\}$ and $\mu_n = \mu|_{\Sigma_n}$, it results that $(A_n, \Sigma_n, \mu_n)$ is a finite measure space. Defining $\varphi_n$ by

$$f_n \in L_p(A_n, \Sigma_n, \mu_n) \longrightarrow \varphi_n(f_n) = \varphi(f_n \mathcal{X}_{A_n}),$$

we have $\varphi_n \in L_p(A_n, \Sigma_n, \mu_n)'$. By the previous case there exists $g_n \in L_{p^*}(A_n, \Sigma_n, \mu_n)$ such that $\varphi_n(f_n) = \int_{A_n} f_n g_n\, d\mu_n$ for each $f_n \in L_p(A_n, \Sigma_n, \mu_n)$. Moreover,

$$\|g_n\|_{L_{p^*}(A_n, \Sigma_n, \mu_n)} = \|\varphi_n\|_{(L_p(A_n, \Sigma_n, \mu_n))'} \le \|\varphi\|_{L_p(X, \Sigma, \mu)'}.$$

Such $g_n$ is unique (we suggest the reader checks this out. Bearing in mind $\mathcal{L}_{p^*}$ instead of $L_{p^*}$, the reader may interpret that $g_n$ is unique up to sets of null measure, that is, another function with the same properties as $g_n$ coincides with $g_n$ up to a set of null measure). We can extend $g_n$ to $X$ by letting $g_n(x) = 0$ for every $x \notin A_n$. As the sequence $(A_n)_{n=1}^\infty$ is increasing, it follows that if $n < m$ then $g_n = g_m$ $\mu$-almost everywhere in $A_n$. It is then well defined and measurable the function $g$ given by $g(x) = g_n(x)$ if $x \in A_n$. By the Monotone Convergence Theorem we conclude that

$$\|g\|_{L_{p^*}(X,\Sigma,\mu)} = \lim_{n\to\infty} \|g_n\|_{L_{p^*}(A_n,\Sigma_n,\mu_n)} \leq \|\varphi\|_{L_p(X,\Sigma,\mu)'},$$

which guarantees that $g \in L_{p^*}(X, \Sigma, \mu)$. For each $f \in L_p(X, \Sigma, \mu)$, by the Dominated Convergence Theorem we know that $\|f\mathcal{X}_{A_n} - f\|_p \longrightarrow 0$, that is, $f\mathcal{X}_{A_n} \longrightarrow f$ in $L_p(X, \Sigma, \mu)$. The continuity of $\varphi$ implies that

$$\varphi(f) = \lim_{n\to\infty} \varphi(f\mathcal{X}_{A_n}) = \lim_{n\to\infty} \varphi_n(f|_{A_n})$$
$$= \lim_{n\to\infty} \int_{A_n} fg_n d\mu_n = \lim_{n\to\infty} \int_{A_n} fg d\mu = \int_X fg\, d\mu,$$

where the next-to-last equality is a consequence of the definition of $\mu_n$ and $g$, and the last equality is another application of the Dominated Convergence Theorem. This proves that $\varphi_g = \varphi$. ∎

## 4.2 The Duals of the Spaces $\ell_p$ and $c_0$

Let $1 \leq p < \infty$ be given. Recall that $\ell_p = L_p(X, \Sigma, \mu)$ where $X = \mathbb{N}$, $\Sigma$ is the set of all subsets of $\mathbb{N}$ and $\mu$ is the counting measure. Note also that the counting measure on $\Sigma$ is $\sigma$-finite. Then we can invoke Theorem 4.1.2. It is immediate that the duality relation, in this case, becomes

$$b = (b_j)_{j=1}^\infty \in \ell_{p^*} \mapsto \varphi_b \in (\ell_p)', \quad \varphi_b((a_j)_{j=1}^\infty) = \sum_{j=1}^\infty a_j b_j \text{ for any } (a_j)_{j=1}^\infty \in \ell_p.$$

(4.1)

Therefore, Theorem 4.1.2 in this setting becomes the following result:

**Proposition 4.2.1** *For $1 \leq p < \infty$, the correspondence $b \in \ell_{p^*} \mapsto \varphi_b \in (\ell_p)'$ establishes an isometric isomorphism between $\ell_{p^*}$ and $(\ell_p)'$ in which the duality relation is given by (4.1).*

**Remark 4.2.2** Knowing that the dual of $\ell_1$ is isometrically isomorphic to $\ell_\infty$, we conclude that the converse of Theorem 3.3.2 is not true, because $\ell_1$ is separable whereas $\ell_\infty$ is non-separable.

The same duality relation above also describes the dual of $c_0$:

## 4.2 The Duals of the Spaces $\ell_p$ and $c_0$

**Proposition 4.2.3** *The spaces $\ell_1$ and $(c_0)'$ are isometrically isomorphic through the duality relation*

$$b = (b_j)_{j=1}^{\infty} \in \ell_1 \mapsto \varphi_b \in (c_0)' , \quad \varphi_b((a_j)_{j=1}^{\infty}) = \sum_{j=1}^{\infty} a_j b_j \text{ for any } (a_j)_{j=1}^{\infty} \in c_0. \tag{4.2}$$

**Proof** We leave to the reader the verification that the correspondence is well defined, is linear and $\|\varphi_b\| \leq \|b\|_1$. We will prove that it is surjective and that $\|\varphi_b\| \geq \|b\|_1$. Let $\varphi \in (c_0)'$ be given. Consider the sequence $b = (\varphi(e_j))_{j=1}^{\infty}$, where $(e_n)_{n=1}^{\infty}$ are the canonical unit vectors (see Example 1.6.4). We must show that $b \in \ell_1$, $\varphi_b = \varphi$ and $\|b\|_1 \leq \|\varphi\|$. For each $n \in \mathbb{N}$ consider the sequence $(\alpha_j)_{j=1}^{\infty}$ where

$$\alpha_j = \frac{\overline{\varphi(e_j)}}{|\varphi(e_j)|} \text{ if } \varphi(e_j) \neq 0 \text{ and } j \leq n , \quad \alpha_j = 0 \text{ otherwise.}$$

It is clear that $|\alpha_j| \leq 1$ for every $j$. Let $j \leq n$ be such that $\varphi(e_j) \neq 0$. Then

$$\alpha_j \varphi(e_j) = \frac{\overline{\varphi(e_j)}}{|\varphi(e_j)|} \cdot \varphi(e_j) = |\varphi(e_j)|.$$

It is also plain that $\alpha_j \varphi(e_j) = |\varphi(e_j)|$ if $\varphi(e_j) = 0$. So $\alpha_j \varphi(e_j) = |\varphi(e_j)|$ for each $j \leq n$. Therefore

$$\sum_{j=1}^{n} |\varphi(e_j)| = \sum_{j=1}^{n} \alpha_j \varphi(e_j) = \varphi\left(\sum_{j=1}^{n} \alpha_j e_j\right) \leq \|\varphi\| \cdot \left\|\sum_{j=1}^{n} \alpha_j e_j\right\|$$

$$= \|\varphi\| \cdot \max\{|\alpha_1|, \ldots, |\alpha_n|\} \leq \|\varphi\|.$$

As this holds for every $n$, letting $n \longrightarrow \infty$ we obtain

$$\|b\|_1 = \sum_{j=1}^{\infty} |\varphi(e_j)| \leq \|\varphi\|,$$

which shows that $b \in \ell_1$ and $\|b\|_1 \leq \|\varphi\|$. Let $x = (a_j)_{j=1}^{\infty} \in c_0$ be given. In Example 1.6.4 we saw, among other things, that $x = \lim_{n \to \infty} \sum_{j=1}^{n} a_j e_j$. It follows that

$$\varphi(x) = \varphi\left(\lim_{n \to \infty} \sum_{j=1}^{n} a_j e_j\right) = \lim_{n \to \infty} \varphi\left(\sum_{j=1}^{n} a_j e_j\right)$$

$$= \lim_{n\to\infty} \sum_{j=1}^{n} a_j \varphi(e_j) = \sum_{j=1}^{\infty} a_j \varphi(e_j) = \varphi_b((a_j)_{j=1}^{\infty}) = \varphi_b(x),$$

proving that $\varphi_b = \varphi$. ∎

Considering what has already been proved in this section and in the previous one, it is usual to write $(L_p(X, \Sigma, \mu))' = L_{p^*}(X, \Sigma, \mu)$, $(\ell_p)' = \ell_{p^*}$ for $1 \leq p < \infty$, and $(c_0)' = \ell_1$, each equality being understood up to the corresponding isometric isomorphism.

## 4.3 Bidual and Reflexive Spaces

For every normed space $E$, we can consider its dual $E'$, which we already know to be a Banach space. We can then consider the dual of $E'$, called the *bidual of $E$* and denoted by $E''$. That is, $E'' = (E')'$. We begin this section by showing that every normed space $E$ can be found within its bidual $E''$:

**Proposition 4.3.1** *For every normed space $E$, the linear operator*

$$J_E \colon E \longrightarrow E'', \quad J_E(x)(\varphi) = \varphi(x) \text{ for any } x \in E \text{ and } \varphi \in E',$$

*is a linear isometry, called the canonical embedding from $E$ to $E''$.*

**Proof** We leave it to the reader to check that $J_E$ is well defined (that is, $J_E(x) \in E''$ for every $x \in E$) and is linear. For each $x \in E$, from

$$\|J_E(x)\| = \sup_{\varphi \in B_{E'}} |J(x)(\varphi)| = \sup_{\varphi \in B_{E'}} |\varphi(x)| = \|x\|,$$

where the last equality follows from the Hahn–Banach Theorem in the form of Corollary 3.1.5, we conclude that $J_E$ is a linear isometry. ∎

**Remark 4.3.2**

(a) If the normed space $E$ contains a subspace isomorphic to the normed space $F$, we say that $E$ contains an *isomorphic copy of $F$*. When $E$ contains a subspace isometrically isomorphic to $F$, we say that $E$ contains an *isometric copy of $F$*. From Proposition 4.3.1 it follows, in particular, that for every normed space $E$, its bidual $E''$ contains an isometric copy of $E$. In this setting, Proposition 3.3.3 states that $\ell_\infty$ contains an isometric copy of every separable space.
(b) Since $E$ and $J_E(E)$ are isometrically isomorphic, it is usual to identify the two spaces and to refer to both simply as $E$. That is, with a slight abuse of notation, we will sometimes write $E$ where we should have written $J_E(E)$.

## 4.3 Bidual and Reflexive Spaces

As a first application of Proposition 4.3.1, we give next a very short proof that every normed space admits a completion. The reader should remember that the proof of the completion theorem in metric spaces is long and laborious. Moreover, adapting it to normed spaces, in the sense of showing that the completion of a normed space is also a normed space (hence a Banach space), adds even more work to the task.

**Proposition 4.3.3** *For every normed space $E$ there exists a Banach space $\widehat{E}$ containing an isometric copy of $E$ which is dense in $\widehat{E}$.*

*Proof* Consider $\widehat{E} = \overline{J_E(E)} \subseteq E''$. ∎

The space $\widehat{E}$ is called the *completion* of the normed space $E$. The reader should not be deceived by the simplicity of the proof above, as it implicitly contains the Hahn–Banach Theorem, used in the proof of Proposition 4.3.1.

For the canonical embedding $J_E$ from $E$ into its bidual $E''$ to be an isometric isomorphism, it only needs to be surjective. This property characterizes an important class of spaces:

**Definition 4.3.4** A normed space $E$ is said to be *reflexive* if the canonical embedding $J_E \colon E \longrightarrow E''$ is surjective, that is, $J_E(E) = E''$. In this case $J_E$ is an isometric isomorphism, and the abuse of notation $E = J_E(E)$ becomes $E = E''$.

The next result shows that there is no loss of generality in defining reflexivity exclusively in Banach spaces:

**Proposition 4.3.5** *Every reflexive normed space is a Banach space.*

*Proof* A reflexive normed space is isometrically isomorphic to its bidual, which is Banach because it is a dual space (Corollary 2.1.5). ∎

**Example 4.3.6**

(a) Finite-dimensional normed spaces are reflexive. In fact, if $E$ has finite dimension $n$, we know from Linear Algebra that its algebraic dual $E^*$ also has dimension $n$. And, by Exercise 2.7.2, the algebraic dual coincides with the topological dual. Repeating the argument, it follows that $E''$ also has dimension $n$ and it easily follows that $J_E$ is surjective, therefore $E$ is reflexive.
(b) Incomplete normed spaces, for example $c_{00}$, are not reflexive (Proposition 4.3.5).
(c) $c_0$ is not reflexive. We already know from Propositions 4.2.1 and 4.2.3 that

$$(c_0)'' = ((c_0)')' = (\ell_1)' = \ell_\infty.$$

It is easy to check that the canonical embedding $J_{c_0} \colon c_0 \longrightarrow (c_0)'' = \ell_\infty$ is in fact the inclusion from $c_0$ to $\ell_\infty$, which obviously is not surjective.

The natural step now is to investigate the reflexivity of the spaces $\ell_p$. We will split this analysis into three steps: $p = 1$, $1 < p < \infty$ and $p = \infty$. Each of them will follow as a consequence of results that are interesting in their own right and that will have other applications throughout the text. We start with the case $p = 1$:

**Proposition 4.3.7** *If $E$ is separable and reflexive, then $E'$ is separable.*

*Proof* Since $E$ is isometrically isomorphic to $E''$, we know that $E''$ is also separable. By Theorem 3.3.2 it then follows that $E'$ is separable. ∎

**Example 4.3.8** $\ell_1$ is not reflexive. In fact, as $\ell_1$ is separable (Example 1.6.4) and its dual $\ell_\infty$ is not separable (Example 1.6.5), by Proposition 4.3.7 it follows that $\ell_1$ is not reflexive.

We seize the study of the reflexivity of spaces $\ell_p$, where $1 < p < \infty$, to introduce the concept of the adjoint operator.

**Definition 4.3.9** Let $E, F$ be normed vector spaces and let $T \in \mathcal{L}(E; F)$ be a continuous linear operator. We define the operator $T': F' \longrightarrow E'$ by

$$T'(\varphi)(x) = \varphi(T(x)) \text{ for all } x \in E \text{ and } \varphi \in F'.$$

The operator $T'$ is called the *adjoint* of $T$.

**Example 4.3.10** Given $1 \leq p < \infty$, consider the backward shift operator

$$T: \ell_p \longrightarrow \ell_p, \quad T((a_1, a_2, a_3, \ldots,)) = (a_2, a_3, \ldots).$$

It is easy to see that $T$ is linear and continuous. By definition, the adjoint operator $T'$ goes from $(\ell_p)'$ to $(\ell_p)'$. It is convenient to identify $(\ell_p)'$ with $\ell_{p^*}$, that is, we will see a functional $\varphi \in (\ell_p)'$ as a sequence $(b_n)_{n=1}^\infty \in \ell_{p^*}$ and the action of the functional $\varphi$ on the elements of $\ell_p$ is given by the duality relation (4.1):

$$\varphi\left((a_n)_{n=1}^\infty\right) = \sum_{j=1}^\infty a_n b_n \text{ for any } (a_n)_{n=1}^\infty \in \ell_p.$$

So, given $\varphi = (b_n)_{n=1}^\infty \in \ell_{p^*}$ and $x = (a_n)_{n=1}^\infty \in \ell_p$,

$$T'(\varphi)(x) = \varphi(T(x)) = \varphi((a_2, a_3, \ldots,)) = b_1 a_2 + b_2 a_3 + \cdots$$
$$= 0 a_1 + b_1 a_2 + b_2 a_3 + \cdots = \psi(x),$$

where $\psi := (0, b_1, b_2, \ldots) \in \ell_{p^*}$. This reveals that $T'(\varphi) = \psi$, that is, the adjoint operator $T'$ is the forward shift operator

$$T': \ell_{p^*} \longrightarrow \ell_{p^*}, \quad T'((b_1, b_2, b_3, \ldots,)) = (0, b_1, b_2, \ldots).$$

## 4.3 Bidual and Reflexive Spaces

It is interesting to note that in the case $p = 2$ we have $T \circ T' = \mathrm{id}_{\ell_2}$ but $T' \circ T \neq \mathrm{id}_{\ell_2}$.

**Proposition 4.3.11** *For every $T \in \mathcal{L}(E, F)$, $T' \in \mathcal{L}(F', E')$ and $\|T'\| = \|T\|$. Furthermore, if $T$ is an (isometric) isomorphism, then $T'$ is also an (isometric) isomorphism.*

*Proof* The reader should check that $T'$ is well defined (that is, $T'(\varphi) \in E'$ for each $\varphi \in F'$) and is linear. For each $\varphi \in F'$ we have

$$\|T'(\varphi)\| = \sup_{x \in B_E} |T'(\varphi)(x)| = \sup_{x \in B_E} |\varphi((T(x))| \leq \|\varphi\| \cdot \sup_{x \in B_E} \|T(x)\| = \|\varphi\| \cdot \|T\|,$$

which shows that $\|T'\| \leq \|T\|$. On the other hand, for each $x \in E$, by Corollary 3.1.5 we have

$$\|T(x)\| = \sup_{\varphi \in B_{F'}} |\varphi(T(x))| = \sup_{\varphi \in B_{F'}} |T'(\varphi)(x)| \leq \|x\| \cdot \sup_{\varphi \in B_{F'}} \|T'(\varphi)\| = \|x\| \cdot \|T'\|,$$

proving that $\|T\| \leq \|T'\|$. Now suppose that $T$ is an isomorphism. To prove that $T'$ is surjective, let $\varphi \in E'$ be given. Consider $\phi \in F'$ given by $\phi(z) = \varphi(T^{-1}(z))$ and notice that $T'(\phi) = \varphi$, proving that $T'$ is surjective. Now we prove that $T'$ is injective. If $\psi \in \ker(T')$, then

$$\psi(T(x)) = T'(\psi)(x) = 0 \text{ for every } x \in E.$$

Since $T$ is surjective, it follows that $\psi = 0$. Assuming that $T$ is an isometric isomorphism, it is enough to show that $\|T'(\varphi)\| = \|\varphi\|$ for every $\varphi \in F'$. As $T$ is an isometric isomorphism, $x \in B_E$ if and only if $T(x) \in B_F$, therefore

$$\|T'(\varphi)\| = \sup_{x \in B_E} |T'(\varphi)(x)| = \sup_{x \in B_E} |\varphi(T(x))|$$

$$= \sup_{T(x) \in B_F} |\varphi(T(x))| = \sup_{z \in B_F} |\varphi(z)| = \|\varphi\|.$$

∎

**Proposition 4.3.12** *For $1 < p < \infty$, the spaces $L_p(X, \Sigma, \mu)$ and $\ell_p$ are reflexive.*

*Proof* It is enough to consider the case of $L_p(X, \Sigma, \mu)$. Let $T: L_{p^*}(X, \Sigma, \mu) \longrightarrow (L_p(X, \Sigma, \mu))'$ and $S: L_p(X, \Sigma, \mu) \longrightarrow (L_{p^*}(X, \Sigma, \mu))'$ be the isometric isomorphisms of Theorem 4.1.2. Therefore $T^{-1}$ is also an isometric isomorphism and, by Proposition 4.3.11, it follows that its adjoint $(T^{-1})': (L_{p^*}(X, \Sigma, \mu))' \longrightarrow (L_p(X, \Sigma, \mu))''$ is an isometric isomorphism as well. Therefore the operator $(T^{-1})' \circ S: L_p(X, \Sigma, \mu) \longrightarrow (L_p(X, \Sigma, \mu))''$ is an isometric isomorphism. It remains to prove that $(T^{-1})' \circ S = J_{L_p(X, \Sigma, \mu)}$. To do so, let $f \in L_p(X, \Sigma, \mu)$ and $\varphi \in (L_p(X, \Sigma, \mu))'$ be given and call $g = T^{-1}(\varphi) \in L_{p^*}(X, \Sigma, \mu)$. Then

$$\left((T^{-1})' \circ S\right)(f)(\varphi) = (T^{-1})'(S(f))(\varphi) = S(f)((T^{-1})(\varphi)) = S(f)(g)$$

$$= \int_X fg\, d\mu = T(g)(f) = \varphi(f) = J_{L_p(X,\Sigma,\mu)}(f)(\varphi),$$

completing the proof. ∎

Finally, for the case $p = \infty$ we will use the following result:

**Proposition 4.3.13** *A Banach space $E$ is reflexive if and only if its dual $E'$ is reflexive.*

**Proof** Suppose that $E$ is reflexive. Then the canonical embedding $J_E : E \longrightarrow E''$ is surjective. We need to prove that the canonical embedding $J_{E'} : E' \longrightarrow E'''$ is also surjective. Let $x''' \in E'''$ be given. By Proposition 4.3.11 we know that $(J_E)' : E''' \longrightarrow E'$ is an isometric isomorphism. Note that $J_{E'}(x') = x'''$, where $x' = (J_E)'(x''') \in E'$. In fact, for each $x \in E$, we have

$$J_{E'}(x')(J_E(x)) = J_E(x)(x') = x'(x) = (J_E)'(x''')(x) = x'''(J_E(x))$$

and, since $J_E(E) = E''$, we conclude that $J_{E'}(x') = x'''$.

Now suppose that $E'$ is reflexive. By contradiction, suppose that $E$ is not reflexive. Then $J_E : E \longrightarrow E''$ is not surjective. As $J_E : E \longrightarrow J_E(E)$ is a linear isometry, it follows that $J_E(E)$ is closed in $E''$ and does not coincide with $E''$. By Proposition 3.3.1, there exists $0 \neq \varphi \in E''' = J_{E'}(E')$ such that $\varphi(J_E(E)) = \{0\}$. As $E'$ is reflexive, $\varphi = J_{E'}(\phi)$ for some $\phi \in E'$. For every $x \in E$,

$$0 = \varphi(J_E(x)) = (J_{E'}(\phi))(J_E(x)) = J_E(x)(\phi) = \phi(x).$$

This proves that $\phi = 0$, therefore $\varphi = J_{E'}(\phi) = 0$, which is a contradiction. ∎

**Example 4.3.14** $\ell_\infty$ is not reflexive. The statement follows by combining the non-reflexivity of $\ell_1$ (Example 4.3.8) with the equality $(\ell_1)' = \ell_\infty$ (Proposition 4.2.1) and Proposition 4.3.13.

Other examples and properties of reflexive spaces will be seen in Sects. 6.4 and 6.5. We will show, in particular, that the spaces $C[a,b]$, $L_1[a,b]$ and $L_\infty[a,b]$ are not reflexive.

## 4.4 Comments and Historical Notes

There is a version of the Riesz representation theorem that describes the continuous linear functionals on the space $(C(K), \|\cdot\|_\infty)$, where $K$ is a compact Hausdorff topological space, as follows: to each functional $\varphi \in C(K)'$ corresponds a Radon measure $\mu$ on the Borel sets of $K$ such that

$$\varphi(f) = \int_K f \, d\mu \text{ for every } f \in C(K).$$

Details can be found in [25, 7.3]. In the case of $C[a, b]$, the linear functionals are represented by functions of bounded variation, and the duality relation is given by the Riemann–Stieltjes integral.

The notion of reflexivity appeared with Helly, who in 1921 noted that the canonical embedding from the space $c$ of convergent sequences with the norm $\|\cdot\|_\infty$ into its bidual is not surjective. With this example, it became necessary to separate the class of spaces for which the canonical embedding is surjective, and this was done by Hahn in 1927. Hahn called such spaces *regular*, an unfortunate term as it appears countless times in other contexts. Fortunately, E. Lorch coined the term *reflexive space* in 1939.

The study we have made of reflexivity is clearly incomplete, for example, we have not touched on the question of whether a closed subspace of a reflexive space is also reflexive (in Exercise 4.5.31 the reader will prove this fact using quotient spaces). The study of reflexivity will be resumed in Chap. 6.

We have seen that if a normed space $E$ is reflexive, then $E$ is isometrically isomorphic to its bidual $E''$ through the canonical embedding $J_E$. If $E$ is isometrically isomorphic to $E''$ through another isometric isomorphism, is $E$ necessarily reflexive? The answer is no: in 1950, R. C. James [33] constructed a space, henceforth known as *James' space*, which is isometrically isomorphic to its bidual but not reflexive. James' space has found many other applications and many other related spaces, with very special properties, have been constructed. The monograph [23] is an excellent reference for James' space, its properties, its applications and its related spaces. In Sect. 2.6 we discussed operators that attain their norms, and in Exercise 2.7.11 we saw that not even continuous linear functionals always attain their norms. On the other hand, in Exercise 4.5.28 we will see that continuous linear functionals on reflexive spaces attain their norms. The converse was proven by the same R. C. James in [34]: a Banach space $E$ is reflexive if and only if every continuous linear functional $\varphi \in E'$ attains its norm. For a proof see [21, Corollary 3.56].

## 4.5 Exercises

**Exercise 4.5.1** Without going through $L_p$ spaces, prove that, for $1 < p < \infty$, the duality relation (4.1) actually establishes an isometric isomorphism between $(\ell_p)'$ and $\ell_{p^*}$.

**Exercise 4.5.2** Let $1 < p < \infty$. Prove, without using reflexivity, that every continuous linear functional on $\ell_p$ attain its norm, that is, for every $\varphi \in (\ell_p)'$ there exists $x \in \ell_p$ such that $\|x\|_p = 1$ and $\|\varphi\| = |\varphi(x)|$.

**Exercise 4.5.3** Let $a = (a_n)_{n=1}^\infty$ be a scalar sequence and let $p \geq 1$ be given. Suppose that, for every sequence $(b_n)_{n=1}^\infty \in \ell_p$, the series $\sum_{n=1}^\infty a_n b_n$ is convergent. Prove that $a \in \ell_\infty$ if $p = 1$ and $a \in \ell_{p*}$ if $p > 1$.

**Exercise 4.5.4** Call $c$ the space of convergent scalar sequences, equipped with the supremum norm $\|\cdot\|_\infty$ (see Exercise 1.8.27). Prove that $c$ is isomorphic to $c_0$, conclude that $c'$ is isomorphic to $\ell_1$ and describe the resulting duality relation.

**Exercise 4.5.5** Given $\varphi \in (\ell_\infty)'$, prove that the series $\sum_{n=1}^\infty |\varphi(e_n)|$ is convergent and that its sum is less than or equal to $\|\varphi\|$.

**Exercise 4.5.6** Prove, without using reflexivity, that $c_0$ and $\ell_p$, $1 < p < \infty$, do not contain a subspace isomorphic to $\ell_1$.

**Exercise 4.5.7** Let $1 < p < \infty$ and let $(E_n)_{n=1}^\infty$ be a sequence of Banach spaces. Define
$$\left(\sum E_n\right)_p = \left\{ (x_n)_{n=1}^\infty : x_n \in E_n \ \forall n \in \mathbb{N} \text{ and } \sum_{n=1}^\infty \|x_n\|^p < \infty \right\}$$
and $\|(x_n)_{n=1}^\infty\|_p = \left( \sum_{n=1}^\infty \|x_n\|^p \right)^{\frac{1}{p}}$. Prove that $\left( \left(\sum E_n\right)_p, \|\cdot\|_p \right)$ is a Banach space and that $\left( \left(\sum E_n\right)_p \right)'$ is isometrically isomorphic to $\left( \left(\sum E'_n\right)_{p*} \right)$.

**Exercise 4.5.8** Let $1 \leq p \leq \infty$ and let $E$ be a Banach space. Define
$$\ell_p(E) = \left\{ (x_n)_{n=1}^\infty : x_n \in E \ \forall n \in \mathbb{N} \text{ and } \|(x_n)_{n=1}^\infty\|_p = \left( \sum_{n=1}^\infty \|x_n\|^p \right)^{\frac{1}{p}} < \infty \right\},$$
with the obvious modification in the case $p = \infty$. Prove that:

(a) $(\ell_p(E), \|\cdot\|_p)$ is a Banach space.
(b) For $1 \leq p < \infty$, $\ell_p(E)'$ is isometrically isomorphic to $\ell_{p*}(E')$.

**Exercise 4.5.9\*** Show that for each $n \in \mathbb{N}$ there is a function $g_n \in L_\infty[0, 1]$ such that
$$p(0) + p'(0) - 3p''(1) = \int_0^1 p(t) g_n(t) dt,$$
for every polynomial $p$ of degree less than or equal to $n$. Also prove that
$$\lim_{n \to \infty} \|g_n\|_{L_1[0,1]} = \infty.$$

## 4.5 Exercises

**Exercise 4.5.10*** Consider the sets

$$L := \{f : [0, 1] \longrightarrow \mathbb{R} : f(0) = 0, \; f(1) = 1 \text{ and } |f(x) - f(y)| \leq 2|x - y|$$
$$\text{for all } x, y \in [0, 1]\},$$

$$S := \left\{ g : [0, 1] \longrightarrow \mathbb{R} : g(t) = \sum_{i=2}^{100} \lambda_i t^{1/i}, \; \lambda_i \geq 0 \; \forall i \text{ and } \sum_{i=2}^{100} \lambda_i = 1 \right\}.$$

Show that there exists a function $\phi \in L_\infty[0, 1]$ such that

$$\sup_{f \in L} \int_0^1 f(t)\phi(t)dt < \inf_{g \in S} \int_0^1 g(t)\phi(t)dt.$$

**Exercise 4.5.11** Let $E$ be a normed space. Prove that the set $J_E(E)$ separates points of $E'$, that is, given $\varphi_1, \varphi_2 \in E'$, $\varphi_1 \neq \varphi_2$, there exists $f \in J_E(E)$ such that $f(\varphi_1) \neq f(\varphi_2)$.

**Exercise 4.5.12** Let $B$ be a subset of a normed space $E$ such that $\varphi(B)$ is bounded for each $\varphi \in E'$. Show that $B$ is bounded.

**Exercise 4.5.13*** Let $F$ and $G$ be closed subspaces of the Banach space $E$ such that $E = F \oplus G$. Prove that:

(a) $E' = F^\perp \oplus G^\perp$. (See the definition of $\perp$ in Exercise 3.6.13)
(b) $E'$ contains complemented isomorphic copies of $F'$ and $G'$.
(c) $E'$ is isometrically isomorphic to $(F' \times G', \|\cdot\|_1)$.

**Exercise 4.5.14** Let $E$ be a normed space, let $\widehat{E}$ be its completion, and let $F$ a Banach space. Prove that the spaces $\mathcal{L}(E, F)$ and $\mathcal{L}(\widehat{E}, F)$ are isometrically isomorphic. In particular, $(\widehat{E})' = E'$.

**Exercise 4.5.15** Let $M$ be a closed subspace of the normed space $E$. Prove that:

(a) $M \subseteq (M^\perp)^\perp$.
(b) If $E$ is reflexive, then $(M^\perp)^\perp = M$.

**Exercise 4.5.16** Determine the adjoints $T' : \ell_\infty \longrightarrow \ell_\infty$ of the following operators $T : \ell_1 \longrightarrow \ell_1$:

(a) $T((a_1, a_2, \ldots,)) = (a_1, \ldots, a_n, 0, 0, \ldots)$, where $n \in \mathbb{N}$ is fixed.
(b) $T((a_n)_{n=1}^\infty) = (b_n a_n)_{n=1}^\infty$, where $(b_n)_{n=1}^\infty \in B_{\ell_\infty}$ is fixed.

**Exercise 4.5.17** Let $1 \leq p < \infty$ and $g \in L_\infty[0, 1]$ be given. Define

$$T_g : L_p[0, 1] \longrightarrow L_p[0, 1], \; T_g(f) = f \cdot g.$$

(a) Prove that $T_g \in \mathcal{L}(L_p[0, 1], L_p[0, 1])$ and $\|T_g\| = \|g\|_\infty$.

(b) Prove that, considering $T'_g: L_{p^*}[0, 1] \longrightarrow L_{p^*}[0, 1]$, we have $T'_g(f) = f \cdot g$ for every $f \in L_{p^*}[0, 1]$.
(c) Conclude that $T'_g = T_g$ when $p = 2$.

**Exercise 4.5.18** Let $E$ and $F$ be normed spaces. Prove that the correspondence $T \in \mathcal{L}(E, F) \mapsto T' \in \mathcal{L}(F', E')$ is an isometric isomorphism onto its range.

**Exercise 4.5.19** Let $E, F, G$ be normed spaces. Given $T \in \mathcal{L}(E, F)$ and $S \in \mathcal{L}(F, G)$, show that $(S \circ T)' = T' \circ S'$.

**Exercise 4.5.20** Let $E, F$ be Banach spaces. Given $T \in \mathcal{L}(E, F)$, prove that $T''$ is an extension of $T$ to $E''$ in the sense that $T'' \circ J_E = J_F \circ T$. In particular, $T''(J_E(E)) \subseteq J_F(F)$ and $T = J_F^{-1} \circ T'' \circ J_E$ where $J_F^{-1}: J_F(F) \longrightarrow F$.

**Exercise 4.5.21** Let $E$ and $F$ be normed spaces with $E$ reflexive and let $T \in \mathcal{L}(E, F)$ be given. Prove that $T''(x) = T(x)$ for every $x \in E$.

**Exercise 4.5.22** Let $E, F$ be normed spaces and let $T \in \mathcal{L}(E, F)$ be given. Prove that if $T$ has a continuous inverse $T^{-1}: F \longrightarrow E$, then $T'$ has a continuous inverse $(T')^{-1}: E' \longrightarrow F'$ and $(T')^{-1} = (T^{-1})'$.

**Exercise 4.5.23** Let $E$ be a normed space. Prove that $T \in \mathcal{L}(E, E)$ is a projection if and only if $T' \in \mathcal{L}(E', E')$ is a projection.

**Exercise 4.5.24*** 

(a) Let $B$ be a subset of the dual $E'$ of the normed space $E$. Prove that
$$^\perp B := \{x \in E : \varphi(x) = 0 \text{ for every } \varphi \in B\}$$
is a closed subspace of $E$.
(b) Let $E$ and $F$ be normed spaces and let $T \in \mathcal{L}(E, F)$ be given. Prove that $\ker(T) = {}^\perp(T'(F'))$ and $\ker(T') = (T(E))^\perp$.

**Exercise 4.5.25*** Let $E$ be a normed space. Prove that:

(a) $(J_E)' \circ J_{E'} = \mathrm{id}_{E'}$.
(b) $(J_E)': E''' \longrightarrow E'$ is surjective.
(c) (Dixmier's Theorem) $E'$ is complemented in $E'''$.

**Exercise 4.5.26** Let $E$ be a normed space. Prove that:

(a) There exists a normed space $F$ such that $E$ is isometrically isomorphic to $F'$ if and only if there exists a Banach space $G$ such that $E$ is isometrically isomorphic to $G'$. In this case, it is said that $E$ is a *dual space*.
(b) If $E$ is a dual space, then $E$ is complemented in its bidual $E''$. Use that $c_0$ is not complemented in $\ell_\infty$ to conclude that $c_0$ is not a dual space.

**Exercise 4.5.27** Prove that every normed space which is isomorphic to a reflexive space is also reflexive.

## 4.5 Exercises

**Exercise 4.5.28** Let $E$ be a reflexive space. Prove that every continuous linear functional on $E$ attains its norm, that is, for every $\varphi \in E'$ there exists $x \in E$, $\|x\| = 1$, such that $\|\varphi\| = |\varphi(x)|$.

**Exercise 4.5.29** Prove that the closed subspace $E = \{f \in C[0, 1] : f(0) = 0\}$ of $C[0, 1]$, studied in Exercise 1.8.34, is not reflexive.

**Exercise 4.5.30** Let $E$ and $F$ be Banach spaces. In the Cartesian product consider any of the usual norms (see Exercise 1.8.12).

(a) Prove that $E' \times F'$ is isomorphic to $(E \times F)'$.
(b) Prove that if $E$ and $F$ are reflexive then $E \times F$ is reflexive.

**Exercise 4.5.31** * Let $M$ be a closed subspace of the Banach space $E$.

(a) For each $\varphi \in M'$ the Hahn–Banach Theorem extends $\varphi$ to a functional $\widetilde{\varphi} \in E'$. Prove that
$$T_M : M' \longrightarrow E'/M^\perp, \quad T_M(\varphi) = [\widetilde{\varphi}],$$
is an isometric isomorphism (as the Hahn–Banach extension is not unique, it is necessary to prove that $T_M$ is well defined).

(b) Let $\pi_M : E \longrightarrow E/M$ be the quotient map. Prove that
$$T^M : (E/M)' \longrightarrow M^\perp, \quad T^M(f) = f \circ \pi_M,$$
is an isometric isomorphism (do not forget to prove that the range of $T^M$ is contained in $M^\perp$).

(c) Conclude that a closed subspace of a reflexive space is reflexive.

**Exercise 4.5.32** Let $E$ and $F$ be normed spaces. Prove that:

(a) $\mathcal{L}(E, F)$ contains an isometric copy of $E'$.
(b) $\bigl(\mathcal{L}(E, F')\bigr)'$ contains an isometric copy of $F$.

**Exercise 4.5.33** Let $E$ and $F$ be Banach spaces. Prove that:

(a) If $\mathcal{L}(E, F)$ is reflexive, then $E$ is reflexive.
(b) If $\mathcal{L}(E, F')$ is reflexive, then $F$ is reflexive.

**Exercise 4.5.34** Let $E$ be a normed space. Just as we defined $E'' = (E')'$, we define $E^{(3)} = (E'')'$, $E^{(4)} = (E^{(3)})'$, ..., $E^{(n)} = (E^{(n-1)})'$, ....

(a) Prove that the following statements are equivalent:
   (i) $E$ is reflexive.
   (ii) $E^{(n)}$ is reflexive for every $n \in \mathbb{N}$.
   (iii) $E^{(n)}$ is reflexive for some $n \in \mathbb{N}$.

(b) If $E$ is not reflexive, then all the natural inclusions
$$E \subseteq E'' \subseteq E^{(4)} \subseteq \cdots \text{ and } E' \subseteq E^{(3)} \subseteq E^{(5)} \subseteq \cdots$$
are strict.

# Chapter 5
# Hilbert Spaces

From Calculus and Analytical Geometry courses, the reader knows that the usual inner product of $\mathbb{R}^n$

$$\langle \cdot, \cdot \rangle \colon \mathbb{R}^n \times \mathbb{R}^n \longrightarrow \mathbb{R} \ , \ (x, y) \mapsto \langle x, y \rangle := \sum_{j=1}^{n} x_j y_j,$$

where $x = (x_1, \ldots, x_n)$ and $y = (y_1, \ldots, y_n)$, is an essential tool in the construction of the theory. It is also known that the Euclidean norm is induced by the inner product in the sense that

$$\|x\|_2 = \sqrt{\langle x, x \rangle} \text{ for every } x \in \mathbb{R}^n.$$

Additionally, from the Linear Algebra course, one certainly remembers the utility of equipping a vector space with an inner product, mainly in the sense that the inner product allows 'geometrizing' the study of vector spaces and linear transformations. We will see in this chapter that in Functional Analysis the gain is also not small when working with spaces with inner product.

It can be said that, on our journey in this book so far, we have been distancing ourselves more and more from the Euclidean spaces $\mathbb{R}^n$. By considering spaces with inner product, we are doing the opposite and getting closer again to the Euclidean spaces. Remembering that these spaces are complete, to get even closer we should consider spaces with inner product that are complete in the norm induced by the inner product. These are the *Hilbert spaces*. Due to this proximity to the Euclidean spaces, Hilbert spaces are, by far, the most used Banach spaces in applications.

## 5.1 Spaces with Inner Product

Let $E$ be a vector space over the field $\mathbb{K} = \mathbb{R}$ or $\mathbb{C}$. An *inner product* on $E$ is a map

$$\langle \cdot, \cdot \rangle \colon E \times E \longrightarrow \mathbb{K}, \ (x, y) \mapsto \langle x, y \rangle,$$

such that for any $x, x_1, x_2, y \in E$ and $\lambda \in \mathbb{K}$:

(P1) $\langle x_1 + x_2, y \rangle = \langle x_1, y \rangle + \langle x_2, y \rangle$,
(P2) $\langle \lambda x, y \rangle = \lambda \langle x, y \rangle$,
(P3) $\langle x, y \rangle = \overline{\langle y, x \rangle}$,
(P4) For every $x \neq 0$, $\langle x, x \rangle$ is a strictly positive real number.

Note that from the properties above it follows that, for all $x, y, y_1, y_2 \in E$ and $\lambda \in \mathbb{K}$:

$$\langle x, 0 \rangle = \langle 0, y \rangle = 0;$$

$$\langle x, x \rangle = 0 \text{ if and only if } x = 0;$$

$$\langle x, y_1 + y_2 \rangle = \langle x, y_1 \rangle + \langle x, y_2 \rangle, \ \langle x, y_1 - y_2 \rangle = \langle x, y_1 \rangle - \langle x, y_2 \rangle; \tag{5.1}$$

$$\langle x, \lambda y \rangle = \overline{\lambda} \langle x, y \rangle \text{ and}$$

$$\text{If } \langle z, y_1 \rangle = \langle z, y_2 \rangle \text{ for every } z \in E, \text{ then } y_1 = y_2. \tag{5.2}$$

Furthermore, it is important to keep in mind that in the real case the inner product is a symmetric bilinear form.

The pair $(E, \langle \cdot, \cdot \rangle)$ is called an *inner product space*. In this case we say that the function

$$\| \cdot \| \colon E \longrightarrow \mathbb{R}, \ \|x\| = \sqrt{\langle x, x \rangle},$$

is the *norm induced by the inner product* $\langle \cdot, \cdot \rangle$. Before proving that the function above is indeed a norm, we shall look at some examples:

**Example 5.1.1**

(a) Given the numerical sequences $x = (x_j)_{j=1}^{\infty}$, $y = (y_j)_{j=1}^{\infty}$, the expression

$$\langle x, y \rangle = \sum_{j=1}^{\infty} x_j \overline{y_j}$$

defines an inner product on $\ell_2$. The convergence of the series above results from the Hölder inequality 1.4.1. The induced norm obviously coincides with the original norm $\| \cdot \|_2$ of $\ell_2$.

## 5.1 Spaces with Inner Product

(b) The expression

$$\langle f, g \rangle = \int_X f\overline{g}\, d\mu$$

defines an inner product on $L_2(X, \Sigma, \mu)$. The existence of the integral above results from the Hölder inequality 1.2.1. The induced norm obviously coincides with the original norm $\|\cdot\|_2$ of $L_2(X, \Sigma, \mu)$.

We need the inequality below to prove that indeed the correspondence $x \in E \mapsto \sqrt{\langle x, x \rangle}$ is a norm on $E$.

**Proposition 5.1.2 (Cauchy–Schwarz Inequality)** *Let $E$ be a vector space with inner product. Then*

$$|\langle x, y \rangle| \leq \|x\| \cdot \|y\|$$

*for any $x, y$ in $E$. Furthermore, equality occurs if and only if the vectors $x$ and $y$ are linearly dependent.*

**Proof** The result is immediate if $x = 0$ or $y = 0$. We can then assume $x \neq 0$ and $y \neq 0$. Let $a = \langle y, y \rangle$ and $b = \langle x, y \rangle$ to conclude that

$$a\overline{b}\langle x, y \rangle = \langle y, y \rangle \overline{b} b = |b|^2 \langle y, y \rangle$$

is a positive real number. Remembering that $a$ is also a positive real number,

$$\begin{aligned}
0 &\leq \langle ax - by, ax - by \rangle \\
&= a\overline{a}\langle x, x \rangle - b\overline{a}\langle y, x \rangle - a\overline{b}\langle x, y \rangle + b\overline{b}\langle y, y \rangle \\
&= a^2 \langle x, x \rangle - 2\operatorname{Re}\left(a\overline{b}\langle x, y \rangle\right) + |b|^2 \langle y, y \rangle \\
&= a^2 \langle x, x \rangle - 2|b|^2 \langle y, y \rangle + |b|^2 \langle y, y \rangle \\
&= a^2 \langle x, x \rangle - |b|^2 \langle y, y \rangle \quad (5.3) \\
&= \langle y, y \rangle \left( \langle y, y \rangle \langle x, x \rangle - |b|^2 \right) \\
&= \langle y, y \rangle \left( \langle y, y \rangle \langle x, x \rangle - |\langle x, y \rangle|^2 \right).
\end{aligned}$$

As $\langle y, y \rangle > 0$ it follows that

$$\|y\|^2 \|x\|^2 - |\langle x, y \rangle|^2 = \langle y, y \rangle \langle x, x \rangle - |\langle x, y \rangle|^2 \geq 0,$$

which proves the desired inequality. Now we move to the second statement. If the vectors $x$ and $y$ are linearly dependent it is easy to see that the equality holds. On the other hand, suppose that the equality is true, that is, $|\langle x, y\rangle| = \|x\| \cdot \|y\|$. In this case, taking again $a = \langle y, y\rangle$ and $b = \langle x, y\rangle$, we get $|b|^2 = \langle x, x\rangle a$. From (5.3) we can conclude that

$$\langle ax - by, ax - by\rangle = a^2 \langle x, x\rangle - |b|^2 \langle y, y\rangle = 0.$$

From axiom (P4) we get $ax - by = 0$. If $y = 0$, then the vectors $x$ and $y$ are obviously linearly dependent. If $y \neq 0$, then $a > 0$, proving that also in this case the vectors $x$ and $y$ are linearly dependent. ∎

Now we can prove that the inner product induces a norm on the space:

**Corollary 5.1.3** *Let E be a vector space with an inner product. The function*

$$\|\cdot\| : E \longrightarrow \mathbb{R}, \|x\| = \sqrt{\langle x, x\rangle}$$

*is a norm on E.*

***Proof*** We will only prove the triangle inequality; the verification of the other axioms is immediate. For all $x, y \in E$, using the Cauchy–Schwarz inequality we get

$$\begin{aligned}
\|x + y\|^2 &= \langle x + y, x + y\rangle = \|x\|^2 + \langle x, y\rangle + \langle y, x\rangle + \|y\|^2 \\
&= \|x\|^2 + \langle x, y\rangle + \overline{\langle x, y\rangle} + \|y\|^2 \\
&= \|x\|^2 + 2\mathrm{Re}(\langle x, y\rangle) + \|y\|^2 \leq \|x\|^2 + 2|\langle x, y\rangle| + \|y\|^2 \\
&\leq \|x\|^2 + 2\|x\| \cdot \|y\| + \|y\|^2 = (\|x\| + \|y\|)^2.
\end{aligned}$$

Extracting the square root, the triangle inequality follows. ∎

Whenever we refer to an inner product space $E$, we are automatically considering on $E$ the induced norm and the consequent topological structure. A second application of the Cauchy–Schwarz inequality ensures that the inner product coexists well with this topological structure:

**Proposition 5.1.4** *Let E be an inner product space and let $y \in E$ be a fixed vector. Then the functions*

$$x \in E \mapsto \langle x, y\rangle \in \mathbb{K} \text{ and } x \in E \mapsto \langle y, x\rangle \in \mathbb{K}$$

*are continuous.*

***Proof*** Let $x_0 \in E$ be given. From

$$|\langle x, y\rangle - \langle x_0, y\rangle| = |\langle x - x_0, y\rangle| \leq \|x - x_0\| \cdot \|y\|,$$

## 5.1 Spaces with Inner Product

the continuity of the first function follows. In view of (5.1), the case of the second function is analogous. ∎

**Definition 5.1.5** An inner product space that is complete in the norm induced by the inner product is called a *Hilbert space*. In particular, a Hilbert space is a Banach space with the norm induced by the inner product.

**Example 5.1.6** As we already know that $\ell_2$ and $L_2(X, \Sigma, \mu)$ are complete in their respective norms, from Example 5.1.1 it follows that these spaces are Hilbert spaces with their respective inner products.

Next we see some useful results that relate the inner product with the norm it induces:

**Proposition 5.1.7** *Let $E$ be a vector space with an inner product. Then, for any $x, y \in E$,*

(i) (Parallelogram Law)

$$\|x + y\|^2 + \|x - y\|^2 = 2\left(\|x\|^2 + \|y\|^2\right).$$

(ii) (Polarization Formula, real case)

$$\langle x, y \rangle = \frac{1}{4}\left(\|x + y\|^2 - \|x - y\|^2\right).$$

(iii) (Polarization Formula, complex case)

$$\langle x, y \rangle = \frac{1}{4}\left[\|x + y\|^2 - \|x - y\|^2 + i\left(\|x + iy\|^2 - \|x - iy\|^2\right)\right].$$

*Proof* Adding the equalities

$$\begin{cases} \|x + y\|^2 = \|x\|^2 + \langle x, y \rangle + \langle y, x \rangle + \|y\|^2 \\ \|x - y\|^2 = \|x\|^2 - \langle x, y \rangle - \langle y, x \rangle + \|y\|^2 \end{cases},$$

we obtain (i); and, in the real case, subtracting them we obtain (ii).

We prove (iii). Subtracting the equalities

$$\begin{cases} \|x + iy\|^2 = \|x\|^2 + \langle x, iy \rangle + \langle iy, x \rangle + \|y\|^2 \\ \qquad = \|x\|^2 - i\langle x, y \rangle + i\langle y, x \rangle + \|y\|^2, \\ \|x - iy\|^2 = \|x\|^2 + \langle x, -iy \rangle + \langle -iy, x \rangle + \|y\|^2 \\ \qquad = \|x\|^2 + i\langle x, y \rangle - i\langle y, x \rangle + \|y\|^2, \end{cases}$$

we obtain

$$\|x + iy\|^2 - \|x - iy\|^2 = -2i\langle x, y \rangle + 2i\langle y, x \rangle.$$

Subtracting again the two equalities used in the proof of items (i) and (ii), it results that

$$\|x+y\|^2 - \|x-y\|^2 = 2\langle x, y\rangle + 2\langle y, x\rangle.$$

Combining these last two equalities we get

$$\|x+y\|^2 - \|x-y\|^2 + i\left(\|x+iy\|^2 - \|x-iy\|^2\right)$$
$$= 2\langle x, y\rangle + 2\langle y, x\rangle + 2\langle x, y\rangle - 2\langle y, x\rangle = 4\langle x, y\rangle,$$

which completes the proof. ∎

## 5.2 Orthogonality

It is said that the inner product geometrizes the study of vector spaces because it is through it that we define the concept of orthogonal vectors. We start by recalling that in $\mathbb{R}^2$ orthogonality is a geometric concept and that two vectors $x = (x_1, x_2)$ and $y = (y_1, y_2)$ of $\mathbb{R}^2$ are orthogonal if and only if

$$x_1 y_1 + x_2 y_2 = 0.$$

In other words, two vectors from $\mathbb{R}^2$ are orthogonal if and only if the inner product between them is equal to zero. This is the key to defining orthogonality in vector spaces with an inner product:

**Definition 5.2.1** Let $E$ be a space with an inner product. We say that the vectors $x$ and $y$ of $E$ are *orthogonal* if $\langle x, y\rangle = 0$. In this case we write $x \perp y$.

To convince yourself that the inner product indeed introduces a geometric view in the study, the reader can easily check the **Pythagorean Theorem**: If $x$ and $y$ are orthogonal vectors in a space with an inner product, then

$$\|x+y\|^2 = \|x\|^2 + \|y\|^2.$$

The theorem below is central in the study of spaces with an inner product. Soon the reader will understand why it is sometimes called the orthogonal projection theorem.

**Theorem 5.2.2** *Let $E$ be a space with an inner product and let $M$ be a complete subspace of $E$. For every $x \in E$ there exists a unique $p \in M$ such that*

$$\|x-p\| = \mathrm{dist}(x, M) := \inf_{y \in M} \|x-y\|.$$

## 5.2 Orthogonality

**Proof** Call $d = \text{dist}(x, M)$. From the definition of infimum, there exists a sequence $(y_n)_{n=1}^\infty$ of vectors from $M$ such that

$$d \leq \|x - y_n\| < d + \frac{1}{n} \tag{5.4}$$

for every $n$. Let $m, n \in \mathbb{N}$ be given. We apply the Parallelogram Law to the vectors $x - y_n$ and $x - y_m$ to obtain

$$2\|x - y_m\|^2 + 2\|x - y_n\|^2 = \|x - y_n + x - y_m\|^2 + \|x - y_n - x + y_m\|^2.$$

Since $M$ is a vector subspace, $\frac{y_n + y_m}{2} \in M$, from the equality above and from (5.4) it follows that

$$\|y_n - y_m\|^2 = 2\|x - y_m\|^2 + 2\|x - y_n\|^2 - \|2x - (y_n + y_m)\|^2$$

$$= 2\|x - y_m\|^2 + 2\|x - y_n\|^2 - 4\left\|x - \frac{y_n + y_m}{2}\right\|^2$$

$$\leq 2\left(d + \frac{1}{m}\right)^2 + 2\left(d + \frac{1}{n}\right)^2 - 4d^2 \longrightarrow 0 \text{ as } m, n \longrightarrow \infty.$$

This proves that the sequence $(y_n)_{n=1}^\infty$ is Cauchy in the complete space $M$, therefore convergent to a certain $p \in M$. As $y_n \longrightarrow p$, letting $n \longrightarrow \infty$ in (5.4) we conclude that $\|x - p\| = d$.

To prove uniqueness, let $q \in M$ be such that $\|x - q\| = d$. Applying the Parallelogram Law to the vectors $x - p$ and $x - q$ we get

$$4d^2 = 2d^2 + 2d^2 = 2\|x - p\|^2 + 2\|x - q\|^2$$

$$= \|x - p + x - q\|^2 + \|x - p - x + q\|^2$$

$$= \|2x - p - q\|^2 + \|p - q\|^2 = 4\left\|x - \frac{p+q}{2}\right\|^2 + \|p - q\|^2.$$

Since $\frac{p+q}{2} \in M$, we have

$$0 \leq \|p - q\|^2 = 4d^2 - 4\left\|x - \frac{p+q}{2}\right\|^2 \leq 4d^2 - 4d^2 = 0,$$

which is sufficient to conclude that $p = q$. ∎

**Definition 5.2.3** Let $E$ be an inner product space and let $A$ be a subset of $E$. We call the subset

$$A^\perp = \{y \in E : \langle x, y \rangle = 0 \text{ for every } x \in A\}$$

the *orthogonal complement* of $A$.

The following properties are easily checked:
- $A \subseteq (A^\perp)^\perp$, $E^\perp = \{0\}$ and $\{0\}^\perp = E$.
- $A^\perp$ is a closed subspace of $E$, even if $A$ is not a subspace (the fact that it is closed comes from Proposition 5.1.4).
- $A \cap A^\perp = \begin{cases} \{0\}, & \text{if } 0 \in A, \\ \emptyset, & \text{if } 0 \notin A. \end{cases}$

**Remark 5.2.4** The notation $A^\perp$ seems to be in conflict with the notation $M^\perp$ from Exercises 3.6.13 and 4.5.15. The upcoming Theorem 5.5.2 will show that these notations are consistent.

**Theorem 5.2.5** *Let $H$ be a Hilbert space and let $M$ be a closed subspace of $H$. Then*

(a) $H = M \oplus M^\perp$, *that is, each $x \in H$ admits a unique representation in the form*

$$x = p + q \text{ with } p \in M \text{ and } q \in M^\perp.$$

*Furthermore,*

$$\|x - p\| = \text{dist}(x, M).$$

*The vector $p$ is called the orthogonal projection of $x$ onto $M$.*

(b) *The operators*

$$P, Q \colon H \longrightarrow H, \ P(x) = p \text{ and } Q(x) = q,$$

*are projections, that is, they are linear, continuous and $P^2 = P$ and $Q^2 = Q$. Moreover, $P(H) = M$, $Q(H) = M^\perp$, $\|P\| = 1$ if $M \neq \{0\}$ and $\|Q\| = 1$ if $M \neq H$. The operator $P$ is called the orthogonal projection of $H$ onto $M$.*

(c) $P \circ Q = Q \circ P = 0.$

***Proof***

(a) Given $x \in H$, by Theorem 5.2.2 there exists a unique vector $p \in M$ such that

$$\|x - p\| = \text{dist}(x, M).$$

Taking $q = x - p$ it immediately follows that $x = p + q$. It remains to prove that $q \in M^\perp$. For every $y \in M$ and any scalar $\lambda$, the vector $p + \lambda y$ belongs to $M$, hence

$$\|q\|^2 = \|x - p\|^2 = \text{dist}(x, M)^2 \leq \|x - (p + \lambda y)\|^2 = \|q - \lambda y\|^2$$
$$= \langle q - \lambda y, q - \lambda y \rangle = \|q\|^2 - \lambda \langle y, q \rangle - \overline{\lambda} \langle q, y \rangle + \lambda \overline{\lambda} \|y\|^2.$$

## 5.2 Orthogonality

We conclude that

$$0 \le |\lambda|^2 \|y\|^2 - 2\operatorname{Re}(\lambda \langle y, q \rangle).$$

Write $\langle y, q \rangle$ in the polar form $|\langle y, q \rangle| e^{i\theta}$ and for each $t \in \mathbb{R}$ call $\lambda = te^{-i\theta}$. From the inequality above it follows that

$$0 \le t^2 \|y\|^2 - 2t |\langle y, q \rangle| \text{ for every } t \in \mathbb{R},$$

and consequently the discriminant of the binomial is less than or equal to zero. This gives $|\langle y, q \rangle| = 0$, therefore $q \in M^\perp$.

To prove uniqueness, suppose that $p + q = p_1 + q_1$ with $p, p_1 \in M$ and $q, q_1 \in M^\perp$. As $M$ and $M^\perp$ are subspaces,

$$p - p_1 = q_1 - q \in M \cap M^\perp = \{0\}.$$

It follows that $p = p_1$ and $q = q_1$.

(b) It follows from (a) and Proposition 3.2.2 that $P$ and $Q$ are linear, continuous and projections. For each $x \in H$, by item (a) we can write $x = p + q$ with $p \perp q$. By the Pythagorean Theorem,

$$\|x\|^2 = \|p + q\|^2 = \|p\|^2 + \|q\|^2,$$

therefore

$$\|P(x)\|^2 = \|p\|^2 \le \|x\|^2 \text{ and } \|Q(x)\|^2 = \|q\|^2 \le \|x\|^2.$$

We conclude that $\|P\| \le 1$ and $\|Q\| \le 1$. The reverse inequalities follow from the fact, already known to us (see Definition 3.2.1), that non-null projections have norms greater than or equal to 1.

(c) These statements are easy to check, so they are left to the reader. ∎

**Remark 5.2.6** It is interesting to note that even item (a) of the previous theorem is not valid, in general, without the hypothesis of $M$ being a complete subspace. As an example consider $H = \ell_2$ and $M = [e_j : j \in \mathbb{N}]$. In Exercise 5.8.4 the reader will check that $M^\perp = \{0\}$, then it is clear that in this case $H \neq M \oplus M^\perp$.

The following corollary is immediate from Theorem 5.2.5. We chose to state it because it is one of the fundamental properties of Hilbert spaces.

**Corollary 5.2.7** *Every non-null closed subspace of a Hilbert space is 1-complemented.*

## 5.3 Orthonormal Sets in Hilbert Spaces

The fact that every vector space has a basis (Proposition A.4) ensures that every vector in a vector space can be written as a linear combination of the basis vectors. In finite dimension, the reader is used to this and surely is convinced of the advantages of having such a representation. However, in infinite dimension, this representation in general is useless. In this section, we will see an extremely useful way of representing vectors in a Hilbert space as sums (not always finite) of multiples of conveniently chosen vectors. This brings Hilbert spaces even closer to Euclidean spaces.

**Definition 5.3.1** Let $E$ be a space with an inner product. A set $S \subseteq E$ is said to be *orthonormal* if for all $x, y \in S$,

$$\langle x, y \rangle = \begin{cases} 0, & \text{if } x \neq y, \\ 1, & \text{if } x = y. \end{cases}$$

An orthonormal set $S$ such that $S^\perp = \{0\}$ is called a *complete orthonormal system*.

**Example 5.3.2** The canonical basis $\{(1, 0, \ldots, 0), \ldots, (0, \ldots, 0, 1)\}$ of $\mathbb{K}^n$ is a complete orthonormal system. The set $\{e_n : n \in \mathbb{N}\}$ of the canonical unit vectors is a complete orthonormal system in $\ell_2$.

The reader will have no difficulty in checking that every orthonormal set in a space with an inner product is linearly independent. In fact, we believe that this has already been done in the Linear Algebra course.

To simplify the notation, we now introduce the *Kronecker delta*: given any set $I$ and $i, j \in I$, we define

$$\delta_{ij} = \begin{cases} 0, & \text{if } i \neq j, \\ 1, & \text{if } i = j. \end{cases}$$

In this way, a set $\{x_i : i \in I\}$ of vectors from a space with an inner product is orthonormal if and only if $\langle x_i, x_j \rangle = \delta_{ij}$ for all $i, j \in I$.

We have already seen in Theorem 5.2.2 that when $M$ is a complete subspace of a space with an inner product $E$, every vector $x$ in $E$ admits a (unique) best approximation in $M$. The following result provides a description of this best approximation of $x$ in a finite-dimensional subspace of a Hilbert space.

**Proposition 5.3.3** *Let $H$ be a Hilbert space and let $\{x_1, \ldots, x_n\}$ be a finite orthonormal set in $H$.*

(a) *If $M = [x_1, \ldots, x_n]$ and $x \in H$, then*

$$\left\| x - \sum_{i=1}^{n} \langle x, x_i \rangle x_i \right\| = \operatorname{dist}(x, M).$$

## 5.3 Orthonormal Sets in Hilbert Spaces

(b) *For every $x$ in $H$, $\sum_{i=1}^{n} |\langle x, x_i \rangle|^2 \leq \|x\|^2$.*

*Proof*

(a) By Theorem 5.2.5 we can choose $p \in M$ and $q \in M^\perp$ such that

$$x = p + q \quad \text{and} \quad \|x - p\| = \text{dist}(x, M).$$

As $p \in M$, there exist scalars $\alpha_1, \ldots, \alpha_n$ such that $p = \alpha_1 x_1 + \cdots + \alpha_n x_n$. And as $x - p = q \in M^\perp$, we have $(x - p) \perp x_j$ for each $j = 1, \ldots, n$. It follows that

$$0 = \langle x - p, x_j \rangle = \langle x, x_j \rangle - \alpha_j,$$

that is, $\alpha_j = \langle x, x_j \rangle$, for each $j = 1, \ldots, n$. Item (a) is proven.

(b) Given $x \in H$, let $\alpha_1, \ldots, \alpha_n$ be as in (a). As $\langle x_i, x_j \rangle = \delta_{ij}$ for all $i, j = 1, \ldots, n$,

$$0 \leq \left\langle x - \sum_{i=1}^{n} \langle x, x_i \rangle x_i, x - \sum_{i=1}^{n} \langle x, x_i \rangle x_i \right\rangle$$

$$= \|x\|^2 - \sum_{i=1}^{n} |\langle x, x_i \rangle|^2 - \sum_{i=1}^{n} |\langle x, x_i \rangle|^2 + \sum_{i=1}^{n} |\langle x, x_i \rangle|^2,$$

and the result follows. ∎

**Remark 5.3.4** The expression $\sum_{i=1}^{n} \langle x, x_i \rangle x_i$ is, for obvious reasons, called the *best approximation* of $x$ in $M = [x_1, \ldots, x_n]$.

We will now move towards establishing two fundamental relations of Hilbert space theory, namely, Bessel's inequality and Parseval's identity. To accomplish these tasks we need the following:

**Lemma 5.3.5** *Let $S = \{x_i : i \in I\}$ be an orthonormal set in a Hilbert space $H$. Then, for each $0 \neq x \in H$, the set*

$$J = \{i \in I : \langle x, x_i \rangle \neq 0\}$$

*is finite or countable.*

*Proof* For each $k \in \mathbb{N}$ we define

$$J_k = \left\{ i \in I : |\langle x, x_i \rangle| > \frac{1}{k} \right\}$$

to obtain $J = \bigcup_{k=1}^{\infty} J_k$. It is enough to show that each $J_k$ is finite. From Proposition 5.3.3(b) we know that for every finite subset $J_0$ of $J$ it is true that

$$\sum_{i \in J_0} |\langle x, x_i \rangle|^2 \leq \|x\|^2.$$

In particular, given a finite number of elements $i_1, \ldots, i_n$ of $J_k$,

$$\frac{n}{k^2} = \frac{1}{k^2} + \cdots + \frac{1}{k^2} < |\langle x, x_{i_1} \rangle|^2 + \cdots + |\langle x, x_{i_n} \rangle|^2 \leq \|x\|^2.$$

Consequently, $n \leq k^2 \|x\|^2$. This means that the number of elements of any finite subset of $J_k$ does not exceed $k^2 \|x\|^2$. It follows that $J_k$ is finite. ∎

**Theorem 5.3.6 (Bessel's Inequality)** *Let $S = \{x_i : i \in I\}$ be an orthonormal set in a Hilbert space $H$. Then, for every $x \in H$,*

$$\sum_{i \in J} |\langle x, x_i \rangle|^2 \leq \|x\|^2,$$

*where $J = \{i \in I : \langle x, x_i \rangle \neq 0\}$.*

*Proof* We know from Lemma 5.3.5 that $J$ is finite or countable. Proposition 5.3.3 solves the case where $J$ is finite. It remains to do the case where $J$ is countable. Since all terms of the series are positive, the order in which we sum the series does not matter. We can then consider any enumeration $i_1, i_2, \ldots$ of the elements of $J$. For each $n \in \mathbb{N}$, Proposition 5.3.3 ensures that

$$\sum_{k=1}^{n} |\langle x, x_{i_k} \rangle|^2 \leq \|x\|^2.$$

Now just let $n$ tend to infinity in this inequality to obtain

$$\sum_{i \in J} |\langle x, x_i \rangle|^2 = \sum_{k=1}^{\infty} |\langle x, x_{i_k} \rangle|^2 \leq \|x\|^2.$$

∎

Our goal is to show that every vector in a Hilbert space can be written as a sum of multiples of vectors in a complete orthonormal system. As in general this sum is not finite, we need to explore the concept of series in normed spaces. This topic will be treated in detail in Sect. 10.1. At the moment we restrict ourselves only to the necessary definitions and an auxiliary result.

## 5.3 Orthonormal Sets in Hilbert Spaces

**Definition 5.3.7** Let $(x_n)_{n=1}^\infty$ be a sequence in the normed space $E$. We say that the series $\sum_{n=1}^\infty x_n$ is *convergent* if there exists $x \in E$ such that the sequence of partial sums $\left(\sum_{j=1}^n x_j\right)_{n=1}^\infty$ converges to $x$. In this case we say that $x$ is the sum of the series and we write

$$x = \sum_{n=1}^\infty x_n.$$

We say that the series $\sum_{n=1}^\infty x_n$ is *unconditionally convergent* if it is convergent in any ordering in which we consider its terms; more precisely, if for every bijective function $\sigma \colon \mathbb{N} \longrightarrow \mathbb{N}$ the series $\sum_{n=1}^\infty x_{\sigma(n)}$ is convergent.

According to these definitions, it is possible for an unconditionally convergent series to converge to different limits considering different orderings of $\mathbb{N}$. It is important to establish at this point that this does not occur:

**Proposition 5.3.8** Let $\sum_{n=1}^\infty x_n$ be an unconditionally convergent series in a normed space $E$. If $\sigma_1, \sigma_2 \colon \mathbb{N} \longrightarrow \mathbb{N}$ are bijective functions, then

$$\sum_{n=1}^\infty x_{\sigma_1(n)} = \sum_{n=1}^\infty x_{\sigma_2(n)}.$$

*Proof* Let $\varphi \in E'$ be given. From the continuity of $\varphi$ it is clear that the numerical series $\sum_{n=1}^\infty \varphi(x_n)$ is unconditionally convergent. Since unconditionally convergent numerical series converge to the same limit regardless of the order of the terms, we have

$$\varphi\left(\sum_{n=1}^\infty x_{\sigma_1(n)}\right) = \sum_{n=1}^\infty \varphi(x_{\sigma_1(n)}) = \sum_{n=1}^\infty \varphi(x_{\sigma_2(n)}) = \varphi\left(\sum_{n=1}^\infty x_{\sigma_2(n)}\right).$$

As this holds for every $\varphi \in E'$, the result follows from the Hahn-Banach Theorem in the form of Corollary 3.1.5 (or, if you prefer, from Exercise 3.6.2). ∎

We want to show that if $\{x_i : i \in I\}$ is a complete orthonormal system in a Hilbert space $H$, then every vector $x \in H$ can be written in the form

$$x = \sum_{i \in I} \langle x, x_i \rangle x_i, \tag{5.5}$$

a fact that highlights the central role played by complete orthonormal systems in the theory of Hilbert spaces. As we know from Lemma 5.3.5 that $\{i \in I : \langle x, x_i \rangle \neq 0\}$ is finite or countable, it is clear that the sum that appears in (5.5) denotes, in fact, a finite sum or a sum over a countable set of indices, that is, a series. However, the convergence of the series and even less its unconditional convergence is not yet clear. The next lemma establishes this fact.

**Lemma 5.3.9** *Let $S = \{x_i : i \in I\}$ be an orthonormal set in a Hilbert space $H$. Then, for each $x \in H$, denoting $I_x = \{i \in I : \langle x, x_i \rangle \neq 0\}$, the series*

$$\sum_{i \in I_x} \langle x, x_i \rangle x_i$$

*is unconditionally convergent.*

**Proof** There is nothing to do when $I_x$ is finite. Suppose that $I_x$ is infinite and take $(y_j)_{j=1}^\infty$ as any enumeration of the set $\{w \in S : \langle x, w \rangle \neq 0\}$. For each $n \in \mathbb{N}$, define $S_n = \sum_{i=1}^n \langle x, y_i \rangle y_i$. From Bessel's inequality we know that the series $\sum_{n=1}^\infty |\langle x, y_n \rangle|^2$ is convergent. As $\langle y_i, y_j \rangle = \delta_{ij}$,

$$\|S_n - S_m\|^2 = \left\| \sum_{i=m+1}^n \langle x, y_i \rangle y_i \right\|^2 = \sum_{i=m+1}^n |\langle x, y_i \rangle|^2$$

for $n > m$, therefore the sequence $(S_n)_{n=1}^\infty$ is Cauchy in $H$, hence convergent. ∎

Having resolved the issue of convergence, we can prove (5.5):

**Theorem 5.3.10** *Let $S = \{x_i : i \in I\}$ be an orthonormal set in a Hilbert space $H$. The following statements are equivalent:*

(a) *For each $x \in H$, $x = \sum_{i \in I} \langle x, x_i \rangle x_i$.*
(b) *$S$ is a complete orthonormal system.*
(c) *$\overline{[S]} = H$.*
(d) *For each $x \in H$, $\|x\|^2 = \sum_{i \in I} |\langle x, x_i \rangle|^2$. (Parseval's Identity)*
(e) *For all $x, y \in H$, $\langle x, y \rangle = \sum_{i \in I} \langle x, x_i \rangle \overline{\langle y, x_i \rangle}$.*

**Proof** (a) $\implies$ (b) Let $x \in S^\perp$ be given. Since $\langle x, x_i \rangle = 0$ for every $i \in I$, from (a) it immediately follows that $x = 0$. Thus $S^\perp = \{0\}$ and $S$ is complete.

(b) $\implies$ (a) Let $x \in H$ be given. Suppose that the set $J := \{i \in I : \langle x, x_i \rangle \neq 0\}$ is infinite, therefore countable. Let $\{i_1, i_2, \ldots\}$ be any enumeration of $J$. From Lemma 5.3.9 and Proposition 5.3.8 we know that

## 5.3 Orthonormal Sets in Hilbert Spaces

$$\sum_{i \in I} \langle x, x_i \rangle x_i = \sum_{j=1}^{\infty} \langle x, x_{i_j} \rangle x_{i_j}.$$

Let $i \in I$ be given. If $i \notin J$, then

$$\left\langle x - \sum_{j=1}^{\infty} \langle x, x_{i_j} \rangle x_{i_j}, x_i \right\rangle = 0.$$

And if $i \in J$, there exists $k \in \mathbb{N}$ such that $i = i_k$. In this case, since $\langle x_{i_j}, x_{i_k} \rangle = \delta_{i_j i_k} = \delta_{jk}$,

$$\left\langle x - \sum_{j=1}^{\infty} \langle x, x_{i_j} \rangle x_{i_j}, x_i \right\rangle = \left\langle x - \sum_{j=1}^{\infty} \langle x, x_{i_j} \rangle x_{i_j}, x_{i_k} \right\rangle$$

$$= \langle x, x_{i_k} \rangle - \sum_{j=1}^{\infty} \langle x, x_{i_j} \rangle \langle x_{i_j}, x_{i_k} \rangle = 0.$$

Since $S$ is complete, we obtain

$$x - \sum_{j \in I} \langle x, x_i \rangle x_i = x - \sum_{j=1}^{\infty} \langle x, x_{i_j} \rangle x_{i_j} = 0.$$

The argument above easily adapts to the case where $J$ is finite.

(b) $\Longrightarrow$ (c) Call $M = \overline{[S]}$. Since $S$ is a subset of $M$, it follows that $M^\perp$ is a subspace of $S^\perp = \{0\}$, therefore $M^\perp = \{0\}$. But $H = M \oplus M^\perp$ by Theorem 5.2.5, and we conclude that $H = M = \overline{[S]}$.

(c) $\Longrightarrow$ (d) Let $x \in H$ and $\varepsilon > 0$ be given. By (c) there exists $y_\varepsilon \in [S]$ such that $\|x - y_\varepsilon\| < \varepsilon$. As $y_\varepsilon \in [S]$, there are a finite subset $J_\varepsilon$ of $I$ and scalars $(a_i)_{i \in J_\varepsilon}$ such that $y_\varepsilon = \sum_{i \in J_\varepsilon} a_i x_i$. From Proposition 5.3.3 we know that the vector $\sum_{i \in J_\varepsilon} \langle x, x_i \rangle x_i$ is the best approximation of $x$ in $[x_i : i \in J_\varepsilon]$, therefore

$$\left\| x - \sum_{i \in J_\varepsilon} \langle x, x_i \rangle x_i \right\| \leq \left\| x - \sum_{i \in J_\varepsilon} a_i x_i \right\| = \|x - y_\varepsilon\| < \varepsilon.$$

Since $\{x_i : i \in J_\varepsilon\}$ is orthonormal,

$$\|x\|^2 - \sum_{i \in J_\varepsilon} |\langle x, x_i \rangle|^2 = \left\langle x - \sum_{i \in J_\varepsilon} \langle x, x_i \rangle x_i, x - \sum_{i \in J_\varepsilon} \langle x, x_i \rangle x_i \right\rangle$$

$$= \left\| x - \sum_{i \in J_\varepsilon} \langle x, x_i \rangle x_i \right\|^2 < \varepsilon^2.$$

Combining this with Bessel's inequality we get

$$\|x\|^2 < \sum_{i \in J_\varepsilon} |\langle x, x_i \rangle|^2 + \varepsilon^2 \le \sum_{i \in I} |\langle x, x_i \rangle|^2 + \varepsilon^2 \le \|x\|^2 + \varepsilon^2.$$

The result follows by letting $\varepsilon \longrightarrow 0^+$.

(d) $\Longrightarrow$ (e) Let $x, y \in H$ and $a$ be a scalar. By (d),

$$\langle ax + y, ax + y \rangle = \|ax + y\|^2 = \sum_{i \in I} |\langle ax + y, x_i \rangle|^2$$

$$= \sum_{i \in I} \langle ax + y, x_i \rangle \overline{\langle ax + y, x_i \rangle},$$

and then,

$$|a|^2 \|x\|^2 + a \langle x, y \rangle + \overline{a} \langle y, x \rangle + \|y\|^2 = \langle ax + y, ax + y \rangle$$

$$= \sum_{i \in I} \left( |a|^2 |\langle x, x_i \rangle|^2 + a \langle x, x_i \rangle \langle x_i, y \rangle + \overline{a} \langle y, x_i \rangle \langle x_i, x \rangle + |\langle x_i, y \rangle|^2 \right)$$

$$\stackrel{(*)}{=} \sum_{i \in I} |a|^2 |\langle x, x_i \rangle|^2 + \sum_{i \in I} a \langle x, x_i \rangle \langle x_i, y \rangle + \sum_{i \in I} \overline{a} \langle y, x_i \rangle \langle x_i, x \rangle + \sum_{i \in I} |\langle x_i, y \rangle|^2$$

$$= |a|^2 \|x\|^2 + \sum_{i \in I} a \langle x, x_i \rangle \langle x_i, y \rangle + \sum_{i \in I} \overline{a} \langle y, x_i \rangle \langle x_i, x \rangle + \|y\|^2.$$

The step $(*)$ is justified because each of the four series is convergent: the first and the last by (d) and the second and the third by Hölder's inequality. Then

$$a \langle x, y \rangle + \overline{a} \langle y, x \rangle = \sum_{i \in I} a \langle x, x_i \rangle \langle x_i, y \rangle + \sum_{i \in I} \overline{a} \langle y, x_i \rangle \langle x_i, x \rangle$$

$$= a \left( \sum_{i \in I} \langle x, x_i \rangle \overline{\langle y, x_i \rangle} \right) + \overline{a} \left( \sum_{i \in I} \langle y, x_i \rangle \overline{\langle x, x_i \rangle} \right).$$

First choosing $a = 1$ and then $a = i$, we obtain

$$\operatorname{Re}\langle x, y\rangle = \operatorname{Re}\left(\sum_{i\in I}\langle x, x_i\rangle\overline{\langle y, x_i\rangle}\right) \quad \text{and} \quad \operatorname{Im}\langle x, y\rangle = \operatorname{Im}\left(\sum_{i\in I}\langle x, x_i\rangle\overline{\langle y, x_i\rangle}\right),$$

and the result follows.

(e) $\Longrightarrow$ (b) Let $x \in S^{\perp}$ be given. Therefore $\langle x, x_i\rangle = 0$ for every $i \in I$. Using (e) with $x = y$ we obtain $\langle x, x\rangle = 0$ and consequently $x$ is the null vector. ∎

## 5.4 Orthogonalization Process and Its Consequences

In the previous section we saw that a complete orthonormal system is something very desirable in a Hilbert space. The aim of this section is to show that every Hilbert space contains complete orthonormal systems.

We start by showing how to transform a linearly independent set into an orthonormal set with good properties of subspace preservation. Any similarity with the Gram–Schmidt process that we all know from Linear Algebra is **not** a mere coincidence.

**Proposition 5.4.1 (Gram–Schmidt Orthogonalization Process)** *Let $E$ be a space with inner product and let $(x_n)_{n=1}^{\infty}$ be a sequence of linearly independent vectors in $E$. There exists an orthonormal sequence $(e_n)_{n=1}^{\infty}$ in $E$ such that for every $n \in \mathbb{N}$,*

$$[x_1, \ldots, x_n] = [e_1, \ldots, e_n].$$

***Proof*** We assume that, from the Linear Algebra course, the reader is already familiar with the construction. Therefore, we will only indicate the steps to be followed. First we define $e_1 = \frac{x_1}{\|x_1\|}$ and write $x_2 = \langle x_2, e_1\rangle e_1 + v_2$ with $v_2 \in [e_1]^{\perp}$. We then define $e_2 = \frac{v_2}{\|v_2\|}$ to obtain $[x_1, x_2] = [e_1, e_2]$. Proceeding by induction, suppose we have constructed orthonormal vectors $e_1, \ldots, e_n$ such that

$$[x_1, \ldots, x_n] = [e_1, \ldots, e_n].$$

We write

$$x_{n+1} = \langle x_{n+1}, e_1\rangle e_1 + \cdots + \langle x_{n+1}, e_n\rangle e_n + v_{n+1}$$

with $v_{n+1} \in [e_1, \ldots, e_n]^{\perp}$. Defining $e_{n+1} = \frac{v_{n+1}}{\|v_{n+1}\|}$ it is true that the vectors $e_1, \ldots, e_{n+1}$ are orthonormal and

$$[x_1, \ldots, x_{n+1}] = [e_1, \ldots, e_{n+1}].$$

∎

**Corollary 5.4.2** *Let $E$ be an inner product space and let $(x_n)_{n=1}^\infty$ be a sequence of linearly independent vectors in $E$. There exists an orthonormal sequence $(e_n)_{n=1}^\infty$ such that*

$$[x_n : n \in \mathbb{N}] = [e_n : n \in \mathbb{N}].$$

***Proof*** Let $(e_n)_{n=1}^\infty$ be the sequence obtained in Proposition 5.4.1. Given $v \in [x_n : n \in \mathbb{N}]$, let $k \in \mathbb{N}$ be such that $v \in [x_1, \ldots, x_k]$. By the construction of the Gram–Schmidt process we know that $[x_1, \ldots, x_k] = [e_1, \ldots, e_k]$. It is clear that $[e_1, \ldots, e_k] \subseteq [e_n : n \in \mathbb{N}]$, therefore $v \in [e_n : n \in \mathbb{N}]$, and consequently $[x_n : n \in \mathbb{N}] \subseteq [e_n : n \in \mathbb{N}]$. The other inclusion is obtained analogously. ∎

We need nothing more to prove that separable Hilbert spaces admit complete orthonormal systems:

**Theorem 5.4.3** *An infinite-dimensional Hilbert space $H$ is separable if and only if there exists in $H$ a countable complete orthonormal system.*

***Proof*** Suppose that $S = \{x_n : n \in \mathbb{N}\}$ is a complete orthonormal system in $H$. By Theorem 5.3.10, every $x$ in $H$ can be written in the form

$$x = \sum_{i=1}^{\infty} \langle x, x_i \rangle x_i.$$

This guarantees, in particular, that $[S]$ is dense in $H$. The separability of $H$ follows from Lemma 1.6.3.

Conversely, suppose that $H$ is separable. Let $D = \{x_n : n \in \mathbb{N}\}$ be a countable dense subset in $H$. We can extract from $D$ a basis $\mathcal{B}$ for $[D]$ (the reader in doubt at this point should consult Proposition A.4 in Appendix A). Suppose that this basis $\mathcal{B} \subseteq D$ is finite, say $\mathcal{B} = \{v_1, \ldots, v_n\}$. In this case, as $[D]$ is also dense in $H$ (for containing $D$ which is dense) and as finite-dimensional subspaces of complete spaces are closed, we have

$$H = \overline{[D]} = \overline{[v_1, \ldots, v_n]} = [v_1, \ldots, v_n].$$

But this cannot happen because $H$ is infinite-dimensional. We conclude that the basis $\mathcal{B}$ of $[D]$ is infinite. But $\mathcal{B} \subseteq D$ and $D$ is countable, so $\mathcal{B}$ is countable as well, say $\mathcal{B} = \{v_n : n \in \mathbb{N}\}$. By the Gram–Schmidt orthogonalization process there is an orthonormal set $S$ in $H$ such that $[S] = [v_n : n \in \mathbb{N}]$. Thus,

$$\overline{[S]} = \overline{[v_n : n \in \mathbb{N}]} = \overline{[D]} = H,$$

and by Theorem 5.3.10 we conclude that $S$ is a complete orthonormal system. ∎

## 5.4 Orthogonalization Process and Its Consequences

The existence of a complete countable orthonormal system so rigidly structures the space that—surprisingly—the only infinite-dimensional separable Hilbert space is, up to isometric isomorphism, our old friend $\ell_2$:

**Theorem 5.4.4 (Riesz–Fischer Theorem)** *Every infinite-dimensional separable Hilbert space is isometrically isomorphic to $\ell_2$.*

*Proof* Let $H$ be an infinite-dimensional separable Hilbert space. By Theorem 5.4.3 there is a complete countable orthonormal system $S = \{x_j : j \in \mathbb{N}\}$ in $H$. From Bessel's inequality we know that, for every $x \in H$, $(\langle x, x_n \rangle)_{n=1}^\infty \in \ell_2$. Then the operator

$$T : H \longrightarrow \ell_2, \quad T(x) = (\langle x, x_n \rangle)_{n=1}^\infty.$$

is well defined. It is easy to check that $T$ is linear. Moreover, by Theorem 5.3.10 we know that every $x$ in $H$ is represented by the series

$$x = \sum_{i=1}^\infty \langle x, x_i \rangle x_i,$$

and so Parseval's identity reveals that $\|T(x)\| = \|x\|$. In particular, $T$ is injective. It remains to prove that $T$ is surjective. To do so, let $(a_j)_{j=1}^\infty \in \ell_2$ be given. We first note that the series $\sum_{j=1}^\infty a_j x_j$ converges in $H$. Indeed, writing for each $k \in \mathbb{N}$, $S_k = \sum_{j=1}^k a_j x_j$, from the Pythagorean Theorem it follows that, for $n > m$,

$$\|S_n - S_m\|^2 = \left\| \sum_{j=m+1}^n a_j x_j \right\|^2 = \sum_{j=m+1}^n \|a_j x_j\|^2 = \sum_{j=m+1}^n |a_j|^2.$$

Since the series $\sum_{j=1}^\infty |a_j|^2$ is convergent, it follows that $(S_n)_{n=1}^\infty$ is a Cauchy sequence in $H$, so it is convergent. Therefore $x := \sum_{j=1}^\infty a_j x_j \in H$ is well defined and from the condition $\langle x_j, x_n \rangle = \delta_{jn}$ it immediately follows that $T(x) = (a_j)_{j=1}^\infty$. ∎

For the non-separable case of the Riesz–Fischer Theorem, see Exercise 5.8.33. With a little more theory, we will prove in the next section that Hilbert spaces are reflexive. But for the separable case we already have enough theory:

**Corollary 5.4.5** *Separable Hilbert spaces are reflexive.*

***Proof*** Let $H$ be a separable Hilbert space. By the Riesz–Fischer Theorem we know that $H$ is isomorphic to $\ell_2$. Since $\ell_2$ is reflexive, from Exercise 4.5.27 it follows that $H$ is reflexive. ∎

To prove the existence of a complete orthonormal system in non-separable Hilbert spaces we need to resort to Zorn's Lemma, the same one used in the proof of the Hahn–Banach Theorem. We reiterate that the reader unfamiliar with this tool should consult Appendix A. The proof below is similar to the proof of the fact that every generating set of a vector space contains a basis of the space (see Proposition A.4).

**Theorem 5.4.6** *Let $H$ be a space with inner product and let $S_0$ be an orthonormal set in $H$. Then there exists a complete orthonormal system in $H$ that contains $S_0$.*

***Proof*** Call $\mathcal{F}$ the family of all orthonormal sets in $H$ that contain $S_0$. Note that $\mathcal{F} \neq \emptyset$ because $S_0 \in \mathcal{F}$. We consider $\mathcal{F}$ with the partial order given by the containment relation: $S_1, S_2 \in \mathcal{F}$, $S_1 \leq S_2$ if and only if $S_1 \subseteq S_2$. Let $\{S_i : i \in I\}$ be a totally ordered subset of $\mathcal{F}$. Now we prove that $\bigcup_{i \in I} S_i \in \mathcal{F}$. Indeed, if $x, y \in \bigcup_{i \in I} S_i$, then there exist $i_1, i_2 \in I$ such that $x \in S_{i_1}$ and $y \in S_{i_2}$. If $x = y$, then $\langle x, y \rangle = \langle x, x \rangle = 1$ because $S_{i_1}$ is orthonormal. Now suppose that $x \neq y$. As $\{S_i : i \in I\}$ is totally ordered, we have $S_{i_1} \subseteq S_{i_2}$ or $S_{i_2} \subseteq S_{i_1}$. Therefore, $x, y \in S_{i_1}$ or $x, y \in S_{i_2}$. In both cases $x$ is orthogonal to $y$, since both $S_{i_1}$ and $S_{i_2}$ are orthogonal. It follows that $\bigcup_{i \in I} S_i$ is an orthonormal set. It is clear that $\bigcup_{i \in I} S_i$ is an upper bound for the set $\{S_i : i \in I\}$, therefore every partially ordered set has an upper bound. Zorn's Lemma guarantees the existence of an orthonormal set $S \in \mathcal{F}$ that is maximal. It remains to show that $S$ is complete. Assuming that $S$ is not complete, there exists $z \in H$ non-null such that $z \in S^\perp$. In this case the set

$$S^1 := S \cup \left\{ \frac{z}{\|z\|} \right\}$$

is orthonormal, distinct from $S$ and $S \leq S^1$. But this contradicts the maximality of $S$, therefore $S$ is complete. ∎

## 5.5 Linear Functionals and the Riesz–Fréchet Theorem

In this section, we will obtain a description of all continuous linear functionals on a Hilbert space. We first see that it is very easy to create such functionals:

**Example 5.5.1** Let $H$ be a Hilbert space. For each $y \in H$, the functional

$$\varphi_y : H \longrightarrow \mathbb{K}, \; \varphi_y(x) = \langle x, y \rangle,$$

## 5.5 Linear Functionals and the Riesz–Fréchet Theorem

is clearly linear. From the Cauchy–Schwarz inequality it follows that $\varphi_y$ is continuous and that $\|\varphi_y\| \leq \|y\|$. For $y = 0$ it is immediate that $\|\varphi_y\| = \|y\|$, and for $y \neq 0$ this also holds because

$$\left|\varphi_y\left(\frac{y}{\|y\|}\right)\right| = \frac{|\varphi_y(y)|}{\|y\|} = \|y\|.$$

In summary, $\varphi_y \in H'$ and $\|\varphi_y\| = \|y\|$.

The next result, a classic due to Riesz and Fréchet, guarantees that all continuous linear functionals on a Hilbert space are of the form described in the previous example.

**Theorem 5.5.2 (Riesz–Fréchet Theorem)** *Let $H$ be a Hilbert space and let $\varphi: H \longrightarrow \mathbb{K}$ be a continuous linear functional. Then there exists a unique $y_0 \in H$ such that*

$$\varphi(x) = \langle x, y_0 \rangle \text{ for every } x \in H.$$

*Furthermore, $\|\varphi\| = \|y_0\|$.*

**Proof** If $\varphi = 0$, we simply choose $y_0 = 0$. Now we consider the case where $\varphi \neq 0$. In this case the kernel of $\varphi$,

$$M := \{x \in H : \varphi(x) = 0\},$$

is a proper closed subspace of $H$. From Theorem 5.2.5(a) we know that $M^\perp \neq \{0\}$, then we can choose $x_0$ in $M^\perp$ of norm 1 (remember that $M^\perp$ is a vector subspace). We prove that $y_0 := \overline{\varphi(x_0)} x_0$ is the sought vector. Given $x \in H$,

$$x = \left(x - \frac{\varphi(x)}{\varphi(x_0)} x_0\right) + \frac{\varphi(x)}{\varphi(x_0)} x_0,$$

where $\left(x - \frac{\varphi(x)}{\varphi(x_0)} x_0\right) \in M$ and $\frac{\varphi(x)}{\varphi(x_0)} x_0 \in M^\perp$. Since $y_0 \in M^\perp$ and $\|x_0\| = 1$,

$$\langle x, y_0 \rangle = \left\langle x - \frac{\varphi(x)}{\varphi(x_0)} x_0, y_0 \right\rangle + \left\langle \frac{\varphi(x)}{\varphi(x_0)} x_0, y_0 \right\rangle$$

$$= 0 + \frac{\varphi(x)}{\varphi(x_0)} \langle x_0, y_0 \rangle = \frac{\varphi(x)}{\varphi(x_0)} \left\langle x_0, \overline{\varphi(x_0)} x_0 \right\rangle = \varphi(x).$$

From the Cauchy–Schwarz inequality we have $\|\varphi\| \leq \|y_0\|$. The equality $|\varphi(x_0)| = \|y_0\|$ ensures the reverse inequality, hence $\|\varphi\| = \|y_0\|$. Uniqueness follows from (5.2). ∎

Let $H$ be a Hilbert space. Combining Example 5.5.1 with Theorem 5.5.2 results in the correspondence

$$y \in H \longleftrightarrow \varphi_y \in H' \tag{5.6}$$

being both bijective and an isometry. It is also linear in the real case, that is, when $\mathbb{K} = \mathbb{R}$. Therefore we have the

**Corollary 5.5.3** *Every real Hilbert space is isometrically isomorphic to its dual through the correspondence (5.6).*

In Exercise 5.8.33 the reader will check that a complex Hilbert space is also isometrically isomorphic to its dual, but not through the correspondence (5.6). For now we stick with a few more interesting consequences of the Riesz–Fréchet Theorem. The following two results are immediate in the case of separable Hilbert spaces; however our context also involves the non-separable case:

**Proposition 5.5.4** *The dual of a Hilbert space is also a Hilbert space.*

*Proof* Let $H$ be a Hilbert space. As $H'$ is complete, it is enough to show that the norm on $H'$ is induced by an inner product. Given $\varphi_1, \varphi_2 \in H'$, by Theorem 5.5.2 there exist unique $y_1, y_2 \in H$ such that $\varphi_1(x) = \langle x, y_1 \rangle$ and $\varphi_2(x) = \langle x, y_2 \rangle$ for every $x \in H$, $\|\varphi_1\| = \|y_1\|$ and $\|\varphi_2\| = \|y_2\|$. It is easy to check that the expression

$$\langle \varphi_1, \varphi_2 \rangle := \langle y_2, y_1 \rangle \tag{5.7}$$

defines an inner product on $H'$ (this is not a typing error; the order of appearance of the $y_j$ in the definition above has to be inverted, because in the natural order the axiom (P2) would not be satisfied in the complex case). From

$$\langle \varphi_1, \varphi_1 \rangle = \langle y_1, y_1 \rangle = \|y_1\|^2 = \|\varphi_1\|^2$$

it follows that this inner product induces the norm on $H'$. ∎

**Corollary 5.5.5** *Hilbert spaces are reflexive.*

*Proof* Let $H$ be a Hilbert space. Given $\Phi \in H''$, as $H'$ is a Hilbert space by Proposition 5.5.4, from the Riesz-Fréchet Theorem, there exists $\varphi \in H'$ such that $\Phi(\psi) = \langle \psi, \varphi \rangle$ for every $\psi \in H'$. A second application of the Riesz-Fréchet Theorem provides a vector $y \in H$ such that $\varphi(x) = \langle x, y \rangle$ for every $x \in H$. Let $\psi \in H'$ be given. Using the Riesz-Fréchet Theorem for the third time, there exists $z \in H$ such that $\psi(x) = \langle x, z \rangle$ for every $x \in H$. From (5.7) we know that $\langle \psi, \varphi \rangle = \langle y, z \rangle$. Thus,

$$J_H(y)(\psi) = \psi(y) = \langle y, z \rangle = \langle \psi, \varphi \rangle = \Phi(\psi).$$

As this holds for every $\psi \in H'$, it follows that $J_H(y) = \Phi$, proving that $H$ is reflexive. ∎

We will see in Sect. 6.6 that Hilbert spaces satisfy a property stronger than reflexivity.

## 5.6 The Lax-Milgram Theorem

The Lax-Milgram Theorem is an important tool in the theory of Partial Differential Equations and in the solution of variational problems. It is a kind of Riesz-Fréchet Theorem in which bilinear forms replace inner products.

**Definition 5.6.1** Let $E$, $F$ be Banach spaces over $\mathbb{K}$. A bilinear form $T: E \times F \longrightarrow \mathbb{K}$ is

(a) *coercive* if $E = F$, $\mathbb{K} = \mathbb{R}$ and there exists a constant $\beta > 0$ such that

$$T(x, x) \geq \beta \|x\|^2$$

for every $x \in E$.
(b) *symmetric* if $E = F$ and $T(x, y) = T(y, x)$ for any $x, y \in E$.
(c) *non-degenerate* if the only vector $x \in E$ such that $T(x, y) = 0$ for every $y \in F$ is the vector $x = 0$; and, analogously, the only vector $y \in F$ such that $T(x, y) = 0$ for every $x \in E$ is the vector $y = 0$.

The continuity of bilinear forms was treated in Corollary 2.3.5 and in Exercise 2.7.25.

**Example 5.6.2** The bilinear form

$$T_1: \ell_2 \times \ell_2 \longrightarrow \mathbb{R}, \ T_1\left((a_j)_{j=1}^\infty, (b_j)_{j=1}^\infty\right) = \sum_{j=1}^\infty a_j b_j,$$

is symmetric, coercive, continuous and non-degenerate. On the other hand, the bilinear form

$$T_2: \ell_2 \times \ell_2 \longrightarrow \mathbb{R}, \ T_2\left((a_j)_{j=1}^\infty, (b_j)_{j=1}^\infty\right) = \sum_{j=1}^\infty a_{2j} b_{2j},$$

is symmetric and continuous but not coercive nor non-degenerate.

The first result of this section is a type of Riesz-Fréchet Theorem for symmetric, continuous, and coercive bilinear forms in Hilbert spaces.

**Proposition 5.6.3** *Let $H$ be a Hilbert space over the field of real numbers and let $T: H \times H \longrightarrow \mathbb{R}$ be a symmetric, continuous, and coercive bilinear form. For every continuous linear functional $\varphi \in H'$ there exists a unique vector $x_0 \in H$ such that $\varphi(x) = T(x, x_0)$ for every $x \in H$.*

**Proof** Note that $T$ is an inner product on $H$. Let $\|\cdot\|_T$ be the norm derived from the inner product $T$, that is,

$$x \in H \mapsto \|x\|_T = \sqrt{T(x,x)}.$$

Let $(x_n)_{n=1}^\infty$ be a Cauchy sequence in $(H, \|\cdot\|_T)$. By the coercivity of $T$ there exists $\beta > 0$ such that

$$\|x_n - x_m\|_T^2 = T(x_n - x_m, x_n - x_m) \geq \beta \|x_n - x_m\|^2,$$

therefore $(x_n)_{n=1}^\infty$ is a Cauchy sequence in $H$ with its original norm. As $H$ is complete, there exists $y \in H$ such that $x_n \longrightarrow y$ in the original norm of $H$. And as $T$ is continuous, by Exercise 2.7.25 there exists $C \geq 0$ such that

$$\|x_n - y\|_T = \sqrt{T(x_n - y, x_n - y)} \leq \sqrt{C \|x_n - y\|^2} = \sqrt{C} \|x_n - y\|,$$

hence $x_n \longrightarrow y$ in $(H, \|\cdot\|_T)$. Then $(H, \|\cdot\|_T)$ is complete and, by the Riesz–Fréchet Theorem, there exists a unique $x_0 \in H$ such that $\varphi(x) = T(x, x_0)$ for every $x \in H$. ∎

The Lax–Milgram Theorem is precisely the previous result without the assumption that the bilinear form is symmetric.

**Theorem 5.6.4 (Lax–Milgram Theorem)** *Let $H$ be a Hilbert space over the field of real numbers and let $T: H \times H \longrightarrow \mathbb{R}$ be a continuous and coercive bilinear form. For every continuous linear functional $\varphi \in H'$ there exists a unique vector $x_0 \in H$ such that $\varphi(x) = T(x, x_0)$ for every $x \in H$.*

**Proof** For each $x \in H$, the map

$$T_x: H \longrightarrow \mathbb{R}, \quad T_x(y) = T(y, x),$$

is a continuous linear functional on $H$. By the Riesz–Fréchet Theorem there exists a unique $w_x \in H$ such that $T(y, x) = T_x(y) = \langle y, w_x \rangle$ for every $y \in H$. It is easy to check that the operator

$$A: H \longrightarrow H, \quad A(x) = w_x,$$

is linear. Furthermore, from the continuity of $T$ there exists $C > 0$ such that

$$\|A(x)\|^2 = \langle w_x, w_x \rangle = \langle A(x), w_x \rangle = T(A(x), x) \leq C \|x\| \cdot \|A(x)\|$$

for every $x \in H$. Therefore $\|A(x)\| \leq C \|x\|$ for every $x \in H$, which proves the continuity of the linear operator $A$. The coercivity of $T$ guarantees the existence of $\beta > 0$ such that

## 5.6 The Lax-Milgram Theorem

$$\beta \|x\|^2 \leq T(x,x) = \langle x, w_x \rangle \leq \|x\| \cdot \|w_x\| = \|x\| \cdot \|A(x)\|,$$

therefore $\beta \|x\| \leq \|A(x)\|$ for every $x \in H$. We are allowed to conclude that $A$ is an isomorphism onto its range, and in particular $A$ is injective. As $H$ is complete and the range of $A$, $R(A)$, is a subspace of $H$ isomorphic to $H$, it follows that $R(A)$ is a closed subspace of $H$. Now we prove that $(R(A))^\perp = \{0\}$: indeed, if $y \in (R(A))^\perp$, we have

$$\beta \|y\|^2 \leq T(y,y) = \langle y, w_y \rangle = \langle y, A(y) \rangle = 0,$$

which guarantees that $y = 0$. From Theorem 5.2.5 it follows that $H = R(A)$, that is, $A$ is surjective. Again by the Riesz–Fréchet Theorem there exists $z \in H$ such that $\varphi(x) = \langle x, z \rangle$ for every $x \in H$. As $A$ is bijective, $z = A(x_0)$ for some $x_0 \in H$. Therefore

$$\varphi(x) = \langle x, A(x_0) \rangle = \langle x, w_{x_0} \rangle = T(x, x_0)$$

for every $x \in H$. Uniqueness follows easily from the coercivity of $T$. ∎

There are versions of the Lax–Milgram Theorem for the complex case and even for bilinear forms in Banach spaces. We highlight one of these versions whose proof, as we will see, invokes the Open Mapping Theorem and the Hahn–Banach Theorem.

**Theorem 5.6.5 (Lax–Milgram Theorem for Banach Spaces)** *Let $E$ and $F$ be Banach spaces over $\mathbb{K} = \mathbb{R}$ or $\mathbb{C}$, with $F$ reflexive. Let $T \colon E \times F \longrightarrow \mathbb{K}$ be a non-degenerate continuous bilinear form. A necessary and sufficient condition for every continuous linear functional $\varphi \in F'$ to have a unique representation of the form*

$$\varphi(y) = T(x_0, y) \text{ for every } y \in F, \tag{5.8}$$

*for some $x_0 \in E$, is the existence of a constant $\beta > 0$ such that*

$$\sup_{\|y\|=1} |T(x,y)| \geq \beta \|x\| \text{ for every } x \in E. \tag{5.9}$$

**Proof** Consider the linear operator

$$A \colon E \longrightarrow F', \quad A(x)(y) = T(x,y).$$

As $T$ is non-degenerate, it follows that $A$ is injective. The continuity of $A$ is a consequence of the continuity of $T$.

Suppose that every continuous linear functional $\varphi \in F'$ has a unique representation in the form (5.8). This means that the operator $A$ is surjective. By the Open Mapping Theorem it follows that the operator $A^{-1}$ is continuous, therefore there is a constant $\beta > 0$ such that $\|A(x)\| \geq \beta \|x\|$ for every $x$ in $E$. But, for every $x \in E$,

$$\|A(x)\| = \sup_{\|y\|=1} \|A(x)(y)\| = \sup_{\|y\|=1} |T(x,y)|,$$

and the result follows.

Conversely, suppose that there exists $\beta > 0$ satisfying (5.9). From

$$\|A(x)\| = \sup_{\|y\|=1} |T(x,y)|$$

for every $x \in E$, it follows that $\|A(x)\| \geq \beta \|x\|$ for every $x$ in $E$. Therefore, $A$ is an isomorphism onto its range in $F'$. Suppose that $A$ is not surjective. In this case consider $\varphi \in F'$, $\varphi \notin R(A)$. It is clear that $\varphi \neq 0$. By the Hahn–Banach Theorem in the form of Corollary 3.4.10, there exists $\psi \in F''$ such that

$$\psi(\varphi) = 1 \text{ and } \psi(R(A)) = \{0\}.$$

Since $F$ is reflexive, there exists $y_0 \in F$ such that $J_F(y_0) = \psi$. Therefore, for every $x \in E$,

$$T(x, y_0) = A(x)(y_0) = J_F(y_0)(A(x)) = 0.$$

Since $T$ is non-degenerate, it follows that $y_0 = 0$. This implies that $\psi = 0$, which is a contradiction because $\psi(\varphi) = 1$. ∎

## 5.7 Comments and Historical Notes

D. Hilbert used in 1906 the closed unit ball of $\ell_2$ as the domain of linear, bilinear, and quadratic forms. That is why the term *Hilbert space* was coined by Schoenflies in 1908. For some time, a Hilbert space was understood to be the space $\ell_2$ or its closed unit ball. The abstract concept of a Hilbert space appeared with von Neumann's axiomatization in 1927.

Von Neumann's definition requires separability, and the reason for this is the Gram–Schmidt process. In finite dimension, this process goes back to Laplace and Cauchy, and the case of infinite dimension came with Gram in 1883 and Schmidt in 1908. But Schmidt's proof only works in the separable case. Only in 1934 did F. Riesz provide a proof that does not require separability, and only from then on were Hilbert spaces considered in the generality we know today.

Schmidt is also remembered, along with F. Riesz, for bringing the language of Euclidean geometry (triangle inequality, Pythagorean Theorem, etc) to the theory of normed spaces and Hilbert spaces.

## 5.7 Comments and Historical Notes

The history of the Cauchy–Schwarz inequality is long and involves several characters. The first versions go back to Lagrange and Cauchy in the case of $\mathbb{R}^2$ and $\mathbb{R}^n$. The version with integrals was proven by Buniakowsky in 1859 and by Schwarz in 1885. The case of $\ell_2$ was proven by Schmidt in 1908 and the abstract version by von Neumann in 1930. Despite Buniakowsky having preceded Schwarz by more than 25 years, Western literature usually forgets his name and consolidates the expression Cauchy–Schwarz inequality.

It is usual that general results keep the name of the discoverer of a particular case, even if the particular case is far from the general version. This is what happens with Parseval's identity (1799) and Bessel's inequality (1828). The fact that every closed subspace of a Hilbert space is complemented (Corollary 5.2.7) characterizes Hilbert spaces up to isomorphisms: J. Lindenstrauss and L. Tzafriri [42] proved in 1971 that when every closed subspace of a Banach space $E$ is complemented, $E$ is isomorphic to a Hilbert space.

Unconditionally convergent series play a central role in the theory of Banach spaces. In Sect. 10.1 we will explore this concept appropriately.

Complete orthonormal systems are also called *Hilbert bases* or *maximal orthonormal systems*. The equality of Theorem 5.3.10(a) justifies the term Hilbert basis and Exercise 5.8.22 justifies the term maximal. The trigonometric systems of Exercises 5.8.5 (real case) and 5.8.6 (complex case) are complete orthonormal systems in $L_2[0, 2\pi]$ (see [58, Theorem 4.6] and [13, Theorem 5.2]). In fact, in the separable case these Hilbert bases are particular cases of the concept of *Schauder basis*, which will be addressed in Sect. 10.3.

Let $\{x_i : i \in I\}$ be a complete orthonormal system in the Hilbert space $H$ and let $x \in H$ be given. The numbers $\langle x, x_i \rangle$ are often called the *Fourier coefficients of* $x$, and the series $x = \sum_{i \in I} \langle x, x_i \rangle x_i$ from Theorem 5.3.10(a) is called the *Fourier expansion of* $x$ or *Fourier series of* $x$.

The theory of compact and self-adjoint operators on Hilbert spaces will be addressed in Chap. 7. Below we suggest readings for some classic topics of Hilbert space theory that are not covered in this book: Stampacchia's Theorem [9, Theorem 5.6], orthogonal polynomials [39, 3.7], normal operators and C*-algebras [13, Chapters VIII and IX], Fourier Analysis ([58, 4.3] and [53, 6.10]), unbounded (discontinuous) linear operators ([39, Chapter 10] and [63, Section 22]), Hilbert–Schmidt operators ([17, Chapter 4] and [66, VI.6]) and the Hille–Yosida Theorem [9, Chapter 7].

The Lax-Milgram theorem, a cornerstone of Functional Analysis and Partial Differential Equations, was formulated by Peter D. Lax and Arthur Milgram in 1954. It is a pivotal result that provides existence and uniqueness of solutions to certain types of boundary value problems with no symmetric assumptions on the operator (see [9, Chapters 8 and 9]). The theorem builds on earlier work in the field of Hilbert spaces and Elliptic Partial Differential Equations, offering a general framework for understanding variational problems. For the complex case of the Lax–Milgram Theorem, see [9, Proposition 11.29].

## 5.8 Exercises

**Exercise 5.8.1** Show that the inner product function (from $E \times E$ to $\mathbb{K}$) is continuous.

**Exercise 5.8.2**

(a) Let $E$ be a real vector space with an inner product. Prove that the converse of the Pythagorean theorem holds: if $\|x+y\|^2 = \|x\|^2 + \|y\|^2$ then $x \perp y$.
(b) Does the same hold in the complex case?

**Exercise 5.8.3** Prove that, for $p \neq 2$, $\ell_p$ is not a space with an inner product.

**Exercise 5.8.4** Show that $S = \{e_j : j \in \mathbb{N}\}$ is a complete orthonormal system in $\ell_2$.

**Exercise 5.8.5** Prove that in the real case the set of functions

$$f_0(t) = \frac{1}{\sqrt{2\pi}}, \quad f_n(t) = \frac{1}{\sqrt{\pi}} \cos(nt), \quad g_n(t) = \frac{1}{\sqrt{\pi}} \sin(nt), \quad n \in \mathbb{N},$$

is an orthonormal system in $L_2[0, 2\pi]$.

**Exercise 5.8.6** Prove that in the complex case the set of functions $f_n(t) = \frac{1}{\sqrt{2\pi}} e^{int}$, $n \in \mathbb{Z}$, is an orthonormal system in $L_2[0, 2\pi]$.

**Exercise 5.8.7**

(a) Let $E$ be a real vector space with an inner product. Prove that the operator

$$T : E \longrightarrow E', \quad T(x)(y) = \langle x, y \rangle \text{ for all } x, y \in E,$$

is well defined (that is, $T(x) \in E'$ for every $x \in E$), is linear, continuous and an isometry.
(b) Does the same hold in the complex case? And if we define $T(x)(y) = \langle y, x \rangle$?

**Exercise 5.8.8** Let $E$ and $F$ be spaces with an inner product and let $T: E \longrightarrow F$ be a linear operator. Prove that $T$ is a linear isometry if and only if $\langle T(x), T(y) \rangle = \langle x, y \rangle$ for all $x, y \in E$.

**Exercise 5.8.9**

(a) Let $E$ be a complex vector space with an inner product and let $T: E \longrightarrow E$ be a linear operator. Prove that if $\langle T(x), x \rangle = 0$ for every $x \in E$ then $T = 0$.
(b) Does the same hold in the real case?

**Exercise 5.8.10** Let $E$ be a normed space. Prove that the norm of $E$ is induced by an inner product if and only if the Parallelogram Law holds.

**Exercise 5.8.11** Prove, without using reflexivity, that the norms of $\ell_1$ and $C[a, b]$ are not induced by inner products.

## 5.8 Exercises

**Exercise 5.8.12** Let $(x_n)_{n=1}^\infty$ and $(y_n)_{n=1}^\infty$ be sequences in the closed unit ball of a Hilbert space. Prove that if $\langle x_n, y_n \rangle \longrightarrow 1$ then $\|x_n - y_n\| \longrightarrow 0$.

**Exercise 5.8.13** Prove that in $(\mathbb{R}^2, \|\cdot\|_\infty)$ there exists a closed subspace $M$ and $x \notin M$ such that the vector $p \in M$ with

$$\|x - p\| = \text{dist}(x, M).$$

is not unique.

**Exercise 5.8.14**

(a) Let $E$ be an inner product space and let $A$ be a non-empty, complete and convex subset of $E$. Prove that for each $x \in E$ there exists a unique $p \in A$ such that

$$\|x - p\| = \text{dist}(x, A).$$

(b) Show that in every non-empty, closed and convex subset of a Hilbert space there exists a unique vector of smallest norm.

**Exercise 5.8.15 (Hellinger–Toeplitz Theorem)** Let $H$ be a Hilbert space and let $T: H \longrightarrow H$ be a linear operator such that $\langle T(x), y \rangle = \langle x, T(y) \rangle$ for all $x, y \in H$. Prove that $T$ is continuous.

**Exercise 5.8.16** Show that an inner product can be defined on every vector space $V$. Moreover, for each Hamel basis $\mathcal{B}$ of $V$ an inner product can be defined on $V$ such that the vectors of $\mathcal{B}$ are orthogonal.

**Exercise 5.8.17** Prove, without using reflexivity, that if $M$ is a closed subspace of a Hilbert space, then $M = (M^\perp)^\perp$.

**Exercise 5.8.18** Consider the vector subspace $M = \{p : p \text{ is a polynomial with real coefficients and degree less than or equal to } 1\}$ of $L_2[-1, 1]$. Determine the best approximation of $f(x) = e^x$ in $M$.

**Exercise 5.8.19\*** If $(y_j)_{j=1}^\infty$ and $(z_j)_{j=1}^\infty$ are enumerations of $\{x_i : i \in I_x\}$ in Lemma 5.3.9, show directly (without using Proposition 5.3.8) that

$$\sum_{i=1}^\infty \langle x, z_i \rangle z_i = \sum_{i=1}^\infty \langle x, y_i \rangle y_i.$$

**Exercise 5.8.20** Let $E$ be an inner product space. Let $S_1 = \{x_n : n \in \mathbb{N}\}$ and $S_2 = \{y_n : n \in \mathbb{N}\}$ be orthonormal sets in $E$ such that $[x_1, \ldots, x_n] = [y_1, \ldots, y_n]$ for each natural number $n$. Show that there exists a sequence $(a_n)_{n=1}^\infty$ of scalars with modulus 1 such that $y_n = a_n x_n$ for every $n$.

**Exercise 5.8.21** Let $E$ be an inner product space and let $S$ be an infinite orthonormal set in $E$. Prove that $S$ is not compact, but is closed and bounded.

**Exercise 5.8.22** Let $S = \{x_i : i \in I\}$ be an orthonormal set in the Hilbert space $H$. Prove that the following statements are equivalent:

(a) $S$ is a complete orthonormal system.
(b) If $\varphi \in H'$ is a functional on which every $x_i$, $i \in I$, vanishes, then $\varphi = 0$.
(c) $S$ is maximal in the sense that no other orthonormal set properly contains it.
(d) If $x \in H$ is orthogonal to each $x_i$, $i \in I$, then $x = 0$.

**Exercise 5.8.23** Let $(x_n)_{n=1}^\infty$ be a complete orthonormal system in the Hilbert space $H$. Prove that for every sequence $(a_n)_{n=1}^\infty \in \ell_2$ there exists $x \in H$ such that

$$a_n = \langle x, x_n \rangle \text{ for every } n \text{ and } \|x\| = \|(a_n)_{n=1}^\infty\|_2.$$

**Exercise 5.8.24** State and prove a result analogous to Exercise 5.8.23 for finite-dimensional Hilbert spaces.

**Exercise 5.8.25** Prove that every Hilbert space of dimension $n$ is isometrically isomorphic to $(\mathbb{K}^n, \|\cdot\|_2)$. Conclude that two finite-dimensional Hilbert spaces over the same field have the same dimension if and only if they are isometrically isomorphic.

**Exercise 5.8.26** Prove that, in the separable case, the representation of Theorem 5.3.10 (a) is unique, that is, if $x = \sum_{n=1}^\infty b_n x_n$, then $b_n = \langle x, x_n \rangle$ for every $n \in \mathbb{N}$.

**Exercise 5.8.27 (Generalized Parallelogram Law)** Let $H$ be a Hilbert space. Prove that for every $n \in \mathbb{N}$ and all $x_1, \ldots, x_n \in H$,

$$\sum_{\varepsilon_j = \pm 1} \left\| \sum_{j=1}^n \varepsilon_j x_j \right\|^2 = 2^n \sum_{j=1}^n \|x_j\|^2.$$

**Exercise 5.8.28** Consider the case $\varphi \neq 0$ in the proof of the Riesz–Fréchet Theorem. The uniqueness of the vector $y_0$ may seem contradictory with the freedom of choice of the vector $x_0$. What is behind this is that $M^\perp$ has dimension 1. Prove this.

**Exercise 5.8.29** Let $H$ be a Hilbert space, let $F$ be a Banach space, let $G$ be a subspace of $H$, and let $\varphi: G \longrightarrow F$ be a bounded linear operator. Show that there exists a linear and continuous operator $\widetilde{\varphi}: H \longrightarrow F$ that extends $\varphi$ and preserves the norm.

**Exercise 5.8.30** Show that the extension of the previous exercise, in general, is not unique.

**Exercise 5.8.31 (The Hahn–Banach Theorem for Hilbert Spaces)** Let $H$ be a Hilbert space, let $F$ be a subspace of $H$ and let $\varphi \in F'$ be given. Prove, without

## 5.8 Exercises

using the Hahn–Banach Theorem, that there exists a *unique* functional $\widetilde{\varphi} \in H'$ that extends $\varphi$ and preserves the norm.

**Exercise 5.8.32**

(a) Suppose $F = \mathbb{K}$ and $G$ is closed in the context of Exercise 5.8.29. Defining

$$\mathcal{F} = \{\widetilde{\varphi} \in H' : \widetilde{\varphi} \text{ is a (continuous) extension of } \varphi\},$$

find an expression (if necessary in terms of the cardinality of $G$) for the cardinality of $\mathcal{F}$.

(b) If $F$ is a Banach space, defining

$$\mathcal{G} = \{\widetilde{\varphi}\colon H \longrightarrow F : \widetilde{\varphi} \text{ is an extension (linear) of } \varphi\} \text{ and}$$

$$\mathcal{H} = \{\widetilde{\varphi}\colon H \longrightarrow F : \widetilde{\varphi} \text{ is an extension (not necessarily linear) of } \varphi\},$$

find an expression (if necessary in terms of the cardinality of $F$ and/or $G$ or related sets) for the cardinalities of $\mathcal{G}$ and $\mathcal{H}$.

**Exercise 5.8.33\*** Let $I$ be any set and let $1 \leq p \leq \infty$. Call $\mathcal{F}$ the collection of all finite subsets of $I$. Given a collection of scalars $(a_j)_{j \in I}$ indexed by $I$, we write

$$\|(a_j)_{j \in I}\|_p = \sup_{A \in \mathcal{F}} \left(\sum_{j \in A} |a_j|^p\right)^{1/p} \text{ if } p < \infty \text{ and } \|(a_j)_{j \in I}\|_\infty = \sup_{i \in I} |a_i|.$$

(a) Prove that

$$\ell_p(I) := \left\{(a_j)_{j \in I} : a_i \in \mathbb{K} \text{ for each } i \in I \text{ and } \|(a_j)_{j \in I}\|_p < \infty\right\}$$

is a vector space in which $\|\cdot\|_p$ is a norm.

(b) Prove that $(\ell_p(I), \|\cdot\|_p)$ is a Banach space.

(c) Prove that, for $1 \leq p < \infty$, $(\ell_p(I))'$ is isometrically isomorphic to $\ell_{p^*}(I)$, with $\frac{1}{p} + \frac{1}{p^*} = 1$ and $1^* = \infty$.

(d) Let $\{x_i : i \in I\}$ be a complete orthonormal system in the Hilbert space $H$. Prove that $H$ is isometrically isomorphic to $\ell_2(I)$.

(e) Prove that every Hilbert space is isometrically isomorphic to its dual.

**Exercise 5.8.34 (Rademacher Functions)** The sign function is defined by

$$\text{sgn}\colon \mathbb{R} \longrightarrow \mathbb{R}, \quad \text{sgn}(x) = \begin{cases} 1, & \text{if } x > 0, \\ 0, & \text{if } x = 0, \\ -1, & \text{if } x < 0. \end{cases}$$

For each $n \in \mathbb{N}$ the $n$-th Rademacher function is the function $r_n(t) = \mathrm{sgn}(\sin(2^n \pi t))$ for every $t \in [0, 1]$. Prove that:

(a) The set $(r_n)_{n=1}^{\infty}$ is orthonormal in $L_2[0, 1]$.
(b) $(r_n)_{n=1}^{\infty}$ is not a complete orthonormal system in $L_2[0, 1]$.

**Exercise 5.8.35** * Let $(r_n)_{n=1}^{\infty}$ be the sequence of Rademacher functions. Prove that for all positive integers $n_1 < n_2 < \cdots < n_k$ and $p_1, \ldots, p_k$, it holds that

$$\int_0^1 r_{n_1}^{p_1}(t) \cdots r_{n_k}^{p_k}(t)\, dt = \begin{cases} 1, & \text{if each } p_j \text{ is even,} \\ 0, & \text{otherwise.} \end{cases}$$

**Exercise 5.8.36** * Let $(X_1, \Sigma_1, \mu_1)$ and $(X_2, \Sigma_2, \mu_2)$ be $\sigma$-finite measure spaces such that $L_2(\mu_1)$ and $L_2(\mu_2)$ are separable. Let $(f_n)_{n=1}^{\infty}$ and $(g_n)_{n=1}^{\infty}$ be complete orthonormal systems in $L_2(\mu_1)$ and $L_2(\mu_2)$, respectively. Prove that by defining $h_{mn}(x, y) = f_m(x) \cdot g_n(y)$, it follows that $(h_{mn})_{m,n=1}^{\infty}$ is a complete orthonormal system in $L_2(\mu_1 \otimes \mu_2)$, where $\mu_1 \otimes \mu_2$ is the product measure.

**Exercise 5.8.37** Prove the uniqueness of the vector $x_0$ in the Lax–Milgram Theorem (Theorem 5.6.4).

# Chapter 6
# Weak Topologies

In Theorem 1.5.4 we proved that the closed unit ball of an infinite-dimensional normed space is not compact. This strong contrast with finite dimension indicates that the existence of compact sets that are not contained in finite-dimensional subspaces is a delicate issue. Examples exist: we saw in Exercise 1.8.35 that the Hilbert cube is compact in $\ell_2$. The issue is not just the existence or non-existence of many compact sets; the central point is that those sets that we would like to be compact, namely the closed balls, are not.

To circumvent this situation, we resort to a classic topological expedient, called *weakening the topology*, described in Sect. 6.1. Next, in Sect. 6.2, we will apply this expedient to the case of normed spaces in general and, in Sect. 6.3, to the case of dual spaces. Also in Sect. 6.3, and also in Sect. 6.4, we will prove that the gain of compact sets is real: in reflexive spaces the closed unit ball is compact in the weak topology and in any dual normed spaces the closed unit ball is compact in the weak-star topology.

In dealing with weak topologies we will use the topological tool known as *nets* (see Appendix B). This choice is justified by the fact that arguments using nets are very similar to arguments using sequences, with which the reader is certainly familiar. In addition, by learning to use nets the students will gain an additional tool for their later studies, since nets have many applications in several areas of Analysis (see, for example, [1, 10, 62]).

In Sect. 6.6 we will study uniformly convex spaces, a class of normed spaces widely used in applications. We will see that uniformly convex spaces form an intermediate class between Hilbert spaces and reflexive spaces, thus enjoying—partially, of course—the properties of Hilbert spaces and the generality of reflexive spaces.

## 6.1 The Topology Generated by a Family of Functions

As this is a typical and classic topic of General Topology, we will omit the proofs of this section. The reader not familiar with the concepts of General Topology should consult Appendix B. We say that one topology has fewer open sets than another if it is properly contained in the other. The expression *fewer open sets* will always be used in this context, without connection to the idea of cardinality.

The basic principle is that the fewer open sets a space has, the greater the chance of a certain set being compact, as there will be fewer of its open coverings. The initial idea then is to consider topologies with fewer open sets, and that is what the expression *weakening the topology* means. The care to be taken is that, considering topologies with fewer open sets in a certain space, while there is a chance of gaining compact sets, the chance of functions defined on it being continuous decreases. And paying the price of losing the continuity of certain functions may not compensate for the (possible) gain with compact sets. One should then choose the functions that should be (or remain) continuous and look for the smallest topology that guarantees the continuity of the previously chosen functions. Let us get to work:

Let $X$ be a set, let $(Y_i)_{i \in I}$ be a family of topological spaces, and let $(f_i)_{i \in I}$ be a family of functions $f_i \colon X \longrightarrow Y_i$ for each $i \in I$. We want to define on $X$ the smallest topology that makes all the functions $f_i$ continuous.

For each $i \in I$ and each open set $A_i$ in $Y_i$ consider the set

$$f_i^{-1}(A_i) = \{x \in X : f_i(x) \in A_i\}.$$

Call $\Phi$ the collection of subsets of $X$ that can be written as finite intersections of sets of the form $f_i^{-1}(A_i)$.

**Proposition 6.1.1** *There exists a topology $\tau$ on $X$ that has $\Phi$ as a basis, that is, the elements of $\tau$ are unions of elements of $\Phi$.*

**Definition 6.1.2** The topology $\tau$ of Proposition 6.1.1 is called the *topology generated by the family of functions* $(f_i)_{i \in I}$.

Despite all the proofs being elementary, some even bordering on tautology, it is convenient to list the properties of the topology we just defined:

**Proposition 6.1.3** *Let $\tau$ be the topology on $X$ generated by the family of functions $(f_i)_{i \in I}$. Then:*

(a) *For each $i \in I$ the function $f_i \colon X \longrightarrow Y_i$ is continuous.*
(b) *$\tau$ is the smallest topology on $X$ such that (a) holds.*
(c) *$\tau$ is the intersection of all topologies on $X$ with respect to which each $f_i$ is continuous.*
(d) *For each $x \in X$, the sets of the form $f_{i_1}^{-1}(A_1) \cap \cdots \cap f_{i_n}^{-1}(A_n)$ where $n \in \mathbb{N}, i_1, \ldots, i_n \in I$ and $A_j$ is a neighborhood of $f_{i_j}(x)$, $j = 1, \ldots, n$, constitute a neighborhood basis for $x$.*

(e) *Let $(x_\lambda)_\lambda$ be a net in X. Then $x_\lambda \longrightarrow x$ in $(X, \tau)$ if and only if $f_i(x_\lambda) \longrightarrow f_i(x)$ in $Y_i$ for every $i \in I$.*
(f) *Let Z be a topological space and let $f: Z \longrightarrow (X, \tau)$ be given. Then f is continuous if and only if $f_i \circ f: Z \longrightarrow Y_i$ is continuous for every $i \in I$.*
(g) *Suppose that each $Y_i$ is a Hausdorff space. Then the topology $\tau$ is Hausdorff if and only if the family $(f_i)_{i \in I}$ separates points of x, that is, for all $x, y \in X$, $x \neq y$, there exists $i \in I$ such that $f_i(x) \neq f_i(y)$.*

## 6.2 The Weak Topology on a Normed Space

Here we apply the construction of the previous section to a normed space $E$. The functions we want to keep continuous are the continuous linear functionals $\varphi \in E'$.

**Definition 6.2.1** The *weak topology* on the normed space $E$, denoted by $\sigma(E, E')$, is the topology generated by the continuous linear functionals $\varphi \in E'$. When a sequence $(x_n)_{n=1}^\infty$ in $E$ converges to $x \in E$ in the weak topology we write $x_n \xrightarrow{w} x$.

The first properties of the weak topology follow, basically, from the general construction we made in the previous section:

**Proposition 6.2.2** *Let E be a normed space. Then:*

(a) *Continuous linear functionals are weakly continuous, that is, for every $\varphi \in E'$, $\varphi: (E, \sigma(E, E')) \longrightarrow \mathbb{K}$ is continuous.*
(b) *For each $x_0 \in E$, the sets of the form*

$$V_{J,\varepsilon} = \{x \in E : |\varphi_i(x) - \varphi_i(x_0)| < \varepsilon \text{ for every } i \in J\},$$

*where J is a finite set, $\varphi_i \in E'$ for all $i \in J$ and $\varepsilon > 0$, form a basis of open neighborhoods of $x_0$ for the weak topology.*
(c) *Let $(x_n)_{n=1}^\infty$ be a sequence in E. Then $x_n \xrightarrow{w} x$ if and only if $\varphi(x_n) \longrightarrow \varphi(x)$ for every $\varphi \in E'$.*
(d) *The weak topology $\sigma(E, E')$ is Hausdorff.*
(e) *Let Z be a topological space and let $f: Z \longrightarrow (E, \sigma(E, E'))$ be a function. Then f is continuous if and only if $\varphi \circ f: Z \longrightarrow \mathbb{K}$ is continuous for every $\varphi \in E'$.*

*Proof* Items (a), (c) and (e) follow immediately from Proposition 6.1.3.
(b) Let $U$ be a neighborhood of $x_0$ in the weak topology. By Proposition 6.1.3(d) there exist a finite set $J$, functionals $\varphi_j \in E'$ and open sets $V_j$ in $\mathbb{K}$ containing $\varphi_j(x_0)$, $j \in J$, such that $\bigcap_{j \in J} \varphi_j^{-1}(V_j)$ is an open set of the weak topology containing $x_0$ and contained in $U$. As $J$ is finite and $\varphi_j(x_0) \in V_j$ for every $j \in J$, there exists $\varepsilon > 0$ such that $B(\varphi_j(x_0); \varepsilon) \subseteq V_j$ for every $j \in J$. It follows that

$$V_{J,\varepsilon} = \{x \in E : |\varphi_j(x) - \varphi_j(x_0)| < \varepsilon \text{ for every } j \in J\}$$
$$= \bigcap_{j \in J} \varphi_j^{-1}(B(\varphi_j(x_0); \varepsilon)) \subseteq \bigcap_{j \in J} \varphi_j^{-1}(V_j) \subseteq U,$$

therefore the sets of the form $V_{J,\varepsilon}$ form a basis of open neighborhoods of $x_0$ for the weak topology.

(d) The fact that $E'$ separates points of $E$ follows easily from the Hahn–Banach Theorem in the form of Corollary 3.1.4. The result then follows from Proposition 6.1.3(g). ∎

**Corollary 6.2.3** *In a normed space $E$, if $x_n \longrightarrow x$ then $x_n \xrightarrow{w} x$.*

**Proof** If $x_n \longrightarrow x$, from the continuity of the functionals of $E'$ it follows that $\varphi(x_n) \longrightarrow \varphi(x)$ for every $\varphi \in E'$. From Proposition 6.2.2(c) it follows that $x_n \xrightarrow{w} x$. ∎

A simple example shows that weak convergence does not imply norm convergence:

**Example 6.2.4** Consider the sequence $(e_n)_{n=1}^{\infty}$ of the canonical unit vectors of $c_0$. Given $\varphi \in (c_0)'$ we know that there exists $(a_j)_{j=1}^{\infty} \in \ell_1$ such that $\varphi\left((b_j)_{j=1}^{\infty}\right) = \sum_{j=1}^{\infty} a_j b_j$ for every sequence $(b_j)_{j=1}^{\infty} \in c_0$. In particular, $\varphi(e_n) = a_n \longrightarrow 0 = \varphi(0)$ because $a_n$ is the general term of a convergent series. From Proposition 6.2.2(c) it follows that $e_n \xrightarrow{w} 0$. As $\|e_{n+1} - e_n\| = 1 \not\longrightarrow 0$, the sequence $(e_n)_{n=1}^{\infty}$ is not Cauchy, therefore it does not converge in norm.

Weakly convergent sequences may not be convergent, but fortunately they are bounded:

**Proposition 6.2.5** *Let $E$ be a normed space.*

(a) *If $x_n \xrightarrow{w} x$ in $E$, then the sequence $(\|x_n\|)_{n=1}^{\infty}$ is bounded and $\|x\| \leq \liminf_n \|x_n\|$.*

(b) *If $x_n \xrightarrow{w} x$ in $E$ and $\varphi_n \longrightarrow \varphi$ in $E'$, then $\varphi_n(x_n) \longrightarrow \varphi(x)$ in $\mathbb{K}$.*

**Proof**

(a) For every $\varphi \in E'$, $\varphi(x_n) \longrightarrow \varphi(x)$ by Proposition 6.2.2(c). Therefore, the sequence $(\varphi(x_n))_{n=1}^{\infty}$ is bounded for every $\varphi \in E'$. From Exercise 4.5.12 it follows that the set $\{x_n : n \in \mathbb{N}\}$ is bounded in $E$, proving the first statement. Let $\varphi \in E'$ be given. From $\varphi(x_n) \longrightarrow \varphi(x)$ and $|\varphi(x_n)| \leq \|\varphi\| \cdot \|x_n\|$ for every $n$, it follows that

$$|\varphi(x)| = \lim_n |\varphi(x_n)| = \liminf_n |\varphi(x_n)| \leq \liminf_n \|\varphi\| \cdot \|x_n\| = \|\varphi\| \cdot \liminf_n \|x_n\|.$$

## 6.2 The Weak Topology on a Normed Space

From the Hahn-Banach Theorem in the form of Corollary 3.1.5 we conclude that

$$\|x\| = \sup_{\|\varphi\|\leq 1} |\varphi(x)| \leq \sup_{\|\varphi\|\leq 1} \left(\|\varphi\| \cdot \liminf_n \|x_n\|\right) = \liminf_n \|x_n\|.$$

(b) Given $\varepsilon > 0$, from the convergences $\varphi_n \longrightarrow \varphi$ and $\varphi(x_n) \longrightarrow \varphi(x)$ there is a natural number $n_0$ such that

$$\|\varphi_n - \varphi\| < \varepsilon \text{ and } |\varphi(x_n) - \varphi(x)| < \varepsilon \text{ for every } n \geq n_0.$$

By item (a) there exists $C > 0$ such that $\|x_n\| \leq C$ for every $n$. Therefore,

$$|\varphi_n(x_n) - \varphi(x)| = |(\varphi_n - \varphi)(x_n) + \varphi(x_n - x)|$$
$$\leq \|\varphi_n - \varphi\| \|x_n\| + |\varphi(x_n) - \varphi(x)|$$
$$\leq C\varepsilon + \varepsilon$$

for every $n \geq n_0$. This proves that $\varphi_n(x_n) \longrightarrow \varphi(x)$. ∎

Example 6.2.4 shows, in particular, that in $c_0$ the norm topology does not coincide with the weak topology. Next we see that this actually characterizes infinite-dimensional spaces:

**Proposition 6.2.6** *Let $E$ be a normed space.*

(a) *The weak topology is contained in the norm topology.*
(b) *The norm and weak topologies coincide if and only if $E$ is finite-dimensional.*

**Proof**

(a) It is enough to observe that the functionals $\varphi \in E'$ are continuous in the norm topology and that the weak topology is the smallest topology on $E$ that keeps these functionals continuous.
(b) Suppose that $E$ is finite-dimensional. From item (a) we already know that every open set in the weak topology is also open in the norm topology. It is enough then to prove that open sets in the norm topology are open in the weak topology. Let $U$ be a non-empty open set in the norm topology and $x_0 \in U$. To prove that $U$ is open in the weak topology it is enough to show that $x_0$ is an interior point of $U$ for the weak topology. In other words, we must find an open set $V$ in the weak topology containing $x_0$ and contained in $U$. To do so, choose $r$ small enough so that $B(x_0; r) \subseteq U$ and fix a basis $\{e_1, \ldots, e_n\}$ of $E$ with $\|e_j\| = 1$ for each $j = 1, \ldots, n$. For each $j = 1, \ldots, n$, consider the functional

$$\varphi_j \colon E \longrightarrow \mathbb{K}, \ x = \sum_{i=1}^n x_i e_i \mapsto \varphi_j(x) = x_j.$$

Note that each $\varphi_j$ is linear and continuous because $E$ is finite-dimensional. By Proposition 6.2.2(b) we know that the set

$$V = \left\{ x \in E : |\varphi_j(x) - \varphi_j(x_0)| < \frac{r}{2n}, \ j = 1, \ldots, n \right\}$$

is an open set in the weak topology containing $x_0$. Moreover, given $x = \sum_{j=1}^{n} x_j e_j \in V$, writing $x_0 = \sum_{j=1}^{n} x_j^{(0)} e_j$,

$$\|x - x_0\| = \left\| \sum_{j=1}^{n} (x_j - x_j^{(0)}) e_j \right\| \leq \sum_{j=1}^{n} \left| x_j - x_j^{(0)} \right|$$

$$= \sum_{j=1}^{n} |\varphi_j(x - x_0)| \leq \sum_{j=1}^{n} \frac{r}{2n} = \frac{r}{2} < r.$$

Therefore $V \subseteq B(x_0, r) \subseteq U$, as we wanted.

Suppose now that $E$ is infinite-dimensional. Let $S = \{x \in E : \|x\| = 1\}$ be the unit sphere of $E$. It is clear that $S$ is closed in the norm topology. We just need to show that $S$ is not closed in the weak topology. To do this we will show that the set $\{x \in E : \|x\| < 1\}$ is contained in the closure of $S$ in the weak topology. Let $x_0 \in E$ such that $\|x_0\| < 1$ and let $V$ be an open set in the weak topology containing $x_0$. To show that $x_0$ belongs to the closure of $S$ in the weak topology, we just need to show that $S \cap V \neq \emptyset$. It is clear that we only need to consider $V$ as an open set from the neighborhood basis of $x_0$. In this case there exist $\varepsilon > 0$ and functionals $\varphi_1, \ldots, \varphi_n$ in $E'$ such that

$$V = \{x \in E : |\varphi_i(x) - \varphi_i(x_0)| < \varepsilon, \ i = 1, \ldots, n\}.$$

Now we show that there exists $y_0 \in E - \{0\}$ such that $\varphi_i(y_0) = 0$ for each $i = 1, \ldots, n$. In fact, otherwise the linear operator

$$T : E \longrightarrow \mathbb{K}^n, \ T(z) = (\varphi_i(z))_{i=1}^{n},$$

would be injective, but this does not occur since $E$ has infinite dimension. We can then consider the function $r(t) = \|x_0 + t y_0\|$ for every $t \in \mathbb{R}$. As $r$ is continuous, $r(0) = \|x_0\| < 1$ and $\lim_{t \to \infty} r(t) = \infty$, the Intermediate Value Theorem guarantees the existence of a real number $t_0$ such that $\|x_0 + t_0 y_0\| = r(t_0) = 1$. Therefore $x_0 + t_0 y_0 \in S \cap V$. ∎

When moving from the norm topology to the weak topology, we do not lose continuous linear functionals by Proposition 6.2.2(a) and we do not gain them by Proposition 6.2.6(a). It is better to record this. By $(E, \sigma(E, E'))'$ we denote the set

## 6.2 The Weak Topology on a Normed Space

of the linear functionals on $E$ that are continuous when $E$ is endowed with the weak topology.

**Corollary 6.2.7** $E' = (E, \sigma(E, E'))'$ *for every normed space* $E$.

With the case of linear functionals settled, we now turn our attention to linear operators. As the weak topology is contained in the norm topology, it is clear that if the linear operator $T: E \longrightarrow F$ is continuous, then $T: E \longrightarrow (F, \sigma(F, F'))$ is also continuous. On the other hand, the identity operator $\mathrm{id}_{c_0}: c_0 \longrightarrow c_0$ is continuous but from Example 6.2.4 we know that $\mathrm{id}_{c_0}: (c_0, \sigma(c_0, \ell_1)) \longrightarrow c_0$ is not continuous. In fact, by Proposition 6.2.6(b) this occurs in any infinite-dimensional space. The remaining case is where we change the topology in both the domain space and the target space. To deal with this case we need to prove Theorem 2.6.1, which we restate as the following lemma:

**Lemma 6.2.8** *Let $X$ and $Y$ be topological spaces with $Y$ Hausdorff. If the function $f: X \longrightarrow Y$ is continuous, then the graph of $f$ is closed in $X \times Y$ with the product topology.*

*Proof* Let $(x_\lambda, f(x_\lambda))_\lambda$ be a net in the graph of $f$ converging to $(x, y) \in X \times Y$. As the projections

$$(t, w) \in X \times Y \mapsto t \in X \text{ and } (t, w) \in X \times Y \mapsto w \in Y$$

are continuous, we have $x_\lambda \longrightarrow x$ in $X$ and $f(x_\lambda) \longrightarrow y$ in $Y$. From the continuity of $f$ it follows that $f(x_\lambda) \longrightarrow f(x)$ in $Y$, and as $Y$ is Hausdorff we have $f(x) = y$, that is, $(x, y)$ belongs to the graph of $f$. ∎

**Proposition 6.2.9** *Let $E$ and $F$ be Banach spaces. A linear operator $T: E \longrightarrow F$ is continuous if and only if $T: (E, \sigma(E, E')) \longrightarrow (F, \sigma(F, F'))$ is continuous.*

*Proof* It is convenient to denote $T: (E, \sigma(E, E')) \longrightarrow (F, \sigma(F, F'))$ by $T_\sigma$ to make the notation more precise. Suppose $T: E \longrightarrow F$ is continuous. For each $\varphi \in F'$, $\varphi \circ T_\sigma \in (E, \sigma(E, E'))'$ by Proposition 6.2.2(a). As this holds for every $\varphi \in F'$, the result follows from Proposition 6.2.2(e).

Conversely, suppose $T_\sigma: (E, \sigma(E, E')) \longrightarrow (F, \sigma(F, F'))$ is continuous. In this case Lemma 6.2.8 reveals that the graph of $T_\sigma$ is closed in $E \times F$ in the product topology of the weak topologies $(E, \sigma(E, E'))$ and $(F, \sigma(F, F'))$. But these weak topologies are contained in the respective norm topologies, and consequently the graph of $T$ is closed in $E \times F$ with the topology induced by the norms. The continuity of $T: E \longrightarrow F$ follows from the Closed Graph Theorem. ∎

**Remark 6.2.10** Proposition 6.2.9 also holds for operators between normed spaces. See [44, Theorem 2.5.11].

From Proposition 6.2.6(b) we know that, in infinite dimension, there are closed sets that are not closed in the weak topology. By imposing an algebraic condition—convexity—this possibility disappears:

**Theorem 6.2.11 (Mazur's Theorem)** *Let E be a normed space and let K be a convex subset of E. Then the closure of K in the norm topology coincides with the closure of K in the weak topology. In particular, a convex set is closed in the weak topology if and only if it is closed in the norm topology.*

**Proof** We consider the real case first. Since the norm topology is described by sequences, from Corollary 6.2.3 it follows that $\overline{K}^{\|\cdot\|} \subseteq \overline{K}^{\sigma(E,E')}$. Suppose there exists $x_0 \in \overline{K}^{\sigma(E,E')} \setminus \overline{K}^{\|\cdot\|}$. In this case $\{x_0\}$ is convex and compact, $\overline{K}^{\|\cdot\|}$ is convex (Exercise 1.8.19) and closed and $\{x_0\} \cap \overline{K}^{\|\cdot\|} = \emptyset$; then, by Theorem 3.4.9, there exist a functional $\varphi \in E'$ and $a \in \mathbb{R}$ such that

$$\varphi(x_0) > a > \varphi(x) \text{ for every } x \in \overline{K}^{\|\cdot\|}. \tag{6.1}$$

However, since $x_0 \in \overline{K}^{\sigma(E,E')}$ there exists a net $(x_\lambda)_\lambda$ in $K$ converging to $x_0$ in the weak topology. From the continuity of the linear functional $\varphi$ in the weak topology it follows that $(\varphi(x_\lambda))_\lambda$ converges to $\varphi(x_0)$. Since each $x_\lambda \in K$, this contradicts (6.1) and completes the proof. The complex case is analogous using Exercise 3.6.26. ∎

We saw in Corollary 6.2.3(c) that sequences converging in norm are weakly convergent, and in Proposition 6.2.6(b) that in infinite dimension the weak and norm topologies never coincide. Since not every topology is described by the convergence of sequences, these two facts do not prevent that, in some infinite-dimensional space, weakly convergent sequences are convergent in norm. In fact, this is exactly what happens in $\ell_1$:

**Theorem 6.2.12 (Schur's Theorem)** *In $\ell_1$ a sequence converges weakly if and only if it converges in the norm topology.*

**Proof** From Corollary 6.2.3 it is enough to prove that every weakly convergent sequence is also convergent in norm. It is clear that it suffices to prove that every sequence weakly converging to zero converges in norm to zero. Let then $\left(z^{(n)}\right)_{n=1}^\infty$ be a sequence in $\ell_1$ that converges to zero in the weak topology. Suppose that $\left(z^{(n)}\right)_{n=1}^\infty$ does not converge to zero in norm. In this case there exist $\varepsilon > 0$ and a subsequence $\left(x^{(n)}\right)_{n=1}^\infty$ of $\left(z^{(n)}\right)_{n=1}^\infty$ such that $\|x^{(n)}\|_1 \geq 5\varepsilon$ for every $n \in \mathbb{N}$. We write $x^{(n)} = \left(x_j^{(n)}\right)_{j=1}^\infty$ for each $n \in \mathbb{N}$. As this subsequence weakly converges to zero, from the duality $(\ell_1)' = \ell_\infty$ we have

$$\lim_{n \to \infty} \sum_{j=1}^\infty b_j x_j^{(n)} = 0 \text{ for every sequence } b = (b_j)_{j=1}^\infty \in \ell_\infty. \tag{6.2}$$

As the canonical unit vectors $e_j$ belong to $\ell_\infty$, it follows that $\lim_{n \to \infty} x_j^{(n)} = 0$ for every $j \in \mathbb{N}$. Define inductively two strictly increasing sequences $(m_k)_{k=1}^\infty$ and $(n_k)_{k=1}^\infty$ of natural numbers in the following way: $m_0 = n_0 = 1$ and, for $k \geq 1$, $n_k$ is the smallest integer greater than $n_{k-1}$ such that

## 6.3 The Weak-Star Topology

$$\sum_{j=1}^{m_{k-1}} \left| x_j^{(n_k)} \right| < \varepsilon;$$

and $m_k$ is the smallest integer greater than $m_{k-1}$ satisfying

$$\sum_{j=m_k}^{\infty} \left| x_j^{(n_k)} \right| < \varepsilon.$$

We define a sequence $(b_j)_{j=1}^{\infty} \in \ell_\infty$ in the following way: for $m_{k-1} < j \leq m_k$, let

$$b_j = \begin{cases} 0, & \text{if } x_j^{(n_k)} = 0 \\ \dfrac{x_j^{(n_k)}}{\left| x_j^{(n_k)} \right|}, & \text{if } x_j^{(n_k)} \neq 0. \end{cases}$$

Note that $|b_j| \leq 1$ for every $j$ and $\sum_{j=m_{k-1}+1}^{m_k} \left( \left| x_j^{(n_k)} \right| - b_j x_j^{(n_k)} \right) = 0$ for every $k$. Hence,

$$5\varepsilon - \left| \sum_{j=1}^{\infty} b_j x_j^{(n_k)} \right| \leq \left| \sum_{j=1}^{\infty} \left| x_j^{(n_k)} \right| - \sum_{j=1}^{\infty} b_j x_j^{(n_k)} \right| \leq \left| \sum_{j=1}^{\infty} \left( \left| x_j^{(n_k)} \right| - b_j x_j^{(n_k)} \right) \right|$$

$$= \left| \sum_{j=1}^{m_{k-1}} \left( \left| x_j^{(n_k)} \right| - b_j x_j^{(n_k)} \right) + \sum_{j=m_k+1}^{\infty} \left( \left| x_j^{(n_k)} \right| - b_j x_j^{(n_k)} \right) \right|$$

$$\leq \sum_{j=1}^{m_{k-1}} \left( \left| x_j^{(n_k)} \right| + \left| b_j x_j^{(n_k)} \right| \right) + \sum_{j=m_k+1}^{\infty} \left( \left| x_j^{(n_k)} \right| + \left| b_j x_j^{(n_k)} \right| \right) \leq 2\varepsilon + 2\varepsilon = 4\varepsilon,$$

for every $k$. Letting $k \longrightarrow \infty$, the inequality above contradicts (6.2) and completes the proof. ∎

## 6.3 The Weak-Star Topology

In the dual $E'$ of a normed space $E$, so far we can consider the topology of the norm and the weak topology $\sigma(E', E'')$, which is the topology generated by the elements of $E''$. It happens that, considering the canonical embedding $J_E \colon E \longrightarrow E''$, the set $J_E(E)$ is also a distinguished set of functions defined on $E'$. We will see in the course of this chapter that it is very fruitful to consider in $E'$ the topology generated by the elements of $J_E(E)$.

**Definition 6.3.1** The *weak-star topology* on the dual $E'$ of the normed space $E$, denoted by $\sigma(E', E)$, is the topology on $E'$ generated by the functions belonging to the set $J_E(E) = \{J_E(x) : x \in E\}$, that is, by the functions $\varphi \in E' \mapsto J_E(x)(\varphi) = \varphi(x) \in \mathbb{K}$, where $x \in E$.

When a sequence $(\varphi_n)_{n=1}^\infty$ in $E'$ converges to $\varphi \in E'$ in the weak-star topology we write $\varphi_n \xrightarrow{w^*} \varphi$.

As in the case of the weak topology, the first properties of the weak-star topology follow from Proposition 6.1.3:

**Proposition 6.3.2** *Let $E$ be a normed space. Then:*

(a) *For every $x \in E$, $J_E(x) \colon (E', \sigma(E', E)) \longrightarrow \mathbb{K}$ is continuous.*
(b) *For each $\varphi_0 \in E'$, the sets of the form*

$$W_{J,\varepsilon} = \{\varphi \in E' : |\varphi(x_i) - \varphi_0(x_i)| < \varepsilon \text{ for every } i \in J\},$$

*where $J$ is a finite set, $x_i \in E$ for all $i \in J$ and $\varepsilon > 0$, form a basis of open neighborhoods of $\varphi_0$ for the weak-star topology.*

(c) *Let $(\varphi_n)_{n=1}^\infty$ be a sequence in $E'$. Then $\varphi_n \xrightarrow{w^*} \varphi$ if and only if $\varphi_n(x) \longrightarrow \varphi(x)$ for every $x \in E$.*
(d) *The weak-star topology $\sigma(E', E)$ is Hausdorff.*
(e) *Let $Z$ be a topological space and let $f \colon Z \longrightarrow (E', \sigma(E', E))$ be given. Then $f$ is continuous if and only if $J_E(x) \circ f \colon Z \longrightarrow \mathbb{K}$ is continuous for every $x \in E$.*

*Proof* Items (a), (c) and (e) follow immediately from Proposition 6.1.3. For item (d) it is enough to combine Exercise 4.5.11 with Proposition 6.1.3(g). Item (b) is obtained from an adaptation of the proof of Proposition 6.2.2(b). ∎

Regarding the convergence of sequences we have:

**Proposition 6.3.3** *Let $E$ be a normed space.*

(a) *If $\varphi_n \xrightarrow{w} \varphi$ in $E'$ then $\varphi_n \xrightarrow{w^*} \varphi$.*
(b) *If $E$ is Banach and $\varphi_n \xrightarrow{w^*} \varphi$ in $E'$ then the sequence $(\|\varphi_n\|)_{n=1}^\infty$ is bounded and $\|\varphi\| \leq \liminf_n \|\varphi_n\|$.*
(c) *If $E$ is Banach, $\varphi_n \xrightarrow{w^*} \varphi$ in $E'$ and $x_n \longrightarrow x$ in $E$, then $\varphi_n(x_n) \longrightarrow \varphi(x)$ in $\mathbb{K}$.*

*Proof*

(a) Follows from Propositions 6.2.2(c) and 6.3.2(c).
(b) Follows from an adaptation of the proof of Proposition 6.2.5(a) using the Banach–Steinhaus Theorem.
(c) Is a consequence of an adaptation of the proof of Proposition 6.2.5(b).

∎

## 6.3 The Weak-Star Topology

Again a simple example shows that weak-star convergence does not imply weak convergence:

**Example 6.3.4** Consider the sequence $(e_n)_{n=1}^\infty$ of the canonical unit vectors of $\ell_1 = (c_0)'$. Given $x = (x_n)_{n=1}^\infty \in c_0$, $e_n(x) = x_n \longrightarrow 0$, which by Proposition 6.3.2(c) guarantees that $e_n \xrightarrow{w^*} 0$ in $\ell_1$. On the other hand, taking $f = (-1, 1, -1, 1, \ldots) \in \ell_\infty = (\ell_1)'$, $f(e_n) = (-1)^n$ for every $n$, so the sequence $(f(e_n))_{n=1}^\infty$ does not converge, therefore by Proposition 6.2.2(c) the sequence $(e_n)_{n=1}^\infty$ does not converge weakly in $\ell_1$.

We denote by $(E', \sigma(E', E))'$ the space formed by the linear functionals from $E'$ to $\mathbb{K}$ that are continuous in the weak-star topology $\sigma(E', E)$. By Proposition 6.3.2(a) we know that $J_E(E) \subseteq (E', \sigma(E', E))'$. To prove the inverse inclusion we need an auxiliary result from Linear Algebra:

**Lemma 6.3.5** *Let $V$ be a vector space and let $\varphi, \varphi_1, \ldots, \varphi_n$ be linear functionals on $V$ such that $\bigcap_{i=1}^{n} \ker(\varphi_i) \subseteq \ker(\varphi)$. Then there exist scalars $a_1, \ldots, a_n$ such that*
$$\varphi = \sum_{i=1}^{n} a_i \varphi_i.$$

***Proof*** Consider the linear transformations
$$T : V \longrightarrow \mathbb{K}^n, \ T(x) = (\varphi_1(x), \ldots, \varphi_n(x)), \text{ and}$$
$$U : T(V) \longrightarrow \mathbb{K}, \ U(\varphi_1(x), \ldots, \varphi_n(x)) = \varphi(x).$$

In order to see that $U$ is well defined, suppose that $(\varphi_1(x), \ldots, \varphi_n(x)) = (\varphi_1(y), \ldots, \varphi_n(y))$. Thus $(\varphi_1(x-y), \ldots, \varphi_n(x-y)) = (0, \ldots, 0)$, and in this case $(x - y) \in \bigcap_{i=1}^{n} \ker(\varphi_i)$. This implies that $(x - y) \in \ker(\varphi)$, therefore $\varphi(x) = \varphi(y)$. The linearity of $U$ is obvious. Let $\widetilde{U} : \mathbb{K}^n \longrightarrow \mathbb{K}$ be a linear extension of $U$. From the matrix representation of linear transformations between finite-dimensional spaces, there exist scalars $a_1, \ldots, a_n$ such that
$$\widetilde{U}(z_1, \ldots, z_n) = \sum_{j=1}^{n} a_j z_j \text{ for all } z_1, \ldots, z_n \in \mathbb{K}.$$

In particular,
$$\varphi(x) = U(\varphi_1(x), \ldots, \varphi_n(x)) = \widetilde{U}(\varphi_1(x), \ldots, \varphi_n(x)) = \sum_{j=1}^{n} a_j \varphi_j(x),$$

for every $x \in V$. ∎

**Proposition 6.3.6** *Let $E$ be a normed space and let $f \colon (E', \sigma(E', E)) \longrightarrow \mathbb{K}$ be a linear and continuous functional. Then there exists $x \in E$ such that $f = J_E(x)$. In other words, $(E', \sigma(E', E))' = J_E(E)$.*

*Proof* Since $f$ is continuous in the weak-star topology $\sigma(E', E)$ and $f(0) = 0$, there exists a neighborhood $V$ of the origin in $E'$ in the weak-star topology such that $|f(\varphi)| < 1$ for every functional $\varphi \in V$. From Proposition 6.3.2(b) we can choose $V$ of the form

$$V = \{\varphi \in E' : |\varphi(x_i)| < \varepsilon,\ i = 1, \ldots, n\},$$

where $\varepsilon > 0$ and $x_1, \ldots, x_n \in E$. If $J_E(x_i)(\varphi) = \varphi(x_i) = 0$ for each $i = 1, \ldots, n$, we clearly have $\varphi \in V$. In this case, for every natural number $k$ it is true that $(k\varphi)(x_i) = 0$ for each $i = 1, \ldots, n$, therefore $k\varphi \in V$. This implies that $|f(k\varphi)| < 1$, that is, $|f(\varphi)| < \frac{1}{k}$ for every natural number $k$, hence $f(\varphi) = 0$. We have just shown that $\bigcap_{i=1}^{n} \ker(J_E(x_i)) \subseteq \ker(f)$. By Lemma 6.3.5 there exist scalars $a_1, \ldots, a_n$ such that $f = \sum_{i=1}^{n} a_i J_E(x_i) = J_E\left(\sum_{i=1}^{n} a_i x_i\right)$. ∎

**Corollary 6.3.7** $(E'', \sigma(E'', E'))' = J_{E'}(E')$ *for every normed space $E$.*

We can now precisely establish the relationship between the weak and weak-star topologies:

**Proposition 6.3.8** *Let $E$ be a normed space.*

(a) *In $E'$, the weak-star topology $\sigma(E', E)$ is contained in the weak topology $\sigma(E', E'')$.*
(b) *The weak $\sigma(E', E'')$ and weak-star $\sigma(E', E)$ topologies coincide on $E'$ if and only if $E$ is reflexive.*

*Proof*

(a) The weak topology $\sigma(E', E'')$ keeps all functionals of $E''$ continuous, in particular it keeps the functionals of the form $J_E(x)$ with $x \in E$ continuous. Since the weak-star topology $\sigma(E', E)$ is the smallest topology that keeps these last functionals continuous, it follows that $\sigma(E', E) \subseteq \sigma(E', E'')$.
(b) If $E$ is reflexive, $\sigma(E', E)$ and $\sigma(E', E'')$ coincide with the smallest topology on $E'$ that keeps the functionals of the form $J_E(x)$ with $x \in E$ continuous, so they coincide. Conversely, suppose that $E$ is not reflexive. In this case there exists $f \in E''$, $f \notin J_E(E)$. Then $f$ is continuous in the norm of $E'$, therefore it is continuous in the weak topology $\sigma(E', E'')$ by Corollary 6.2.7. But by Proposition 6.3.6 we know that $f$ is not continuous in the weak-star topology $\sigma(E', E)$, which proves that the two topologies do not coincide. ∎

## 6.3 The Weak-Star Topology

It is time to show that with these weak topologies we gain the compactness of closed balls. We now reverse the path: we start with the weak-star topology and then arrive at the weak topology. Recall that, given a normed space $E$, $B_E$ represents the closed unit ball of $E$, that is,

$$B_E = \{x \in E : \|x\| \leq 1\}.$$

In stark contrast to the fact that in infinite dimension the ball $B_E$ is never compact in the topology of the norm, we have the

**Theorem 6.3.9 (Banach–Alaoglu–Bourbaki Theorem)** *For every normed space $E$, the ball $B_{E'}$ is compact in the weak-star topology $\sigma(E', E)$ of $E'$.*

**Proof** Consider the generalized Cartesian product $Y := \prod_{x \in E} \mathbb{K}$, that is, the elements of $Y$ are of the form $\omega = (\omega_x)_{x \in E}$ with $\omega_x \in \mathbb{K}$ for every $x \in E$. For each $x \in E$ consider the function

$$\pi_x : Y \longrightarrow \mathbb{K}, \quad \pi_x(\omega) = \omega_x.$$

In a simplified way, we can say that $\pi_x$ is the projection on the $x$-th coordinate. We equip $Y$ with the product topology, which is the smallest topology on $Y$ that makes the functions $(\pi_x)_{x \in E}$ continuous. Consider the function

$$\Psi : (E', \sigma(E', E)) \longrightarrow Y, \quad \Psi(\varphi) = (\varphi(x))_{x \in E}.$$

For each $x \in E$, $\pi_x \circ \Psi = J_E(x)$, which is continuous in the weak-star topology. Therefore, from Proposition 6.1.3(f), we conclude that $\Psi$ is continuous. It is evident that $\Psi$ is injective. To prove that $\Psi$ is a homeomorphism onto its range, it remains to show that $\Psi^{-1} : \Psi(E') \longrightarrow (E', \sigma(E', E))$ is continuous. To do so, let $x \in E$ be given. From

$$J_E(x) \circ \Psi^{-1}((\varphi(y))_{y \in E}) = J_E(x)\left(\Psi^{-1}((\varphi(y))_{y \in E})\right) = J_E(x)(\varphi) = \varphi(x),$$

we conclude that $J_E(x) \circ \Psi^{-1}$ is the restriction of $\pi_x$ to $\Psi(E')$, therefore it is continuous. Thus $J_E(x) \circ \Psi^{-1}$ is continuous for every $x \in E$, which ensures, by Proposition 6.3.2(e), that $\Psi^{-1}$ is continuous. We have thus proved that $\Psi : (E', \sigma(E', E)) \longrightarrow \Psi(E')$ is a homeomorphism. In particular, $(B_{E'}, \sigma(E', E))$ is homeomorphic to $\Psi(B_{E'})$ with the product topology. It is then enough to prove that $\Psi(B_{E'})$ is compact in $\Psi(E')$ in the product topology. It is clear that it is enough to show that $\Psi(B_{E'})$ is compact in $Y$. Since $|\varphi(x)| \leq \|x\|$ for all $x \in E$ and $\varphi \in B_{E'}$, we have $\Psi(B_{E'}) \subseteq \prod_{x \in E} B_{\mathbb{K}}[0, \|x\|]$. Each closed ball $B_{\mathbb{K}}[0, \|x\|]$ is compact in $\mathbb{K}$, so $\prod_{x \in E} B_{\mathbb{K}}[0, \|x\|]$ is compact in $Y$ by Tychonoff's Theorem. It remains to prove that $\Psi(B_{E'})$ is closed in $Y$. This is, however, an easy task. Consider a net $(\Psi(\varphi_\lambda))_\lambda$

in $\Psi(B_{E'})$, that is, $\varphi_\lambda \in B_{E'}$ for every $\lambda$, which converges in the product topology to $w = (\omega_x)_{x \in E} \in Y$. By Proposition 6.1.3(e) we know that

$$\varphi_\lambda(x) = \pi_x((\varphi_\lambda(y))_{y \in E}) = \pi_x(\Psi(\varphi_\lambda)) \xrightarrow{\lambda} \pi_x(w) = \omega_x,$$

for every $x \in E$. We define

$$\varphi \colon E \longrightarrow \mathbb{K}, \ \varphi(x) = \omega_x.$$

The linearity of $\varphi$ follows from the continuity of the vector space operations. For the continuity of $\varphi$, as $|\varphi_\lambda(x)| \leq \|x\|$ for every $\lambda$ and every $x \in E$, from the continuity of the modulus it follows that

$$|\varphi(x)| = |\omega_x| = \lim_\lambda |\varphi_\lambda(x)| \leq \|x\|,$$

for every $x \in E$. This proves that $\varphi \in B_{E'}$. It is clear that $\omega = \Psi(\varphi) \in \Psi(B_{E'})$. ∎

## 6.4 Weak Compactness and Reflexivity

One of the main results involving compactness in Functional Analysis is the fact that the closed unit ball of a reflexive space is compact in the weak topology. This result is an immediate consequence of the Lemma below and the Banach–Alaoglu–Bourbaki Theorem.

**Lemma 6.4.1** *Let $E$ be a Banach space. The canonical embedding $J_E$ is a homeomorphism from $(E, \sigma(E, E'))$ onto its range $J_E(E)$ with the topology induced by the weak-star topology of $E''$. That is, the function*

$$J_E \colon (E, \sigma(E, E')) \longrightarrow J_E(E) \subseteq (E'', \sigma(E'', E'))$$

*is a homeomorphism.*

**Proof** It is clear that the function is bijective. The fact that it is continuous and has a continuous inverse follows from the fact that, for every net $(x_\lambda)_\lambda$ in $E$,

$$x_\lambda \xrightarrow{w} x \text{ in } E \iff \varphi(x_\lambda) \longrightarrow \varphi(x) \text{ for every functional } \varphi \in E'$$
$$\iff J_E(x_\lambda)(\varphi) \longrightarrow J_E(x)(\varphi) \text{ for every functional } \varphi \in E'$$
$$\iff J_E(x_\lambda) \xrightarrow{w^*} J_E(x) \text{ in } E''$$
$$\iff J_E(x_\lambda) \xrightarrow{w^*} J_E(x) \text{ in } J_E(E).$$

## 6.4 Weak Compactness and Reflexivity

The first equivalence follows from an application of Proposition 6.1.3(e) for the weak topology, the second is obvious, the third follows from one more application of Proposition 6.1.3(e), now for the weak-star topology, and the last from the fact that in $J_E(E)$ we are considering the topology induced by the weak-star topology of $E''$. ∎

**Theorem 6.4.2** *For every reflexive space $E$, the ball $B_E$ is compact in the weak topology $\sigma(E, E')$.*

**Proof** Since $E$ is reflexive, $J_E(E) = E''$, so $J_E \colon (E, \sigma(E, E')) \longrightarrow (E'', \sigma(E'', E'))$ is a homeomorphism by Lemma 6.4.1. The Banach–Alaoglu–Bourbaki Theorem guarantees that $B_{E''}$ is compact in the topology $\sigma(E'', E')$, therefore $B_E = J_E^{-1}(B_{E''})$ is compact in the topology $\sigma(E, E')$. ∎

We will now work towards proving that the converse of the result above holds true, that is, only in reflexive spaces is the unit ball compact in the weak topology. Some preparatory results are required.

**Lemma 6.4.3 (Helly's Lemma)** *Let $E$ be a real Banach space, let $\varphi_1, \ldots, \varphi_n \in E'$ be fixed functionals, and let $a_1, \ldots, a_n \in \mathbb{R}$ be fixed scalars. The following conditions are equivalent:*

(a) *For every $\varepsilon > 0$ there exists $x \in B_E$ such that $|\varphi_j(x) - a_j| < \varepsilon$ for every $j = 1, \ldots, n$.*

(b) *For all scalars $b_1, \ldots, b_n$, $\left|\sum_{j=1}^{n} b_j a_j\right| \leq \left\|\sum_{j=1}^{n} b_j \varphi_j\right\|.$*

**Proof** (a) $\Longrightarrow$ (b) Let $b_1, \ldots, b_n \in \mathbb{R}$ be fixed scalars. Given $\varepsilon > 0$, take $x \in B_E$ according to (a). Then

$$\left|\sum_{j=1}^{n} b_j a_j\right| \leq \left|\sum_{j=1}^{n} b_j a_j - \sum_{j=1}^{n} b_j \varphi_j(x)\right| + \left|\sum_{j=1}^{n} b_j \varphi_j(x)\right|$$

$$\leq \sum_{j=1}^{n} (|b_j| \cdot |\varphi_j(x) - a_j|) + \left|\left(\sum_{j=1}^{n} b_j \varphi_j\right)(x)\right|$$

$$\leq \varepsilon \left(\sum_{j=1}^{n} |b_j|\right) + \left\|\sum_{j=1}^{n} b_j \varphi_j\right\|,$$

because $\|x\| \leq 1$. As this is true for every $\varepsilon > 0$, it is enough to make $\varepsilon \longrightarrow 0$ to obtain (b).

(b) $\Longrightarrow$ (a) Consider the continuous linear operator

$$T \colon E \longrightarrow \mathbb{R}^n, \ T(x) = (\varphi_1(x), \ldots, \varphi_n(x)).$$

Note that (a) is equivalent to saying that $(a_1, \ldots, a_n) \in \overline{T(B_E)}$. Suppose that (a) is false, that is, $(a_1, \ldots, a_n) \notin \overline{T(B_E)}$. In this case, in $\mathbb{R}^n$ the convex and compact set $\{(a_1, \ldots, a_n)\}$ is disjoint from the convex and closed set $\overline{T(B_E)}$. By the geometric form of the Hahn–Banach Theorem (Theorem 3.4.9), there exist a functional $\varphi \in (\mathbb{R}^n)'$ and a real number $r$ such that

$$\varphi(T(x)) < r < \varphi((a_1, \ldots, a_n)) \text{ for every } x \in B_E.$$

Consider $b_1, \ldots, b_n \in \mathbb{R}$ such that

$$\varphi(x_1, \ldots, x_n) = \sum_{j=1}^{n} b_j x_j$$

for every $(x_1, \ldots, x_n) \in \mathbb{R}^n$. Hence,

$$\sum_{j=1}^{n} b_j \varphi_j(x) < r < \sum_{j=1}^{n} b_j a_j \text{ for every } x \in B_E.$$

For each $x \in B_E$, as $-x \in B_E$,

$$-\sum_{j=1}^{n} b_j \varphi_j(x) = \sum_{j=1}^{n} b_j \varphi_j(-x) < r.$$

Therefore

$$\left| \sum_{j=1}^{n} b_j \varphi_j(x) \right| < r < \sum_{j=1}^{n} b_j a_j \text{ for every } x \in B_E.$$

Taking the supremum over $x \in B_E$ we arrive at

$$\left\| \sum_{j=1}^{n} b_j \varphi_j \right\| \leq r < \sum_{j=1}^{n} b_j a_j \leq \left| \sum_{j=1}^{n} b_j a_j \right|.$$

The proof is complete as this contradicts (b). ∎

**Theorem 6.4.4 (Goldstine's Theorem)** *Let $E$ be a Banach space and let $J_E \colon E \longrightarrow E''$ be the canonical embedding. Then $J_E(B_E)$ is dense in $B_{E''}$ in the weak-star topology.*

**Proof** We have to show that $\overline{J_E(B_E)}^{\sigma(E'', E')} = B_{E''}$. As $J_E$ is an isometry, $J_E(B_E) \subseteq B_{E''}$. By the Banach–Alaoglu–Bourbaki Theorem we know that $B_{E''}$ is

## 6.4 Weak Compactness and Reflexivity

compact in the weak-star topology, therefore it is closed in the weak-star topology, as this topology is Hausdorff. Thus, $\overline{J_E(B_E)}^{\sigma(E'',E')} \subseteq \overline{B_{E''}}^{\sigma(E'',E')} = B_{E''}$.

We will prove the reverse inclusion only in the real case. For the complex case the reader will have to wait until Chap. 8, where we will give a proof that includes the complex case. Let $f \in B_{E''}$. We must prove that $J_E(B_E)$ intersects every neighborhood of $f$ in the weak-star topology. It is clear that it is enough to prove that this occurs for every basic neighborhood. Let $V$ be such a basic neighborhood of $f$ in the weak-star topology. By Proposition 6.3.2(b) there exist $\varepsilon > 0$ and functionals $\varphi_1, \ldots, \varphi_n \in E'$ such that

$$V = \{g \in E'' : |g(\varphi_i) - f(\varphi_i)| < \varepsilon, \ i = 1, \ldots, n\}.$$

For all $b_1, \ldots, b_n \in \mathbb{R}$,

$$\left|\sum_{i=1}^n b_i f(\varphi_i)\right| = \left|f\left(\sum_{i=1}^n b_i \varphi_i\right)\right| \leq \|f\| \cdot \left\|\sum_{i=1}^n b_i \varphi_i\right\| \leq \left\|\sum_{i=1}^n b_i \varphi_i\right\|.$$

By Lemma 6.4.3 there exists $x \in B_E$ such that

$$|J_E(x)(\varphi_i) - f(\varphi_i)| = |\varphi_i(x) - f(\varphi_i)| < \varepsilon$$

for each $i = 1, \ldots, n$. This proves that $J_E(x) \in J_E(B_E) \cap V$ and finally completes the proof. ∎

We are now in a position to characterize reflexive spaces as those in which the closed unit ball is weakly compact:

**Theorem 6.4.5 (Kakutani's Theorem)** *A Banach space $E$ is reflexive if and only if the closed unit ball $B_E$ is compact in the weak topology $\sigma(E, E')$.*

*Proof* One of the implications has already been proven in Theorem 6.4.2. For the reverse implication, suppose that $B_E$ is compact in the weak topology. As a continuous function takes compact sets to compact sets, from Lemma 6.4.1 it follows that $J_E(B_E)$ is compact in the weak-star topology. But the weak-star topology is Hausdorff (Proposition 6.3.2(d)), therefore $J_E(B_E)$ is closed in the weak-star topology since compact sets in Hausdorff spaces are closed. This means that $J_E(B_E) = \overline{J_E(B_E)}^{\sigma(E'',E')}$, and from Goldstine's Theorem it follows that $J_E(B_E) = B_{E''}$. From the linearity of $J_E$ we conclude that $J_E(E) = E''$. ∎

**Corollary 6.4.6** *If $E$ is reflexive, then every closed subspace of $E$ is reflexive.*

*Proof* Let $F$ be a closed subspace of $E$. It is not difficult to check (cf. Exercise 6.8.4) that the weak topology $\sigma(F, F')$ of $F$ is precisely the topology induced on $F$ by the weak topology $\sigma(E, E')$ of $E$. As $B_F$ is convex and closed in $F$, by Theorem 6.2.11 $B_F$ is closed in the weak topology $\sigma(F, F')$. Then $B_F$ is a subset of $B_E$ closed in the weak topology $\sigma(E, E')$. As $B_E$ is compact in this topology $\sigma(E, E')$ by Theorem 6.3.9 and as closed sets within compact sets are compact, it follows that $B_F$ is also compact in the weak topology $\sigma(E, E')$ and consequently in the topology $\sigma(F, F')$. Theorem 6.4.5 reveals that $F$ is reflexive. ∎

## 6.5 Metrizability and Separability

We have already seen that in infinite dimension the weak and weak-star topologies never coincide with the norm topology. The most we can hope for these weak topologies is that they are metrizable. The aim of this section is to show that this possibility is closely related to the separability of the space.

**Proposition 6.5.1** *If $E$ is separable then $(B_{E'}, \sigma(E', E))$ is metrizable.*

**Proof** Let $Y = \{y_n : n \in \mathbb{N}\}$ be an countable and dense subset in $E$. Define $X = \{x_n \in Y : x_n \in B_E\}$, note that $X$ is dense in $B_E$ and consider

$$d: B_{E'} \times B_{E'} \longrightarrow [0, \infty) \, , \, d(\varphi, \psi) = \sum_{n=1}^{\infty} \frac{1}{2^n} |\varphi(x_n) - \psi(x_n)| .$$

It is easy to see that $d$ is well defined and that it is a metric. To show that the topology on $B_{E'}$ induced by the metric $d$ coincides with the weak-star topology we must prove that for each $\varphi_0 \in B_{E'}$, every neighborhood of $\varphi_0$ in the weak-star topology is contained in an open ball centered at $\varphi_0$ according to the metric $d$ and vice versa. To do so, let $\varphi_0 \in B_{E'}$ be given. We start with a neighborhood $V$ of $\varphi_0$ in the weak-star topology. By Proposition 6.3.2(b) there exist $\varepsilon > 0$ and $z_1, \ldots, z_m \in E$ such that

$$V = \{\varphi \in B_{E'} : |\varphi(z_i) - \varphi_0(z_i)| < \varepsilon, i = 1, \ldots, m\} .$$

It is clear that we can assume that the vectors $z_i$ are non-zero.

Let $M = \max\{\|z_1\|, \ldots, \|z_m\|\}$, $\delta = \frac{\varepsilon}{M}$ and $y_i = \frac{z_i}{\|z_i\|}$ for $i = 1, \ldots, m$. Then

$$V' := \{\varphi \in B_{E'} : |\varphi(y_i) - \varphi_0(y_i)| < \delta, i = 1, \ldots, m\}$$

is also a neighborhood of $\varphi_0$ in the weak-star topology, $V' \subseteq V$ and $\|y_i\| = 1$ for $i = 1, \ldots, m$. For each $i = 1, \ldots, m$, from the denseness of $X$ in $B_E$ there exists $n_i \in \mathbb{N}$ such that $\|x_{n_i} - y_i\| < \frac{\delta}{4}$. Choose $r > 0$ such that $r < \min\left\{\frac{\delta}{2^{n_i+1}} : i = 1, \ldots, m\right\}$. In this way, $r \cdot 2^{n_i} < \frac{\delta}{2}$ for each $i = 1, \ldots, m$. If $\varphi \in B_{E'}$ and $d(\varphi, \varphi_0) < r$, then

$$\frac{1}{2^{n_i}} |\varphi(x_{n_i}) - \varphi_0(x_{n_i})| < r \text{ for all } 1 \leq i \leq m.$$

In this case,

## 6.5 Metrizability and Separability

$$|\varphi(y_i) - \varphi_0(y_i)| = |\varphi(y_i - x_{n_i}) - \varphi_0(y_i - x_{n_i}) + \varphi(x_{n_i}) - \varphi_0(x_{n_i})|$$

$$\leq \|\varphi - \varphi_0\| \cdot \|y_i - x_{n_i}\| + |\varphi(x_{n_i}) - \varphi_0(x_{n_i})|$$

$$< (\|\varphi\| + \|\varphi_0\|)\frac{\delta}{4} + r \cdot 2^{n_i} \leq 2 \cdot \frac{\delta}{4} + \frac{\delta}{2} = \delta$$

for each $i = 1, \ldots, m$, therefore $\varphi \in V'$. We have then proved that

$$B(\varphi_0, r) = \{\varphi \in B_{E'} : d(\varphi, \varphi_0) < r\} \subseteq V' \subseteq V.$$

Conversely, consider the open ball $B(\varphi_0, r_0) = \{\varphi \in B_{E'} : d(\varphi, \varphi_0) < r_0\}$, where $\varphi_0 \in B_{E'}$ and $r_0 > 0$. As the series $\sum_{i=1}^{\infty} \frac{1}{2^i}$ converges, we can choose $m \in \mathbb{N}$ large enough so that

$$2\left(\sum_{i=m+1}^{\infty} \frac{1}{2^i}\right) < \frac{r_0}{2}.$$

Now choose $0 < \varepsilon < \frac{r_0}{2}$. Then

$$W := \{\varphi \in B_{E'} : |\varphi(x_i) - \varphi_0(x_i)| < \varepsilon, i = 1, \ldots, m\}$$

is a neighborhood of $\varphi_0$ in the weak-star topology. It remains to prove that $W \subseteq B(\varphi_0, r_0)$ to complete the proof. But this is true because, given $\varphi \in W$,

$$d(\varphi, \varphi_0) = \sum_{i=1}^{m} \frac{1}{2^i}|\varphi(x_i) - \varphi_0(x_i)| + \sum_{i=m+1}^{\infty} \frac{1}{2^i}|\varphi(x_i) - \varphi_0(x_i)|$$

$$< \varepsilon\left(\sum_{i=1}^{m} \frac{1}{2^i}\right) + \|\varphi - \varphi_0\|\left(\sum_{i=m+1}^{\infty} \frac{1}{2^i}\|x_i\|\right) < r_0.$$

∎

We will discuss the (non) metrizability of the weak topology in $E$ and the weak-star topology in $E'$ in Sect. 6.7.

**Corollary 6.5.2** *If $E$ is separable, then $E'$ is separable in the weak-star topology.*

*Proof* For each natural number $n$, in the weak-star topology the set $nB_{E'}$ is compact by Theorem 6.3.9 and metrizable by Proposition 6.5.1. As compact subsets of metric spaces are separable, it follows that $nB_{E'}$ is separable in the weak-star topology for every $n$. Therefore $E' = \bigcup_{n=1}^{\infty} nB_{E'}$ is separable in the weak-star topology as the countable union of separable sets in this topology. ∎

**Proposition 6.5.3** *If $E'$ is separable, then $(B_E, \sigma(E, E'))$ is metrizable.*

**Proof** As a subset of a separable metric space is also separable, we can consider a set $\{\varphi_i : i \in \mathbb{N}\}$ dense in the set $\{\varphi \in E' : \|\varphi\| = 1\}$. Now we just proceed as in the proof of Proposition 6.5.1 with the metric

$$d: B_E \times B_E \longrightarrow [0, \infty) \, , \, d(x, y) = \sum_{n=1}^{\infty} \frac{1}{2^n} |\varphi_n(x - y)|.$$

∎

The following result is quite useful. Its converse shall be discussed in Sect. 6.7.

**Theorem 6.5.4** *In a reflexive space, every bounded sequence has a weakly convergent subsequence.*

**Proof** Let $(x_n)_{n=1}^{\infty}$ be a bounded sequence in the reflexive space $E$. Call $F$ the closure of the subspace generated by the elements of the sequence $(x_n)_{n=1}^{\infty}$. So $F$ is separable by Lemma 1.6.3 and reflexive by Corollary 6.4.6. It follows from Theorem 4.3.7 that $F'$ is separable. Thus, in the weak topology, $B_F$ is compact by Theorem 6.4.5 and metrizable by Proposition 6.5.3. Taking $L$ such that $\|x_n\| \leq L$ for every $n$, the sequence $(\frac{x_n}{L})_{n=1}^{\infty}$ is contained in the compact metric space $(B_F, \sigma(F, F'))$. As in metric spaces sequences in compact sets admit convergent subsequences, it follows that $(\frac{x_n}{L})_{n=1}^{\infty}$ has a subsequence that converges in the weak topology $\sigma(F, F')$, therefore it also converges in the weak topology $\sigma(E, E')$. The return to the original sequence $(x_n)_{n=1}^{\infty}$ is trivial. ∎

Next we look at another very interesting consequence of Proposition 6.5.1. In Proposition 3.3.3 we proved that $\ell_\infty$ contains isometric copies of all separable spaces. But $\ell_\infty$ is not separable, which leads to questioning whether there is a separable space that contains an isometric copy of every separable space. We will see next that such a universal separable space exists and is one of our old acquaintances.

**Theorem 6.5.5 (Banach–Mazur Theorem)** *Let $E$ be a separable normed space. Then $E$ is isometrically isomorphic to a subspace of $C[0, 1]$.*

**Proof** Since $E$ is separable, $(B_{E'}, \sigma(E', E))$ is compact by Theorem 6.3.9 and metrizable by Proposition 6.5.1. By the Hahn–Mazurkiewicz Theorem there exists a function

$$f: [0, 1] \longrightarrow (B_{E'}, \sigma(E', E))$$

that is surjective and continuous. Define

$$g: E \longrightarrow C[0, 1] \, , \, g(x)(t) = f(t)(x).$$

Now we prove that, for each $x \in E$, $g(x) \in C[0, 1]$. Indeed, if $t_n \longrightarrow t$ in $[0, 1]$, from the continuity of $f$ it follows that $f(t_n) \xrightarrow{w^*} f(t)$, therefore

$$g(x)(t_n) = f(t_n)(x) = J_E(x)(f(t_n)) \longrightarrow J_E(x)(f(t)) = f(t)(x) = g(x)(t).$$

It is immediate that $g$ is linear. Using the surjectivity of $f$ and Corollary 3.1.5 we conclude that

$$\|g(x)\| = \sup_{t\in[0,1]} |g(x)(t)| = \sup_{t\in[0,1]} |f(t)(x)| = \sup_{\varphi\in B_{E'}} |\varphi(x)| = \|x\|$$

for every $x \in E$. This proves that $g$ is an isometric isomorphism onto its range. ∎

**Corollary 6.5.6** *The spaces $C[0,1]$, $L_1[0,1]$ and $L_\infty[0,1]$ are not reflexive.*

*Proof* Since $c_0$ is separable, $C[0,1]$ contains a copy of $c_0$ by Theorem 6.5.5. But $c_0$ is not reflexive (Example 4.3.6(c)), so $C[0,1]$ is not reflexive by Corollary 6.4.6. But $C[0,1]$ is a closed subspace of $L_\infty[0,1]$, so this same Corollary 6.4.6 ensures that $L_\infty[0,1]$ is not reflexive. Since $L_\infty[0,1] = (L_1[0,1])'$ (Theorem 4.1.2), the non-reflexivity of $L_1[0,1]$ follows from Proposition 4.3.13. ∎

## 6.6 Uniformly Convex Spaces

Our geometric intuition clearly tells us that the closed unit ball of the Euclidean space $(\mathbb{R}^2, \|\cdot\|_2)$ is round, while those of the spaces $(\mathbb{R}^2, \|\cdot\|_1)$ and $(\mathbb{R}^2, \|\cdot\|_\infty)$ are not round at all; in fact, they are square. These last two examples are extreme, as their unit spheres contain line segments.

To prevent the closed unit sphere $S_E = \{x \in E : \|x\| = 1\}$ of a space $E$ from containing line segments, it is enough to ask that if $x, y \in B_E$, then the midpoint of the segment connecting $x$ to $y$ is inside the ball, that is,

$$\left\|\frac{x+y}{2}\right\| < 1 \text{ whenever } \|x\| \leq 1 \text{ and } \|y\| \leq 1.$$

But the unit sphere of $(\mathbb{R}^2, \|\cdot\|_2)$, that is, the circumference, satisfies a much stronger property than simply not containing line segments. Fix a distance $\varepsilon > 0$ and take any two points $x$ and $y$ on the circumference that are at least $\varepsilon$ apart. Appealing again to geometric intuition, we realize that, regardless of where the points $x$ and $y$ are located on the circumference, the midpoint of the segment that joins them remains inside the circumference and, more importantly, at a safe distance from the circumference. That is, we can move the points $x$ and $y$ on the circumference, just keeping them at a distance $\varepsilon$, that there is no risk of the midpoint of the interval that joins them approaching dangerously close to the circumference. In this section, we will study the normed spaces that share this uniform behavior of the points of the unit ball with the Euclidean spaces.

**Definition 6.6.1** We say that a normed space $E$ is *uniformly convex* or, more precisely, that its norm is *uniformly convex* if, for each $\varepsilon > 0$ there exists $\delta = \delta(\varepsilon) > 0$ such that

$$\left\|\frac{x+y}{2}\right\| \leq 1 - \delta \text{ whenever } x, y \in B_E \text{ and } \|x - y\| \geq \varepsilon.$$

Since two points in $B_E$ are at most 2 apart, it is sufficient to show the condition above for $0 < \varepsilon \leq 2$.

Some more recent books use the term *uniformly rotund space*, which indeed better translates the underlying geometric idea.

**Example 6.6.2** Every Hilbert space $H$ is uniformly convex. Indeed, given $x, y \in B_H$ and $0 < \varepsilon \leq 2$, from the Parallelogram Law (Proposition 5.1.7(a)), it follows that if $\|x - y\| \geq \varepsilon$ then

$$\left\|\frac{x+y}{2}\right\|^2 = \frac{\|x\|^2}{2} + \frac{\|y\|^2}{2} - \frac{\|x-y\|^2}{4} \leq 1 - \frac{\varepsilon^2}{4}.$$

It is then enough to choose $\delta = 1 - \left(1 - \varepsilon^2/4\right)^{\frac{1}{2}} > 0$.

**Example 6.6.3** The spaces $\left(\mathbb{R}^2, \|\cdot\|_\infty\right)$ and $\left(\mathbb{R}^2, \|\cdot\|_1\right)$ are not uniformly convex. In fact, given $0 < \varepsilon < 1$, it is enough to take two vectors that are on the same line segment of the unit sphere and are $\varepsilon$ apart from each other. By adding zeroes in the other coordinates, we transform these ordered pairs into sequences, which proves that the spaces $c_0$, $\ell_1$ and $\ell_\infty$ are not uniformly convex.

**Remark 6.6.4** As $\left(\mathbb{R}^2, \|\cdot\|_2\right)$ is Hilbert, the examples above show, in particular, that a space isomorphic to a uniformly convex space is not always uniformly convex.

We will now work towards proving that uniformly convex spaces are reflexive, which, on the one hand, guarantees a good property for uniformly convex spaces and, on the other hand, provides many examples of spaces that are not uniformly convex.

Let $(I, \leq)$ be a directed set. In $I \times I$ we consider the natural direction given by $(\alpha_1, \alpha_2) \leq (\beta_1, \beta_2)$ if and only if $\alpha_1 \leq \beta_1$ and $\alpha_2 \leq \beta_2$.

**Lemma 6.6.5** *Let $E$ be a uniformly convex space. If $(x_\alpha)_{\alpha \in I}$ is a net in $B_E$ such that $\lim\limits_{(\alpha,\beta)} \left\|\frac{x_\alpha + x_\beta}{2}\right\| = 1$, then $(x_\alpha)_{\alpha \in I}$ is a Cauchy net, that is, for every $\varepsilon > 0$ there exists $\alpha_0 \in I$ such that $\|x_\alpha - x_\beta\| < \varepsilon$ for all $\alpha, \beta \in I$ with $\alpha, \beta \geq \alpha_0$.*

**Proof** Suppose that the net $(x_\alpha)_{\alpha \in I}$ is not Cauchy. In this case there exists $0 < \varepsilon \leq 2$ such that for every index $\alpha_0 \in I$ there correspond indices $\alpha, \beta \geq \alpha_0$ such that $\|x_\alpha - x_\beta\| \geq \varepsilon$. Consider $\delta > 0$ corresponding to this $\varepsilon > 0$ according to the definition of uniformly convex space. We have, in particular, that

$$\left\|\frac{x_\alpha + x_\beta}{2}\right\| \leq 1 - \delta.$$

## 6.6 Uniformly Convex Spaces

The result follows, as this contradicts the convergence $\lim_{(\alpha,\beta)} \left\| \frac{x_\alpha + x_\beta}{2} \right\| = 1$. ∎

**Theorem 6.6.6 (Milman–Pettis Theorem)** *Uniformly convex Banach spaces are reflexive.*

*Proof* Let $E$ be a uniformly convex Banach space. Given a functional $f \in E''$, $\|f\| = 1$, by Theorem 6.4.4 there exists a net $(x_\alpha)_{\alpha \in I}$ in $B_E$ such that $J_E(x_\alpha) \xrightarrow{w^*} f$. We shall check that, with the natural direction in $I \times I$,

$$J_E\left(\frac{1}{2}(x_\alpha + x_\beta)\right) \xrightarrow{w^*} f. \tag{6.3}$$

To do so, consider a neighborhood $W$ of $f$ in the weak-star topology. In Exercise 6.8.14 the reader will prove that there is a neighborhood $W_0$ of the origin in the weak-star topology such that $W = f + W_0$. Let

$$V = \{g \in E'' : |g(\varphi_i)| < \delta, \ i = 1, \ldots, m\}$$

be a basic neighborhood of the origin in the weak-star topology contained in $W_0$, where $\delta > 0$ and $\varphi_1, \ldots, \varphi_m \in E'$. Then

$$V_0 := \left\{g \in E'' : |g(\varphi_i)| < \frac{\delta}{2}, \ i = 1, \ldots, m\right\}$$

is a neighborhood of the origin in the weak-star topology and $V_0 + V_0 \subseteq V \subseteq W_0$. As $\frac{f}{2} + V_0$ is a neighborhood of $\frac{f}{2}$ in the weak-star topology and $J_E\left(\frac{x_\alpha}{2}\right) \xrightarrow{w^*} \frac{f}{2}$, there exists $\alpha_1 \in I$ such that

$$J_E\left(\frac{x_\alpha}{2}\right) \in \frac{f}{2} + V_0 \quad \text{whenever } \alpha \geq \alpha_1.$$

Hence,

$$J_E\left(\frac{x_\alpha}{2}\right) + J_E\left(\frac{x_\beta}{2}\right) \in \left(\frac{f}{2} + V_0\right) + \left(\frac{f}{2} + V_0\right) \subseteq f + W_0 = W$$

whenever $(\alpha, \beta) \geq (\alpha_1, \alpha_1)$. This proves (6.3). Given $\varepsilon > 0$, from the definition of $\|f\|$ there exists $\varphi_0 \in E'$ such that $\|\varphi_0\| = 1$ and $|f(\varphi_0)| > \|f\| - \varepsilon$. From (6.3) it follows that

$$\lim_{(\alpha,\beta)} \left| J_E\left(\frac{1}{2}(x_\alpha + x_\beta)\right)(\varphi_0) \right| = |f(\varphi_0)| > \|f\| - \varepsilon,$$

therefore there exists $(\alpha_\varepsilon, \beta_\varepsilon) \in I \times I$ such that

$$\left| J_E \left( \frac{1}{2} (x_\alpha + x_\beta) \right) (\varphi_0) \right| > \|f\| - \varepsilon \text{ whenever } (\alpha, \beta) \geq (\alpha_\varepsilon, \beta_\varepsilon).$$

Since $\left\| \frac{1}{2}(x_\alpha + x_\beta) \right\| \leq 1$ and $J_E$ is an isometry,

$$1 \geq \left\| J_E \left( \frac{1}{2} (x_\alpha + x_\beta) \right) \right\| \geq \left| J_E \left( \frac{1}{2} (x_\alpha + x_\beta) \right) (\varphi_0) \right| > \|f\| - \varepsilon = 1 - \varepsilon$$

whenever $(\alpha, \beta) \geq (\alpha_\varepsilon, \beta_\varepsilon)$. It follows then that

$$\left\| \frac{1}{2} (x_\alpha + x_\beta) \right\| = \left\| J_E \left( \frac{1}{2} (x_\alpha + x_\beta) \right) \right\| \longrightarrow 1.$$

It follows from Lemma 6.6.5 that $(x_\alpha)_\alpha$ is a Cauchy net in $E$. As $E$ is a Banach space, there exists $x_0 \in E$ such that $x_\alpha \longrightarrow x_0$ (see Exercise 6.8.24), therefore $J_E(x_\alpha) \longrightarrow J_E(x_0)$. But norm convergence implies weak-star convergence, hence $J_E(x_\alpha) \xrightarrow{w^*} J_E(x_0)$. The fact that the weak-star topology is Hausdorff implies that $J_E(x_0) = f$, proving that $E$ is reflexive. ∎

Before moving on, we shall note another interesting consequence of Lemma 6.6.5:

**Proposition 6.6.7** *Let $E$ be a uniformly convex Banach space and let $(x_n)_{n=1}^\infty$ be a sequence in $E$. If $x_n \xrightarrow{w} x$ and $\|x_n\| \longrightarrow \|x\|$, then $x_n \longrightarrow x$.*

**Proof** There is nothing to do in the case where $x = 0$. Suppose then $x \neq 0$. In this case, from $\|x_n\| \longrightarrow \|x\| > 0$ we can choose $n_0 \in \mathbb{N}$ such that $\|x_n\| > 0$ for every $n \geq n_0$. Let $y = \frac{x}{\|x\|}$ and $y_n = \frac{x_n}{\|x_n\|}$ for every $n \geq n_0$. It is immediate that $y_n \xrightarrow{w} y$. As $y \neq 0$, by the Hahn–Banach Theorem in the form of Corollary 3.1.4 there exists $\varphi \in E'$ with $\|\varphi\| = 1$ and $\varphi(y) = \|y\| = 1$. For all $n, m \geq n_0$,

$$\left| \frac{\varphi(y_n)}{2} + \frac{\varphi(y_m)}{2} \right| = \left| \varphi \left( \frac{y_n + y_m}{2} \right) \right| \leq \left\| \frac{y_n + y_m}{2} \right\| \leq \frac{\|y_n\|}{2} + \frac{\|y_m\|}{2} = 1.$$

From this inequality and the convergence $\varphi(y_n) \longrightarrow \varphi(y) = 1$ it follows that $\lim_{m,n} \left\| \frac{y_n + y_m}{2} \right\| = 1$. By Lemma 6.6.5 it follows that that $(y_n)_{n=1}^\infty$ is a Cauchy sequence, hence convergent in the norm topology, therefore in the weak topology, to a certain $z$. As the weak topology is Hausdorff, it is clear that $z = y$. The result follows because

$$\|x_n - x\| = \| \|x_n\| y_n - \|x\| y \| \leq \| \|x_n\| y_n - \|x_n\| y \| + \| \|x_n\| y - \|x\| y \|$$
$$= \|x_n\| \cdot \|y_n - y\| + |\|x_n\| - \|x\|| \cdot \|y\| \longrightarrow 0.$$

∎

## 6.6 Uniformly Convex Spaces

From Theorem 6.6.6 and Corollary 5.5.6 we conclude that $L_1(X, \Sigma, \mu)$ and $L_\infty(X, \Sigma, \mu)$ are not uniformly convex. The goal now is to show that the spaces $L_p(X, \Sigma, \mu)$ are uniformly convex for $1 < p < \infty$. First we have to prove Clarkson's inequalities.

**Lemma 6.6.8** *Let $2 \leq p < \infty$ and $a, b \in \mathbb{R}$ be given. Then*

$$|a+b|^p + |a-b|^p \leq 2^{p-1}\left(|a|^p + |b|^p\right).$$

*Proof* Use the Hölder inequality (Proposition 1.4.1) to obtain

$$\left(|a+b|^p + |a-b|^p\right)^{\frac{1}{p}} \leq \left((a+b)^2 + (a-b)^2\right)^{\frac{1}{2}} = (2a^2 + 2b^2)^{\frac{1}{2}}$$

$$= 2^{\frac{1}{2}}\left(|a|^2 + |b|^2\right)^{\frac{1}{2}} = 2^{\frac{1}{2}}\left(|a|^2 \cdot 1 + |b|^2 \cdot 1\right)^{\frac{1}{2}}$$

$$\leq 2^{\frac{1}{2}}\left[\left(\left(|a|^2\right)^{\frac{p}{2}} + \left(|b|^2\right)^{\frac{p}{2}}\right)^{\frac{2}{p}} \cdot (1+1)^{\frac{p-2}{p}}\right]^{\frac{1}{2}}$$

$$= 2^{\frac{1}{2}} \cdot 2^{\frac{1}{2}-\frac{1}{p}}\left(|a|^p + |b|^p\right)^{\frac{1}{p}} = 2^{\frac{p-1}{p}}\left(|a|^p + |b|^p\right)^{\frac{1}{p}}.$$

Raise both sides to $p$ to obtain the result. ∎

**Proposition 6.6.9 (First Clarkson's Inequality)** *Let $2 \leq p < \infty$ and $f, g \in L_p(X, \Sigma, \mu)$ be given. Then*

$$\|f+g\|_p^p + \|f-g\|_p^p \leq 2^{p-1}\left(\|f\|_p^p + \|g\|_p^p\right).$$

*Proof* Just make $a = f(t)$ and $b = g(t)$ in Lemma 6.6.8 and integrate. ∎

The case $1 < p < 2$ is a bit more laborious. Just like in the previous case we need an arithmetic lemma. For simplicity, we will denote by $q$ the conjugate of the number $p > 1$, that is, $\frac{1}{p} + \frac{1}{q} = 1$.

**Lemma 6.6.10** *Let $1 < p \leq 2$ be given. Then*

$$|a+b|^q + |a-b|^q \leq 2\left(|a|^p + |b|^p\right)^{q-1} \text{ for all } a, b \in \mathbb{R}.$$

*Proof* Consider the function $f: [0, 1] \times [0, 1] \longrightarrow \mathbb{R}$ given by

$$f(\alpha, t) = (1 + \alpha^{1-q}t)(1 + \alpha t)^{q-1} + (1 - \alpha^{1-q}t)(1 - \alpha t)^{q-1}.$$

Note that $(1 - \alpha^{-q}) \leq 0$ and $(1 + \alpha t)^{q-2} - (1 - \alpha t)^{q-2} \geq 0$ because $q \geq 2$. Therefore

$$\frac{\partial f}{\partial \alpha} = (q-1)t(1-\alpha^{-q})\left[(1+\alpha t)^{q-2} - (1-\alpha t)^{q-2}\right] \leq 0.$$

As $t^{p-1} \leq 1$,

$$(1+t)^q + (1-t)^q = f(1,t) \leq f(t^{p-1}, t) = 2\left(1+t^p\right)^{q-1},$$

proving that

$$(1+t)^q + (1-t)^q \leq 2\left(1+t^p\right)^{q-1} \quad \text{for every } 0 \leq t \leq 1. \tag{6.4}$$

Let $a, b \in \mathbb{R}$ be given. Since the inequality we want to prove is symmetric in $a$ and $b$, we can assume $a \leq b$. The case $a = b$ is immediate, so we can actually assume $a < b$. We split into cases.

Case $a, b > 0$. Just apply (6.4) for $t = \frac{a}{b}$ to obtain the desired inequality.
Case $a, b < 0$. Apply the previous case for $-b$ and $-a$.
Case $a < 0$ and $b > 0$. Apply the first case for $-a$ and $b$ (or $b$ and $-a$). ∎

**Proposition 6.6.11 (Second Clarkson's Inequality)** *Let $1 < p \leq 2$ and $f, g \in L_p(X, \Sigma, \mu)$ be given. Then*

$$\|f+g\|_p^q + \|f-g\|_p^q \leq 2\left(\|f\|_p^p + \|g\|_p^p\right)^{q-1}.$$

***Proof***

$$\begin{aligned}
\|f+g\|_p^q + \|f-g\|_p^q &= \left\||f+g|^q\right\|_{p-1} + \left\||f-g|^q\right\|_{p-1} \\
&\leq \left\||f+g|^q + |f-g|^q\right\|_{p-1} \\
&= \left(\int_X \left(|f+g|^q + |f-g|^q\right)^{p-1} d\mu\right)^{\frac{1}{p-1}} \\
&\leq 2\left(\int_X \left(|f|^p + |g|^p\right)^{(p-1)(q-1)} d\mu\right)^{\frac{1}{p-1}} \\
&= 2\left(\int_X |f|^p \, d\mu + \int_X |g|^p \, d\mu\right)^{\frac{1}{p-1}} \\
&= 2\left(\|f\|_p^p + \|g\|_p^p\right)^{q-1}.
\end{aligned}$$

In the first step we used the equality $\|h\|_p^q = \||h|^q\|_{p-1}$ because $q = \frac{p}{p-1}$; in the second we use the reverse Minkowski inequality (Theorem 1.7.2) because $0 < p - 1 < 1$; in the third we use the definition of $\|\cdot\|_{p-1}$; in the fourth we use Lemma 6.6.10 for $a = f(t)$ and $b = g(t)$; in the fifth and sixth we use that $(p-1)(q-1) = 1$. ∎

## 6.6 Uniformly Convex Spaces

**Theorem 6.6.12** $L_p(X, \Sigma, \mu)$ *is uniformly convex for every* $1 < p < \infty$.

***Proof*** Let $0 < \varepsilon \leq 2$ and let $f, g \in B_{L_p(X,\Sigma,\mu)}$ be such that $\|f - g\|_p \geq \varepsilon$. We first consider the case $2 \leq p < \infty$. From Proposition 6.6.9 it follows that

$$\|f + g\|_p^p \leq 2^{p-1}\left(\|f\|_p^p + \|g\|_p^p\right) - \|f - g\|_p^p \leq 2^{p-1}(1+1) - \varepsilon^p = 2^p - \varepsilon^p.$$

Therefore

$$\left\|\frac{f+g}{2}\right\|_p \leq \left(1 - \left(\frac{\varepsilon}{2}\right)^p\right)^{\frac{1}{p}}.$$

It suffices to choose $\delta = 1 - \left(1 - \left(\frac{\varepsilon}{2}\right)^p\right)^{\frac{1}{p}}$. For $1 < p < 2$, from Proposition 6.6.11 it follows that

$$\|f + g\|_p^q \leq 2(1+1)^{q-1} - \|f - g\|_p^q \leq 2^q - \varepsilon^q.$$

Similarly to the previous case, it suffices to settle $\delta = 1 - \left(1 - \left(\frac{\varepsilon}{2}\right)^q\right)^{\frac{1}{q}}$. ∎

As promised in Chap. 4, we will now present an alternative proof of Theorem 4.1.2 for the case of real scalars and $1 < p < \infty$. At that time, we obtained the reflexivity of the spaces $L_p(X, \Sigma, \mu)$, $1 < p < \infty$, from Theorem 4.1.2. Note that we have just proven, without using Theorem 4.1.2 or any of its consequences, that $L_p(X, \Sigma, \mu)$, $1 < p < \infty$, is uniformly convex, therefore reflexive. We thus have a proof of the reflexivity of such spaces that does not depend on Theorem 4.1.2. This is exactly what will allow us to now give a new proof of Theorem 4.1.2 for $1 < p < \infty$ in the real case:

**Theorem 6.6.13** *Let* $\mathbb{K} = \mathbb{R}$, $1 < p < \infty$ *and let* $(X, \Sigma, \mu)$ *be a measure space. Then the correspondence* $g \mapsto \varphi_g$ *establishes an isometric isomorphism between* $L_{p^*}(X, \Sigma, \mu)$ *and* $L_p(X, \Sigma, \mu)'$ *in which the duality relation is given by*

$$\varphi_g(f) = \int_X fg\, d\mu \ \text{for every}\ f \in L_p(X, \Sigma, \mu).$$

***Proof*** By Example 4.1.1 it is enough to show that the linear operator

$$T: L_{p^*}(X, \Sigma, \mu) \longrightarrow L_p(X, \Sigma, \mu)',\ T(g)(f) = \int_X fg\, d\mu,$$

is surjective. The range of $T$ is closed because it is isometrically isomorphic to the Banach space $L_{p^*}(X, \Sigma, \mu)$. Therefore, it is sufficient to show that $T(L_{p^*}(X, \Sigma, \mu))$ is dense in $L_p(X, \Sigma, \mu)'$. Arguing by contradiction, suppose it is not dense. In this case there exists $\varphi \in L_p(X, \Sigma, \mu)' - \overline{T(L_{p^*}(X, \Sigma, \mu))}$.

By Corollary 3.4.10 there exists $\psi \in L_p(X, \Sigma, \mu)''$ such that $\|\psi\| = 1$ and $\psi(h) = 0$ for every $h \in \overline{T(L_{p^*}(X, \Sigma, \mu))}$. In particular, $\psi(T(g)) = 0$ for every $g \in L_{p^*}(X, \Sigma, \mu)$. Since $L_p(X, \Sigma, \mu)$ is reflexive, there exists $f \in L_p(X, \Sigma, \mu)$ such that $\psi = J_{L_p(X,\Sigma,\mu)}(f)$. Then

$$\int_X fg\, d\mu = T(g)(f) = J_{L_p(X,\Sigma,\mu)}(f)(T(g)) = \psi(T(g)) = 0$$

for every $g \in L_{p^*}(X, \Sigma, \mu)$. Taking $g = |f|^{p-2} \cdot f \in L_{p^*}(X, \Sigma, \mu)$, it follows that

$$0 = \left(\int_X f \cdot |f|^{p-2} \cdot f\, d\mu\right)^{\frac{1}{p}} = \left(\int_X |f|^p\, d\mu\right)^{\frac{1}{p}} = \|f\|_p$$
$$= \|J_{L_p(X,\Sigma,\mu)}(f)\| = \|\psi\| = 1.$$

This contradiction completes the proof. ∎

## 6.7 Comments and Historical Notes

The topology generated by a family of functions is also called the *induced topology* (or *weakly induced topology*) by the family of functions. This is because if we consider a subset $Y$ of the topological space $X$, the topology induced on $Y$ by $X$ coincides with the topology generated by the inclusion $Y \hookrightarrow X$. The term *initial topology* is used by the Bourbaki group [8] and its followers. As we saw in the proof of Theorem 6.3.9, this topology is closely related to the product topology in the generalized Cartesian product. A good discussion about this can be found in [67, Section 8].

Weakly convergent sequences were first studied by Hilbert in $L_2[0, 1]$, by Riesz in $L_p[0, 1]$ and by Schur in $\ell_1$, but the understanding that it was a topology — and a very useful one — came only with von Neumann in the 1930s.

Theorem 6.2.12 was proved by I. Schur in 1921. Because of this theorem, a space in which every weakly convergent sequence is norm convergent is called a *Schur space* (sometimes it is said that the space has the *Schur property*). Among the infinite-dimensional spaces with which we have been working in this book, only $\ell_1$ is Schur. The other natural candidate would be $L_1[0, 1]$, which is not Schur because it contains a copy of $\ell_2$ (see [17, Theorem 1.2]), and according to Exercise 6.8.21 spaces that contain infinite-dimensional reflexive subspaces are not Schur. Chapter 11 of [40] is entirely dedicated to the study of reflexive subspaces of $L_1(\mu)$.

In 1928 Banach proved that the unit ball of the dual of a separable normed vector space is sequentially compact in the weak-star topology. The general form of Theorem 6.3.9 was proved by L. Alaoglu in 1940. Theorem 6.4.5 was also initially

## 6.7 Comments and Historical Notes

proved by Banach for the separable case and sequential compactness. More abstract versions were successively proved by S. Kakutani, V. I. Smulian and W. F. Eberlein.

In Theorem 6.2.12 we saw that, even in an infinite-dimensional space, it can happen that sequences convergent in norm coincide with weakly convergent sequences. In independent works published in 1975, B. Josefson and A. Nissenzweig proved that this never occurs with the weak-star topology:

**Theorem 6.7.1 (Josefson–Nissenzweig Theorem)** *Let $E$ be an infinite-dimensional Banach space. In $E'$ there is a sequence $(\varphi_n)_{n=1}^\infty$ such that $\varphi_n \xrightarrow{w^*} 0$ and $\|\varphi_n\| = 1$ for every $n$.*

For the proof see [15, Chapter XII].

Proposition 6.5.1 leads to questioning about the metrizability of the weak-star topology in $E'$ and also of the weak topology in $E$. In finite dimension the answer is obviously true, unlike in infinite dimension:

**Proposition 6.7.2 ([44, Propositions 2.5.14 and 2.6.12])**

(a) *The weak topology on a normed space $E$ is metrizable if and only if $E$ is finite-dimensional.*
(b) *The weak-star topology on the dual $E'$ of a Banach space $E$ is metrizable if and only if $E$ is finite-dimensional.*

In metric spaces, particularly in the norm topology of a normed space, a set $K$ is compact if and only if every sequence in $K$ admits a convergent subsequence in $K$. The understanding that the weak topology enjoys this property began in 1940 when V. L. Smulian proved that every sequence in a weakly compact subset $K$ of a Banach space admits a weakly convergent subsequence in $K$. In 1947 W. F. Eberlein proved the converse:

**Theorem 6.7.3 (Eberlein–Smulian Theorem)** *A subset $K$ of a Banach space is weakly compact if and only if every sequence in $K$ admits a weakly convergent subsequence in $K$.*

The proof can be found in [15, Chapter III]. Combining this result with Theorem 6.4.5, we obtain the converse of Theorem 6.5.4. The analogous result to the Eberlein–Smulian Theorem for the weak-star topology is not true (see [15, Exercise III.1]).

The notion of uniformly convex space was introduced by J. A. Clarkson in 1936, in the same work in which he proved Theorem 6.6.12. Theorem 6.6.6 was proved, independently, by D. P. Milman (1939), B. J. Pettis (1939) and S. Kakutani (1939).

Neither of the implications

$$E \text{ uniformly convex} \iff E' \text{ uniformly convex}$$

is true. The following example was communicated to the authors by V. Ferenczi, to whom we publicly thank. Call $C$ the subset of $\mathbb{R}^2$ formed by the intersection of the open disc centered at 1 and radius 2 with the open disc centered at -1 and radius 2. Calling $\|\cdot\|_C$ the Minkowski functional of $C$ (see Definition 3.4.4 and Exercise 3.6.23), $(\mathbb{R}^2, \|\cdot\|_C)$ is uniformly convex but its dual is not. The dual notion to the concept of uniformly convex space is the notion of *uniformly smooth space*. For more information see Chapter 5 of [44].

The examples we have seen of reflexive spaces that are not uniformly convex are all finite-dimensional (Example 6.6.3). It is not an easy task to exhibit an infinite-dimensional reflexive space that is not uniformly convex. Examples can be found in [21, p. 294], [48, Exercise 16.205] or [64, Example 3, p. 166].

Proposition 6.6.7 was first proven for the spaces $L_p(X, \Sigma, \mu)$ by J. Radon in 1913 and F. Riesz in 1928. Therefore, it is said that a space in which the result holds has the *Radon–Riesz property*. That is, Proposition 6.6.7 proves that uniformly convex Banach spaces have the Radon–Riesz property. In the literature this property is also called the *Kadets–Klee property* and the *property (H)*. The concept of strictly convex space was mentioned—without drawing attention to the terminology—at the beginning of Sect. 6.6. This concept will be formally introduced and explored in the exercises. For an example of a strictly convex space that is not reflexive see [44, Example 5.1.8].

It is said that a property of Banach spaces is an *analytic property* if it is invariant by isomorphisms. A *geometric property* is one that is invariant only by isometric isomorphisms. That is, a geometric property has to do with the norm of the space while an analytic property has to do only with the topology induced by the norm. For example, reflexivity is an analytic property (Exercise 4.5.27) and uniform convexity is a geometric property (Remark 6.6.4 and Exercise 6.8.25). In recent decades, the *geometry of Banach spaces*, that is, the study of the geometric properties of Banach spaces, has been one of the most effervescent areas within Functional Analysis, culminating with the advances described in the Introduction of this book. The activity in the area can be evidenced by the two (heavy) volumes of the Handbook of the Geometry of Banach Spaces [36, 37].

## 6.8 Exercises

**Exercise 6.8.1** Prove Proposition 6.1.1.

**Exercise 6.8.2** Prove Proposition 6.1.3.

**Exercise 6.8.3** Let $F$ be a subspace of the normed space $E$, $(x_n)_{n=1}^{\infty}$ be a sequence in $F$ and $x \in F$. Prove that $x_n \xrightarrow{w} x$ in $F$ if and only if $x_n \xrightarrow{w} x$ in $E$.

**Exercise 6.8.4** Let $F$ be a subspace of the normed space $E$. Prove that the weak topology $\sigma(F, F')$ of $F$ is the topology induced in $F$ by the weak topology $\sigma(E, E')$ of $E$.

**Exercise 6.8.5** Discuss the weak convergence of the sequence $(e_n)_{n=1}^{\infty}$ formed by the canonical unit vectors in the following spaces: $c_{00}, \ell_1, \ell_p$ with $1 < p < \infty$ and $\ell_\infty$.

## 6.8 Exercises

**Exercise 6.8.6** Prove that every orthonormal sequence in a Hilbert space weakly converges to zero.

**Exercise 6.8.7**

(a) Prove that if $x_n \xrightarrow{w} x$ and $y_n \longrightarrow y$ in a space with inner product, then $\langle x_n, y_n \rangle \longrightarrow \langle x, y \rangle$.
(b) Give an example where $x_n \xrightarrow{w} x$, $y_n \xrightarrow{w} y$ but $(\langle x_n, y_n \rangle)_{n=1}^{\infty}$ is not convergent.

**Exercise 6.8.8** Is it possible to adapt the proof of Theorem 6.2.12 to prove that weakly convergent nets in $\ell_1$ are convergent in norm?

**Exercise 6.8.9** Prove that every non-empty open set in the weak topology of an infinite-dimensional space is unbounded. Use this to give another proof that the weak and norm topologies never coincide on infinite-dimensional spaces.

**Exercise 6.8.10** Let $(x_n)_{n=1}^{\infty}$ be a sequence in the Banach space $E$ such that $x_n \xrightarrow{w} x \in E$. Prove that there exist convex combinations $(y_n)_{n=1}^{\infty}$ of the set $\{x_1, x_2, \ldots\}$ such that $y_n \longrightarrow x$. (A convex combination of a subset $A$ of a vector space is a vector of the form $\sum_{j=1}^{m} a_j z_j$ where $m \in \mathbb{N}$, $a_1, \ldots, a_n \geq 0$, $a_1 + \cdots + a_m = 1$ and $z_1, \ldots, z_m \in A$.)

**Exercise 6.8.11** Prove item (b) of Proposition 6.3.2.

**Exercise 6.8.12** Prove items (b) and (c) of Proposition 6.3.3.

**Exercise 6.8.13** Consider the linear functional

$$\varphi: \ell_1 \longrightarrow \mathbb{K}, \ \varphi((a_j)_{j=1}^{\infty}) = \sum_{j=1}^{\infty} a_j.$$

Show that it is continuous in norm but not continuous in the weak-star topology of $\ell_1 = (c_0)'$. Conclude that Proposition 6.2.9 does not hold for the weak-star topology.

**Exercise 6.8.14** Let $E$ be a normed space. Prove that if $W$ is a neighborhood of $f \in E''$ in the weak-star topology, then there exists a neighborhood $W_0$ of the origin in the weak-star topology such that $W = f + W_0$.

**Exercise 6.8.15** A sequence $(x_n)_{n=1}^{\infty}$ in a normed space $E$ is called a *weakly Cauchy sequence* if for each $\varphi \in E'$, the sequence $(\varphi(x_n))_{n=1}^{\infty}$ is a Cauchy sequence in $\mathbb{K}$. It is said that $E$ is *weakly sequentially complete* if every weakly Cauchy sequence in $E$ is weakly convergent.

(a) Show that every weakly Cauchy sequence is bounded.
(b) Prove that reflexive spaces are weakly sequentially complete.

**Exercise 6.8.16** Prove that, for $1 < p < \infty$, $\ell_p$ does not contain an isomorphic copy of any of the following spaces: $c_0, \ell_\infty$ and $\ell_1$.

**Exercise 6.8.17** Let $F$ be a closed subspace of a reflexive Banach space $E$. Prove that $E/F$ is reflexive.

**Exercise 6.8.18** As discussed in Sect. 4.4, a very delicate theorem due to R. C. James states that in every non-reflexive Banach space $E$ there exists $\varphi_0 \in E'$ such that $\varphi_0(x) < \|\varphi_0\| \cdot \|x\|$ for every $x \in X$. Prove the converse of Theorem 6.5.4 using this result due to James.

**Exercise 6.8.19\*** Let $E$ be a Banach space and $1 \leq p < \infty$. For the definition of the space $\ell_p(E)$ see Exercise 4.5.8. Prove that:

(a) $\ell_p(E)$ contains 1-complemented isometric copies of $\ell_p$ and of $E$.
(b) For every Banach space $E$, $\ell_1(E)$ is not reflexive.
(c) For a given Banach space $E$, the following statements are equivalent:

  (i) $E$ is reflexive.
  (ii) $\ell_p(E)$ is reflexive for every $1 < p < \infty$.
  (iii) $\ell_p(E)$ is reflexive for some $1 < p < \infty$.

**Exercise 6.8.20** Prove that the function $d$ from the proof of Proposition 6.5.1 is well defined and is a metric.

**Exercise 6.8.21** Recall that a normed space $E$ is Schur if weakly convergent sequences in $E$ converge in norm. Prove that:

(a) Subspace of Schur space is Schur.
(b) Space that is isomorphic to a Schur space is also Schur.
(c) A reflexive normed space is Schur if and only if it is finite-dimensional.
(d) A space that contains a reflexive infinite-dimensional subspace is not Schur.

**Exercise 6.8.22** Prove that $\ell_1$ does not have an infinite-dimensional reflexive subspace.

**Exercise 6.8.23\*** A Banach space $E$ is *weakly compactly generated* if there exists a weakly compact subset $K$ of $E$ such that $E = \overline{[K]}$. Prove that:

(a) $E$ is weakly compactly generated if and only if there exists a subset $K$ of $E$ that is convex, symmetric (that is, $-x \in K$ if $x \in K$) and weakly compact such that $E = \overline{[K]}$.
(b) Reflexive spaces are weakly compactly generated.
(c) Separable spaces are weakly compactly generated.

**Exercise 6.8.24** Prove that every Cauchy net (see definition in the statement of Lemma 6.6.5) in a Banach space is convergent.

**Exercise 6.8.25** If a uniformly convex space contains an isometric copy of the normed space $E$, prove that $E$ is uniformly convex.

**Exercise 6.8.26** Prove that a normed space $E$ is uniformly convex if and only if $E''$ is uniformly convex.

## 6.8 Exercises

**Exercise 6.8.27** Let $E$ be a normed space. Prove that the implications

$$E \text{ uniformly convex} \implies E' \text{ uniformly convex, and}$$

$$E' \text{ uniformly convex} \implies E \text{ uniformly convex,}$$

are equivalent, so a counterexample to one of them is enough to conclude that both are false.

**Exercise 6.8.28** Prove that every normed space that is isomorphic to a uniformly convex Banach space is reflexive.

**Exercise 6.8.29\*** Show that the following statements are equivalent for a Banach space $E$:

(a) $E$ is uniformly convex.
(b) If $x_n, y_n \in E$ for every $n \in \mathbb{N}$, $\lim_{n \to \infty} \left(2 \|x_n\|^2 + 2 \|y_n\|^2 - \|x_n + y_n\|^2\right) = 0$, and $(x_n)_{n=1}^\infty$ is bounded, then $\lim_{n \to \infty} \|x_n - y_n\| = 0$.
(c) If $x_n, y_n \in B_E$ for every $n \in \mathbb{N}$ and $\lim_{n \to \infty} \|x_n + y_n\| = 2$, then $\lim_{n \to \infty} \|x_n - y_n\| = 0$.

**Exercise 6.8.30\*** Prove that the following statements are equivalent for a normed space $E$:

(a) If $x, y \in E$, $\|x\| = \|y\| = 1$ and $\|x + y\| = 2$, then $x = y$.
(b) If $x, y \in E$ and $\|x + y\|^2 = 2\|x\|^2 + 2\|y\|^2$, then $x = y$.
(c) If $x, y \in E$, $x \neq y$ and $\|x + y\| = \|x\| + \|y\|$, then $x$ is a positive multiple of $y$.

We say that $E$ is *strictly convex* when it satisfies the equivalent conditions above.

**Exercise 6.8.31** Show that every uniformly convex space is strictly convex.

**Exercise 6.8.32** Show that the spaces $(\mathbb{R}^2, \|\cdot\|_\infty)$, $(\mathbb{R}^2, \|\cdot\|_1)$, $c_0$, $\ell_1$ and $\ell_\infty$ are not strictly convex.

**Exercise 6.8.33 (A Strictly Convex Space That Is Not Uniformly Convex)** For $f \in C[0, 1]$, define

$$\|f\|_0 = \left(\|f\|_\infty^2 + \|f\|_2^2\right)^{\frac{1}{2}},$$

where $\|\cdot\|_\infty$ is the usual sup norm of $C[0, 1]$ and $\|\cdot\|_2$ is the usual norm of $L_2[0, 1]$.

(a) Prove that $\|\cdot\|_0$ is a norm on $C[0, 1]$ equivalent to the usual norm $\|\cdot\|_\infty$.
(b) Prove that $(C[0, 1], \|\cdot\|_0)$ is strictly convex.
(c) Prove that $(C[0, 1], \|\cdot\|_0)$ is not uniformly convex.

# Chapter 7
# Spectral Theory of Compact Self-adjoint Operators

Just as in Linear Algebra, in Functional Analysis the study of eigenvalues and eigenvectors of a linear and continuous operator is very useful. The following concepts are certainly familiar to the reader:

**Definition** Let $E$ be a vector space and let $T \colon E \longrightarrow E$ be a linear operator. An *eigenvalue* of $T$ is a scalar $\lambda$ for which there exists a non-zero vector $x \in E$ such that $T(x) = \lambda x$. The subspace

$$V_\lambda = \{x \in E : T(x) = \lambda x\}$$

is called the *eigenspace* associated with the eigenvalue $\lambda$ and its elements are called *eigenvectors* of $T$ associated with the eigenvalue $\lambda$.

For simplicity, in this chapter we will denote the identity operator on a vector space by $I$. Suppose that $V$ is finite-dimensional. In this case,

$$\lambda \text{ is an eigenvalue of } T \iff \ker(T - \lambda I) \neq \{0\}$$
$$\iff T - \lambda I \text{ is not injective}$$
$$\iff T - \lambda I \text{ is not bijective}$$
$$\iff \text{there is no } (T - \lambda I)^{-1} \colon E \longrightarrow E.$$

That is, in finite dimension the eigenvalues are exactly the scalars $\lambda$ for which $T - \lambda I$ is not invertible. In infinite dimension this argument does not work and, moreover, when $(T - \lambda I)^{-1}$ exists, we must take into account its continuity. This need will lead to the concept of *spectrum* of a linear operator.

In this chapter we will study the eigenvalues, the eigenvectors and the spectrum of two important classes of operators, namely, the compact operators between normed spaces and the self-adjoint operators between Hilbert spaces.

## 7.1 Spectrum of a Continuous Linear Operator

Let $E$ be a normed space, let $T \in \mathcal{L}(E, E)$ and let $\lambda$ be a scalar that is *not* an eigenvalue of $T$. Then $\ker(T - \lambda I) = \{0\}$ and it makes sense to consider the operator

$$(T - \lambda I)^{-1} : (T - \lambda I)(E) \subseteq E \longrightarrow E,$$

which is injective and linear. However, we know nothing about the continuity of $(T - \lambda I)^{-1}$ and the surjectivity of $(T - \lambda I)$. These concerns lead to the following definition:

**Definition 7.1.1** The scalar $\lambda$ is a *regular value* of the operator $T$ if $(T - \lambda I)$ is bijective and

$$(T - \lambda I)^{-1} : E \longrightarrow E$$

is continuous. The set of regular values of $T$ is called the *resolvent set of $T$* and denoted by $\rho(T)$. Its complement $(\mathbb{K} - \rho(T))$ is called the *spectrum of $T$* and denoted by $\sigma(T)$.

If $E$ is a Banach space, from the Open Mapping Theorem we know that

$$\rho(T) = \{\lambda \in \mathbb{K} : (T - \lambda I) \text{ is bijective}\}.$$

In normed spaces, the continuity of the inverse $(T - \lambda I)^{-1}$ does not always follow from the continuity of $(T - \lambda I)$ (see Example 2.4.3).

From the definition it follows that every eigenvalue of $T$ belongs to the spectrum of $T$. The following example shows that the spectrum does not always coincide with the set of eigenvalues.

**Example 7.1.2** In Example 4.3.10 we studied the linear and continuous operator

$$T : \ell_2 \longrightarrow \ell_2, \quad T((a_1, a_2, a_3, \ldots,)) = (0, a_1, a_2, \ldots).$$

Note that $T$ does not have eigenvalues. On the other hand, $T$ is injective but not bijective, therefore $0 \in \sigma(T)$ even though it is not an eigenvalue of $T$.

The main purpose of this section is to prove the compactness of the spectrum of a continuous linear operator on a Banach space. The following result will be of great help. For $T \in \mathcal{L}(E, E)$ and $n \in \mathbb{N}$, we write $T^n = T \circ \overset{(n)}{\cdots} \circ T$ and $T^0 = I$.

**Proposition 7.1.3** *Let $E$ be a Banach space and let $T \in \mathcal{L}(E, E)$ be such that $\|T\| < 1$. Then $1 \in \rho(T)$ and, in particular, $(I - T)$ has a continuous inverse and*

$$(I - T)^{-1} = \sum_{j=0}^{\infty} T^j \in \mathcal{L}(E, E).$$

## 7.1 Spectrum of a Continuous Linear Operator

**Proof** As $\|T^j\| \leq \|T\|^j$ for every $j$ and $\sum_{j=0}^{\infty} \|T\|^j < \infty$ because $\|T\| < 1$, it follows that $\sum_{j=0}^{\infty} \|T^j\| < \infty$. Given natural numbers $m > n$, from

$$\left\| \sum_{j=0}^{m} T^j - \sum_{j=0}^{n} T^j \right\| = \left\| \sum_{j=n+1}^{m} T^j \right\| \leq \sum_{j=n+1}^{m} \|T^j\| \longrightarrow 0 \text{ if } m, n \longrightarrow \infty,$$

we conclude that the sequence $\left( \sum_{j=0}^{n} T_j \right)_{n=1}^{\infty}$ is Cauchy in the Banach space $\mathcal{L}(E, E)$, hence convergent. This means that the series $\sum_{j=0}^{\infty} T^j$ converges, say $S = \sum_{j=0}^{\infty} T^j \in \mathcal{L}(E, E)$. A simple calculation reveals that

$$(I - T) \circ (I + T + \cdots + T^n) = I - T^{n+1} = (I + T + \cdots + T^n) \circ (I - T)$$

for every $n \in \mathbb{N}$. Letting $n \longrightarrow \infty$, as $T^{n+1} \longrightarrow 0$ since $\|T\| < 1$, we get

$$(I - T) \circ S = I = S \circ (I - T).$$

This shows that $(I - T)$ is invertible and that $(I - T)^{-1} = S$, proving the result. ∎

**Theorem 7.1.4** *Let $E$ be a Banach space and let $T \in \mathcal{L}(E, E)$ be given. Then the spectrum of $T$ is a compact set contained in the disk (complex case) or interval (real case) $\{\lambda \in \mathbb{K} : |\lambda| \leq \|T\|\}$.*

**Proof** Let $\lambda \in \mathbb{K}$ with $|\lambda| > \|T\|$. Then $\|\frac{1}{\lambda}T\| < 1$, therefore from Proposition 7.1.3 we have $1 \in \rho\left(\frac{1}{\lambda}T\right)$. It follows that $\lambda \in \rho(T)$, more precisely,

$$(T - \lambda I)^{-1} = -\frac{1}{\lambda}\left(I - \frac{1}{\lambda}T\right)^{-1} = -\frac{1}{\lambda}\left(\sum_{j=0}^{\infty} \left(\frac{1}{\lambda}T\right)^j\right) \in \mathcal{L}(E, E).$$

We have proved that $\{\lambda \in \mathbb{K} : |\lambda| > \|T\|\} \subseteq \rho(T)$, hence $\sigma(T) \subseteq \{\lambda \in \mathbb{K} : |\lambda| \leq \|T\|\}$. In particular the resolvent set is non-empty and the spectrum is bounded.

To complete the proof, it is enough to prove that $\sigma(T)$ is closed. To do so, we will show that $\rho(T)$ is open. We already know that $\rho(T) \neq \emptyset$. Let $\lambda_0 \in \rho(T)$. Then $(T - \lambda_0 I)^{-1} \in \mathcal{L}(E, E)$, and in particular $\|(T - \lambda_0 I)^{-1}\| > 0$. Let $\lambda \in \mathbb{K}$ with $|\lambda - \lambda_0| < \frac{1}{\|(T-\lambda_0 I)^{-1}\|}$. From

$$T - \lambda I = T - \lambda_0 I - (\lambda - \lambda_0)I = (T - \lambda_0 I) \circ \left(I - (\lambda - \lambda_0)(T - \lambda_0 I)^{-1}\right),$$

calling $U = I - (\lambda - \lambda_0)(T - \lambda_0 I)^{-1}$ we have $T - \lambda I = (T - \lambda_0 I) \circ U$. As

$$\|(\lambda - \lambda_0)(T - \lambda_0 I)^{-1}\| = |\lambda - \lambda_0| \cdot \|(T - \lambda_0 I)^{-1}\| < 1,$$

Proposition 7.1.3 guarantees that $U$ has a continuous inverse defined on $E$. Thus

$$U^{-1} \circ (T - \lambda_0 I)^{-1} = ((T - \lambda_0 I) \circ U)^{-1} = (T - \lambda I)^{-1}$$

exists and is continuous. We can conclude that $\lambda \in \rho(T)$. It follows that $\left\{\lambda \in \mathbb{K} : |\lambda - \lambda_0| < \frac{1}{\|(T - \lambda_0 I)^{-1}\|}\right\}$ is a neighborhood of $\lambda_0$ entirely contained in $\rho(T)$, proving that $\rho(T)$ is open. ∎

## 7.2 Compact Operators

Let $E$ be an infinite-dimensional normed space and let $T \colon E \longrightarrow F$ be a continuous linear operator. We know that $B_E$ is not compact in $E$, so if $T(B_E)$ is compact in $F$, the operator $T$ will have the great advantage of 'fixing' the non-compactness of the unit ball. Although always bounded, $T(B_E)$ may not be closed, so it is reasonable to only ask that its closure be compact:

**Definition 7.2.1** A linear operator $T \colon E \longrightarrow F$ between normed spaces is said to be *compact* if $\overline{T(B_E)}$ is compact in $F$.

The next result contains some examples and counterexamples.

**Proposition 7.2.2**

(a) *Every compact operator is continuous.*
(b) *Every continuous linear operator of finite rank is compact.*
(c) *A normed space $E$ is finite-dimensional if and only if the identity on $E$ is a compact operator.*

*Proof*

(a) Let $T \colon E \longrightarrow F$ be a compact operator. As it is a subset of the compact $\overline{T(B_E)}$, the set $T(B_E)$ is bounded, therefore $T$ is continuous by Proposition 2.1.1(f).
(b) Closed and bounded sets in finite-dimensional spaces are compact (Proposition 1.5.1).
(c) Follows from Theorem 1.5.4. ∎

The following characterizations are simple but very useful.

## 7.2 Compact Operators

**Proposition 7.2.3** *Let E and F be normed spaces. The following statements are equivalent for a linear operator $T: E \longrightarrow F$.*

(a) *T is compact.*
(b) $\overline{T(A)}$ *is compact in F for every bounded A in E.*
(c) *For every bounded sequence $(x_n)_{n=1}^{\infty}$ in E, the sequence $(T(x_n))_{n=1}^{\infty}$ has a convergent subsequence in F.*

**Proof**

(a) $\Longrightarrow$ (b) Let $A$ be bounded in $E$, say $\|x\| \leq C$, with $C > 0$ for every $x \in A$. Then $\frac{1}{C}A \subseteq B_E$, therefore $\frac{1}{C}\overline{T(A)} = \overline{\frac{1}{C}T(A)} \subseteq \overline{T(B_E)}$. It follows that $\frac{1}{C}\overline{T(A)}$ is compact as it is a closed set within a compact set. The compactness of $\overline{T(A)}$ is now immediate.

(b) $\Longrightarrow$ (c) This follows from the fact that in metric spaces, in particular in normed spaces, every sequence in a compact set has a convergent subsequence.

(c) $\Longrightarrow$ (a) Let $(y_n)_{n=1}^{\infty}$ be a sequence in $\overline{T(B_E)}$. For each $n$ we can find $x_n \in B_E$ such that

$$\|T(x_n) - y_n\| < \frac{1}{n}.$$

By assumption $(T(x_n))_{n=1}^{\infty}$ has a convergent subsequence, say $T(x_{n_j}) \longrightarrow y \in F$. From

$$\|y_{n_j} - y\| \leq \|y_{n_j} - T(x_{n_j})\| + \|T(x_{n_j}) - y\| \leq \frac{1}{n_j} + \|T(x_{n_j}) - y\| \overset{j \to \infty}{\longrightarrow} 0,$$

it follows that $y_{n_j} \longrightarrow y$. This proves that $\overline{T(B_E)}$ is compact. ∎

The following example is widely used in applications.

**Example 7.2.4 (Integral Operators)** Let $K: [a,b] \times [c,d] \longrightarrow \mathbb{C}$ be a continuous function. Consider the operator

$$T: C[a,b] \longrightarrow C[c,d], \quad T(f)(t) = \int_a^b K(s,t) f(s)\, ds \text{ for every } t \in [c,d].$$

We leave to the reader the task of proving that $T$ is well defined, that is, $T(f) \in C[c,d]$ for every $f \in C[a,b]$, and that $T$ is linear. We want to show that the operator $T$ is compact, that is, $\overline{T(B_{C[a,b]})}$ is compact in $C[c,d]$. We shall use the Ascoli Theorem (Theorem B.7 of Appendix B). We must prove that $\overline{T(B_{C[a,b]})}$ satisfies conditions (a) and (b) of Theorem B.7.

(a) Let $t_0 \in [c,d]$ and $\varepsilon > 0$ be given. Since $[a,b] \times [c,d]$ is compact and $K$ is continuous, we conclude that $K$ is uniformly continuous. So, there exists $\delta > 0$ such that

$$|K(s, t_1) - K(s, t_2)| < \frac{\varepsilon}{b-a}$$

for all $s \in [a, b]$ and $t_1, t_2 \in [c, d]$ with $|t_1 - t_2| < \delta$. Thus, if $f \in B_{C[a,b]}$ and $|t - t_0| < \delta$, then

$$|T(f)(t) - T(f)(t_0)| = \left| \int_a^b K(s, t) f(s) - K(s, t_0) f(s) \, ds \right|$$

$$\leq \int_a^b |K(s, t) - K(s, t_0)| \cdot |f(s)| \, ds$$

$$\leq \int_a^b |K(s, t) - K(s, t_0)| \, ds \leq \frac{\varepsilon}{b-a}(b-a) = \varepsilon.$$

This proves that $T(B_{C[a,b]})$ is equicontinuous in $C[c, d]$.

(b) Let $t_0 \in [c, d]$ and $f \in C[a, b]$, $\|f\|_\infty \leq 1$ be given. Then

$$|T(f)(t_0)| \leq \int_a^b |K(s, t_0)| \cdot |f(s)| \, ds \leq \int_a^b |K(s, t_0)| \, ds.$$

It follows that the operator $T$ is compact. The function $K$ is called the *kernel* of the integral operator $T$.

We denote by $\mathcal{K}(E, F)$ the set of compact operators from $E$ to $F$.

**Proposition 7.2.5** *Let $E$ and $F$ be normed spaces. Then $\mathcal{K}(E, F)$ is a vector subspace of $\mathcal{L}(E, F)$. If $F$ is a Banach space then $\mathcal{K}(E, F)$ is closed in $\mathcal{L}(E, F)$.*

**Proof** We leave to the reader to prove that $\mathcal{K}(E, F)$ is a vector subspace of $\mathcal{L}(E, F)$. Suppose $F$ is complete. We shall prove that $\mathcal{K}(E, F)$ is closed in $\mathcal{L}(E, F)$. To do so, let $(T_n)_{n=1}^\infty$ be a sequence of compact operators such that $T_n \longrightarrow T$ in $\mathcal{L}(E, F)$. The proof will be complete when we prove that $T$ is compact. Let $(x_m)_{m=1}^\infty$ be a bounded sequence in $E$, say $\|x_m\| \leq C$ for every $m$ and some $C > 0$. By Theorem 7.2.3 it is enough to show that $(T(x_m))_{m=1}^\infty$ has a convergent subsequence in $F$. As $T_1$ is compact, $(x_m)_{m=1}^\infty$ has a subsequence $(x_{1,m})_{m=1}^\infty$ such that $(T_1(x_{1,m}))_{m=1}^\infty$ is Cauchy in $F$. For the same reasons, the sequence $(x_{1,m})_{m=1}^\infty$ has a subsequence $(x_{2,m})_{m=1}^\infty$ such that $(T_2(x_{2,m}))_{m=1}^\infty$ is Cauchy in $F$. In each step of this process, the constructed sequence is obviously bounded and the next operator is compact, so the process can be repeated indefinitely. We can thus consider the diagonal sequence

$$(y_m)_{m=1}^\infty = (x_{1,1}, x_{2,2}, x_{3,3}, \ldots),$$

which is a subsequence of $(x_m)_{m=1}^\infty$ and satisfies the condition that $(T_n(y_m))_{m=1}^\infty$ is a Cauchy sequence in $F$ for every natural $n$. We have, in particular, that $\|y_m\| \leq C$. Let $\varepsilon > 0$. As $T_n \longrightarrow T$, there is $n_0 \in \mathbb{N}$ such that

## 7.2 Compact Operators

$$\|T - T_{n_0}\| < \frac{\varepsilon}{3C}.$$

Since $(T_{n_0}(y_m))_{m=1}^\infty$ is Cauchy, there exists $n_1 \in \mathbb{N}$ such that

$$\|T_{n_0}(y_j) - T_{n_0}(y_k)\| < \frac{\varepsilon}{3} \text{ for all } j, k > n_1.$$

Finally,

$$\|T(y_j) - T(y_k)\| \leq \|T(y_j) - T_{n_0}(y_j)\| + \|T_{n_0}(y_j) - T_{n_0}(y_k)\|$$
$$+ \|T_{n_0}(y_k) - T(y_k)\|$$
$$\leq \|T - T_{n_0}\| \cdot \|y_j\| + \frac{\varepsilon}{3} + \|T_{n_0} - T\| \cdot \|y_k\|$$
$$< \frac{\varepsilon}{3C} C + \frac{\varepsilon}{3} + \frac{\varepsilon}{3C} C = \varepsilon$$

for all $j, k > n_1$. This shows that $(T(y_m))_{m=1}^\infty$ is a Cauchy sequence, therefore convergent in $F$. ∎

**Proposition 7.2.6 (Ideal Property)** *Let $E_0, E, F, F_0$ be normed spaces and let $S: E_0 \longrightarrow E$, $T: E \longrightarrow F$ and $U: F \longrightarrow F_0$ be linear operators with $S$ and $U$ continuous and $T$ compact. Then $U \circ T \circ S$ is compact.*

**Proof** We will prove the compactness of $U \circ T \circ S$ using condition (c) of Proposition 7.2.3. Let $(x_n)_{n=1}^\infty$ be a bounded sequence in $E_0$. As continuous linear operators transform bounded sets into bounded sets, the sequence $(S(x_n))_{n=1}^\infty$ is bounded in $E$. From the compactness of $T$, it follows that $(T(S(x_n)))_{n=1}^\infty$ has a convergent subsequence in $F$, say $T(S(x_{n_k})) \longrightarrow y \in F$. The continuity of $U$ reveals that $U(T(S(x_{n_k}))) \longrightarrow U(y) \in F_0$, proving the compactness of $U \circ T \circ S$. ∎

**Theorem 7.2.7 (Schauder's Theorem)** *Let $E$ and $F$ be Banach spaces and let $T \in \mathcal{L}(E, F)$ be given. Then $T$ is compact if and only if $T': F' \longrightarrow E'$ is compact.*

**Proof** Suppose $T$ is compact. Given a sequence $(\varphi_n)_{n=1}^\infty$ in $B_{F'}$ we must prove that the sequence $(T'(\varphi_n))_{n=1}^\infty$ has a convergent subsequence in $E'$. Calling $K = \overline{T(B_E)}$ we have from the compactness of $T$ that $K$ is a compact metric space. For each $n$ call $f_n$ the restriction of $\varphi_n$ to $K$, $f_n = \varphi_n|_K \in C(K)$. Then $A := \{f_n : n \in \mathbb{N}\}$ is a subset of $C(K)$. We shall prove that conditions (a) and (b) of Ascoli's Theorem (Theorem B.7 of Appendix B) are satisfied:

(a) Given $t_0 \in K$ and $\varepsilon > 0$, take $\delta = \varepsilon$. If $t \in K$ is such that $\|t - t_0\| < \delta$, then

$$\sup_n |f_n(t) - f_n(t_0)| = \sup_n |\varphi_n(t - t_0)| \leq \sup_n \|\varphi_n\| \cdot \|t - t_0\| \leq \|t - t_0\| < \delta = \varepsilon.$$

(b) For every $t \in K$,

$$\sup_n |f_n(t)| = \sup_n |\varphi_n(t)| \leq \sup_n \|\varphi_n\| \cdot \|t\| \leq \|t\|.$$

From Ascoli's Theorem it follows that $\overline{A}$ is compact in $C(K)$, therefore there exist a subsequence $(f_{n_k})_{k=1}^\infty$ of $(f_n)_{n=1}^\infty$ and $f \in C(K)$ such that $\varphi_{n_k}|_K = f_{n_k} \longrightarrow f$ in $C(K)$. This convergence means that

$$\sup_{x \in B_E} |T'(\varphi_{n_k})(x) - f(T(x))| = \sup_{x \in B_E} |\varphi_{n_k}(T(x)) - f(T(x))|$$
$$\leq \sup_{t \in \overline{T(B_E)}} |\varphi_{n_k}(t) - f(t)|$$
$$= \|f_{n_k} - f\|_\infty \longrightarrow 0 \text{ as } k \longrightarrow \infty.$$

From the triangle inequality we conclude that

$$\|T'(\varphi_{n_k}) - T'(\varphi_{n_j})\| = \sup_{x \in B_E} |T'(\varphi_{n_k})(x) - T'(\varphi_{n_j})(x)| \longrightarrow 0 \text{ if } k, j \longrightarrow \infty,$$

proving that the subsequence $(T'(\varphi_{n_k}))_{k=1}^\infty$ is Cauchy in the Banach space $E'$, hence convergent.

Conversely, suppose $T'$ is compact. From what we have just proved, we know that $T'' \colon E'' \longrightarrow F''$ is compact. By Exercise 4.5.20, $T''(J_E(E)) \subseteq J_F(F)$, therefore we can consider the restriction of $T''$ to $J_E(E)$ taking values in $J_F(F)$, which we will call $T_0 \colon J_E(E) \subseteq E'' \longrightarrow J_F(F) \subseteq F''$. As $T''$ is compact and $J_F(F)$ is closed in $F''$ since $F$ is Banach, it follows that $T_0$ is compact. Considering $J_F^{-1} \colon J_F(F) \longrightarrow F$, we have $T = J_F^{-1} \circ T_0 \circ J_E$, hence $T$ is compact by Proposition 7.2.6.

■

We finish this section by characterizing compact operators defined on reflexive spaces as exactly those that transform weakly convergent sequences into convergent sequences.

**Proposition 7.2.8** *Let $E$ and $F$ be normed spaces and let $T \in \mathcal{L}(E, F)$ be given.*

(a) *If $T$ is compact, then the implication*

$$x_n \xrightarrow{w} x \text{ in } E \implies T(x_n) \longrightarrow T(x) \text{ in } F \tag{7.1}$$

*holds.*

(b) *If $E$ is reflexive, then $T$ is compact if and only if (7.1) holds.*

**Proof**

(a) From the assumption $x_n \xrightarrow{w} x$ and the implication of Proposition 6.2.9 that does not require completeness, it follows that $T(x_n) \xrightarrow{w} T(x)$. Suppose that $(T(x_n))_{n=1}^\infty$ does not converge in norm to $T(x)$. In this case there exist $\varepsilon > 0$ and a subsequence $(T(x_{n_k}))_{k=1}^\infty$ such that

$$\|T(x_{n_k}) - T(x)\| \geq \varepsilon \text{ for every } k. \tag{7.2}$$

From $x_{n_k} \xrightarrow{w} x$ it follows that $(x_{n_k})_{k=1}^{\infty}$ is bounded, therefore, by the compactness of $T$, $(T(x_{n_k}))_{k=1}^{\infty}$ admits a convergent subsequence $(T(x_{n_{k_j}}))_{j=1}^{\infty}$, say

$$T(x_{n_{k_j}}) \longrightarrow y \in F.$$

A fortiori $T(x_{n_{k_j}}) \xrightarrow{w} y$ and, as the weak topology is Hausdorff, we conclude that $T(x) = y$. Then $T(x_{n_{k_j}}) \longrightarrow T(x)$, which contradicts (7.2).

(b) Suppose $E$ is reflexive and that (7.1) holds. Given a bounded sequence $(x_n)_{n=1}^{\infty}$ in $E$, by Theorem 6.5.4, we can extract a weakly convergent subsequence, say $x_{n_k} \xrightarrow{w} x \in E$. By hypothesis $T(x_{n_k}) \longrightarrow T(x)$, therefore $T$ is compact by Proposition 7.2.3. ∎

## 7.3 Spectral Theory of Compact Operators

The aim of this section is to describe the spectrum of a compact linear operator. We start by showing that the eigenspaces associated with the non-zero eigenvalues of a compact operator are all finite-dimensional.

**Proposition 7.3.1** *Let $E$ be a Banach space, let $T : E \longrightarrow E$ be a compact operator and let $\lambda$ be a non-zero scalar. Then:*

(a) $V_\lambda := \ker(T - \lambda I)$ *is finite-dimensional.*
(b) $(T - \lambda I)(E)$ *is closed in $E$.*

*Proof*

(a) By Theorem 1.5.4 it is enough to show that the closed unit ball $B_{V_\lambda}$ is compact. To do so, let $(x_n)_{n=1}^{\infty}$ be a sequence in $B_{V_\lambda}$. From the compactness of $T$ we can extract from $(T(x_n))_{n=1}^{\infty}$ a convergent subsequence, say $T(x_{n_k}) \longrightarrow x \in E$. As each $x_n \in \ker(T - \lambda I)$, $T(x_n) = \lambda x_n$, therefore

$$x_{n_k} = \frac{1}{\lambda} T(x_{n_k}) \longrightarrow \frac{1}{\lambda} x.$$

Since $B_{V_\lambda}$ is closed, it is clear that $\frac{1}{\lambda} x \in B_{V_\lambda}$. It follows that the ball $B_{V_\lambda}$ is compact.

(b) By item (a) we know that $V_\lambda$ is finite-dimensional, so it is complemented in $E$ (see Example 3.2.4). From Proposition 3.2.2 there exists a closed subspace $M$ of $E$ such that $E = V_\lambda \oplus M$. Consider the continuous linear operator

$$S: M \longrightarrow E, \quad S(x) = T(x) - \lambda x.$$

As $M \cap \ker(T - \lambda I) = M \cap V_\lambda = \{0\}$, it follows that $S$ is injective. The operator $(T - \lambda I)$ is null on $V_\lambda$, so $S(M) = (T - \lambda I)(E)$, therefore it is enough to prove that $S(M)$ is closed. We shall prove that there exists $r > 0$ such that

$$r\|x\| \leq \|S(x)\| \quad \text{for every } x \in M. \tag{7.3}$$

Suppose that (7.3) is not valid. In this case, for each $n \in \mathbb{N}$ consider $0 \neq z_n \in M$ such that $\|S(z_n)\| < \frac{1}{n}\|z_n\|$. Calling $x_n = \frac{z_n}{\|z_n\|}$ we have $\|S(x_n)\| < \frac{1}{n}$ for every $n$. So, $\|x_n\| = 1$ for every $n$ and $S(x_n) \longrightarrow 0$. From the compactness of $T$ we can extract from $(T(x_n))_{n=1}^\infty$ a convergent subsequence, say $T(x_{n_k}) \longrightarrow x_0 \in E$. Then

$$\lambda x_{n_k} = T(x_{n_k}) - S(x_{n_k}) \longrightarrow x_0. \tag{7.4}$$

As $M$ is a closed subspace, we conclude that $x_0 \in M$. Furthermore,

$$S(x_0) = \lim_k S(\lambda x_{n_k}) = \lambda \lim_k S(x_{n_k}) = 0.$$

From the injectivity of $S$ we get $x_0 = 0$. On the other hand, (7.4) provides

$$\|x_0\| = \lim_k \|\lambda x_{n_k}\| = |\lambda| > 0,$$

because $\|x_n\| = 1$ for every $n$. This contradiction proves the existence of $r > 0$ satisfying (7.3). Let now $(S(y_n))_{n=1}^\infty$ be a Cauchy sequence in $S(M)$. From

$$\|y_m - y_n\| \leq \frac{1}{r}\|S(y_m - y_n)\| = \frac{1}{r}\|S(y_m) - S(y_n)\|,$$

we conclude that the sequence $(y_n)_{n=1}^\infty$ is a Cauchy sequence in $M$, therefore convergent, say $y_n \longrightarrow y \in M$. From the continuity of $S$ it follows that $S(y_n) \longrightarrow S(y) \in S(M)$. This proves that $S(M)$ is complete, therefore closed. ∎

**Remark 7.3.2** One should not expect Proposition 7.3.1(a) to work for $\lambda = 0$. An extreme example is the identically null operator on an infinite-dimensional space, which is compact and for which $V_0$ is infinite-dimensional because it coincides with the whole space.

The following elementary lemma will be useful for the proof of Lemma 7.3.4, which is the main ingredient for the proof of Fredholm's Alternative:

**Lemma 7.3.3** *Let $A$ be any set and let $f: A \longrightarrow A$ be an injective function that is not surjective. Then $f(f(A))$ is a proper subset of $f(A)$.*

## 7.3 Spectral Theory of Compact Operators

**Proof** It is clear that $f(f(A)) \subseteq f(A)$. Since $f$ is not surjective, let $x \in A - f(A)$. It is clear that $f(x) \in f(A)$. Suppose that $f(x) \in f(f(A))$. In this case there exists $y \in A$ such that $f(x) = f(f(y))$. Since $f$ is injective it follows that $x = f(y) \in f(A)$. This contradiction shows that $f(x) \notin f(f(A))$, therefore $f(f(A))$ is a proper subset of $f(A)$. ∎

**Lemma 7.3.4** *Let $E$ be a Banach space, let $T: E \longrightarrow E$ be a compact operator and let $\lambda \in \mathbb{K}$, $\lambda \neq 0$. Then the operator $(T - \lambda I)$ is injective if and only if it is surjective.*

**Proof** If $E$ is finite-dimensional, the result is immediate. Now we deal with the case where $E$ is infinite-dimensional. Suppose that $(T - \lambda I)$ is injective but not surjective. In this case $E_1 := (T - \lambda I)(E)$ is a proper subspace of $E$. From Proposition 7.3.1(b) we know that $E_1$ is a Banach space. Moreover, given $x \in E_1$ there exists $y \in E$ such that $x = T(y) - \lambda y$. Calling $z = T(y) \in E$,

$$T(x) = T(T(y)) - \lambda T(y) = T(z) - \lambda z \in (T - \lambda I)(E) = E_1,$$

proving that $T(E_1) \subseteq E_1$. It is clear that the restriction of a compact operator is a compact operator, therefore the restriction of $T$ to $E_1$,

$$T|_{E_1}: E_1 \longrightarrow E_1,$$

is a compact operator defined on the Banach space $E_1$. Then $E_2 := (T - \lambda I)(E_1) = (T - \lambda I)^2(E)$ is a closed subspace of $E_1$ by Proposition 7.3.1(b). As $(T - \lambda I)$ is injective, by Lemma 7.3.3 we have $E_2 \neq E_1$. Proceeding in this way we construct a strictly decreasing sequence $(E_n)_{n=1}^\infty$ of closed subspaces of $E$ where $E_n = (T - \lambda I)^n(E)$ for every $n$. By the Riesz Lemma (Lemma 1.5.3) we can construct a sequence $(x_n)_{n=1}^\infty$ such that each $x_n \in E_n$, $\|x_n\| = 1$ and

$$\operatorname{dist}(x_n, E_{n+1}) \geq \frac{1}{2}.$$

On the one hand, the sequence $(T(x_n))_{n=1}^\infty$ admits a convergent subsequence because $T$ is compact. On the other hand, for $n > m$ we have $(T(x_n) - \lambda x_n) \in E_{n+1} \subseteq E_{m+1}$, $(T(x_m) - \lambda x_m) \in E_{m+1}$ and $\lambda x_n \in E_n \subseteq E_{m+1}$, therefore

$$\frac{1}{\lambda}[(T(x_n) - \lambda x_n) - (T(x_m) - \lambda x_m) + \lambda x_n] \in E_{m+1}.$$

It follows that

$$\|T(x_n) - T(x_m)\| = \|(T(x_n) - \lambda x_n) - (T(x_m) - \lambda x_m) + \lambda x_n - \lambda x_m\|$$

$$= \left\|\lambda\left(\frac{1}{\lambda}[(T(x_n) - \lambda x_n) - (T(x_m) - \lambda x_m) + \lambda x_n] - x_m\right)\right\|$$

$$= |\lambda| \cdot \left\| \frac{1}{\lambda} [(T(x_n) - \lambda x_n) - (T(x_m) - \lambda x_m) + \lambda x_n] - x_m \right\|$$

$$\geq |\lambda| \cdot \operatorname{dist}(x_m, E_{m+1}) \geq \frac{|\lambda|}{2}$$

for all $n > m$. This clearly contradicts the fact that $(T(x_n))_{n=1}^{\infty}$ has a convergent subsequence and completes the proof that the injectivity of $(T - \lambda I)$ implies its surjectivity.

Conversely suppose $(T - \lambda I)$ is surjective, that is $(T - \lambda I)(E) = E$. As the adjoint of the identity on $E$ is the identity on $E'$,

$$\ker(T' - \lambda I) = \ker((T - \lambda I)') = ((T - \lambda I)(E))^{\perp} = E^{\perp} = \{0\}$$

(see Exercise 4.5.24). We conclude that $(T' - \lambda I)$ is injective. But $T'$ is compact by Theorem 7.2.7; then the first part of the proof shows that $(T - \lambda I)' = (T' - \lambda I)$ is surjective. Using again Exercise 4.5.24,

$$\ker(T - \lambda I) = {}^{\perp}((T - \lambda I)')(E') = {}^{\perp}(E') = \{0\},$$

which proves that $(T - \lambda I)$ is injective. ∎

Now we apply Lemma 7.3.4 for the scalar $\lambda = 1$. The condition $(T - I)$ being injective means that $T$ does not have a non-zero fixed point, and the condition $(T - I)$ being surjective means that the equation $T(x) - x = y$ has a solution for every $y \in E$. We then have the

**Theorem 7.3.5 (Fredholm Alternative)** *Let $E$ be a Banach space. If $T: E \longrightarrow E$ is a compact operator, then one and only one of the possibilities below occurs:*

(a) *$T$ has a non-zero fixed point.*
(b) *The equation $T(x) - x = y$ has a solution for every $y \in E$.*

We will now use Lemma 7.3.4 to prove that, with the possible exception of the origin, the spectrum of a compact operator is made up only of eigenvalues.

**Theorem 7.3.6** *Let $E$ be an infinite-dimensional Banach space and let $T: E \longrightarrow E$ be a compact operator. Then*

$$\sigma(T) = \{0\} \bigcup \{\lambda \in \mathbb{K} : \lambda \text{ is an eigenvalue of } T\}.$$

*Proof* It is clear that $0 \in \sigma(T)$ because, if $0 \in \rho(T)$, then $T^{-1} \in \mathcal{L}(E, E)$, which implies that $I = T \circ T^{-1}$ is compact by Proposition 7.2.6, which is a contradiction by Proposition 7.2.2(c). The fact that every eigenvalue belongs to the spectrum follows from the definitions. Conversely, suppose $\lambda \in \sigma(T)$, $\lambda \neq 0$. Then the operator $(T - \lambda I)$ is not injective or not surjective. But, if it is injective, by Lemma 7.3.4 it will also be surjective. It remains only the possibility that $(T - \lambda I)$ is not injective, and this exactly means that $\lambda$ is an eigenvalue of $T$. ∎

## 7.3 Spectral Theory of Compact Operators

The last result of this section essentially informs that the spectrum of a compact operator behaves like a sequence whose only convergent subsequences, if they exist, converge to zero.

**Theorem 7.3.7** *The spectrum of a compact operator $T: E \longrightarrow E$ on a Banach space $E$ is countable or finite, and the only possible accumulation point is zero.*

*Proof* In view of Theorem 7.3.6 it is enough to show that for every $\varepsilon > 0$, the set of eigenvalues $\lambda$ such that $|\lambda| \geq \varepsilon$ is finite. Suppose this does not happen. In this case there exists $\varepsilon_0 > 0$ such that the set of eigenvalues $\lambda$ such that $|\lambda| \geq \varepsilon_0$ is infinite. We can then consider a sequence $(\lambda_n)_{n=1}^{\infty}$ formed by distinct eigenvalues of $T$ such that $|\lambda_n| \geq \varepsilon_0$ for every $n$. Therefore, for each $n$ there exists $x_n \neq 0$ such that $T(x_n) = \lambda_n x_n$. From the Linear Algebra course the reader knows that eigenvectors associated with distinct eigenvalues are linearly independent. Therefore the set $\{x_1, x_2, \ldots\}$ is linearly independent. For each $n$ consider the $n$-dimensional space $M_n = [x_1, \ldots, x_n]$. Then each $z \in M_n$ has a unique representation in the form

$$z = \sum_{j=1}^{n} a_j x_j, \text{ where } a_1, \ldots, a_n \in \mathbb{K}.$$

Thus

$$(T - \lambda_n I)(z) = \sum_{j=1}^{n} a_j T(x_j) - \sum_{j=1}^{n} \lambda_n a_j x_j = \sum_{j=1}^{n} a_j \lambda_j x_j - \sum_{j=1}^{n} \lambda_n a_j x_j$$

$$= \sum_{j=1}^{n-1} a_j (\lambda_j - \lambda_n) x_j,$$

and consequently

$$(T - \lambda_n I)(z) \in M_{n-1} \text{ for every } z \in M_n. \tag{7.5}$$

Each $M_n$ is obviously closed; therefore using the Riesz Lemma (Lemma 1.5.3) we construct a sequence $(y_n)_{n=1}^{\infty}$ such that for every $n$,

$$y_n \in M_n, \ \|y_n\| = 1 \text{ and } \operatorname{dist}(y_n, M_{n-1}) \geq \frac{1}{2}.$$

In a very similar way to what we did in the proof of Proposition 7.3.4 we will show that

$$\|T(y_n) - T(y_m)\| \geq \frac{\varepsilon_0}{2} \text{ for all } n > m,$$

which will be in clear contradiction with the compactness of $T$. Start by observing that $T(M_n) \subseteq M_n$ because each $x_j$ is an eigenvalue of $T$. Let $m, n$ be positive integers with $n > m$. As $y_m \in M_m \subseteq M_{n-1}$,

$$T(y_m) \in T(M_{n-1}) \subseteq M_{n-1}.$$

As $y_n \in M_n$, from (7.5) it follows that

$$\lambda_n y_n - T(y_n) = -(T - \lambda_n I)(y_n) \in M_{n-1}.$$

Therefore $(\lambda_n y_n - T(y_n)) + T(y_m) \in M_{n-1}$. It then follows that

$$\|T(y_m) - T(y_n)\| = \|\lambda_n y_n - T(y_n) + T(y_m) - \lambda_n y_n\|$$
$$= \left\| \lambda_n \left( \frac{1}{\lambda_n}[(\lambda_n y_n - T(y_n)) + T(y_m)] - y_n \right) \right\|$$
$$= |\lambda_n| \cdot \left\| \frac{1}{\lambda_n}[(\lambda_n y_n - T(y_n)) + T(y_m)] - y_n \right\|$$
$$\geq |\lambda_n| \cdot \operatorname{dist}(y_n, M_{n-1}) \geq \frac{\varepsilon_0}{2},$$

for all $n > m$. The proof is complete. ∎

## 7.4 Self-adjoint Operators on Hilbert Spaces

Let $H$ be a Hilbert space. We call $u_H : H \longrightarrow H'$ the map given by

$$u_H(x)(y) = \langle y, x \rangle \text{ for all } x, y \in H.$$

From the Riesz–Fréchet Theorem (Theorem 5.5.2) we know that $u_H$ is bijective and satisfies the condition $\|u_H(x)\| = \|x\|$ for every $x \in H$. In particular, $u_H$ is continuous. In the real case $u_H$ is an isometric isomorphism (Corollary 5.5.3), but in the complex case $u_H$ is not linear because

$$u_H(\lambda x) = \overline{\lambda} \cdot u_H(x). \tag{7.6}$$

If $H_1$ and $H_2$ are real Hilbert spaces and $T \in \mathcal{L}(H_1, H_2)$, we know that the adjoint operator $T' : H_2' \longrightarrow H_1'$ is linear continuous and has the same norm as $T$ (Proposition 4.3.11). We are about to see that the chain

$$H_2 \xrightarrow{u_{H_2}} H_2' \xrightarrow{T'} H_1' \xrightarrow{u_{H_1}^{-1}} H_1$$

## 7.4 Self-adjoint Operators on Hilbert Spaces

shows how to recognize the adjoint of $T$ as an operator from $H_2$ to $H_1$:

**Proposition 7.4.1** *Let $H_1$ and $H_2$ be Hilbert spaces and let $T \in \mathcal{L}(H_1, H_2)$ be given. Defining $T^* := u_{H_1}^{-1} \circ T' \circ u_{H_2}$ it is true that:*

(a) $T^* \in \mathcal{L}(H_2, H_1)$.
(b) $\langle T(x), y \rangle = \langle x, T^*(y) \rangle$ *for all $x \in H_1$ and $y \in H_2$.*
(c) $T^*$ *is the only operator in $\mathcal{L}(H_2, H_1)$ satisfying (b).*
(d) *The correspondence $T \in \mathcal{L}(H_1, H_2) \mapsto T^* \in \mathcal{L}(H_2, H_1)$ is bijective and $\|T\| = \|T^*\|$. In the real case this correspondence is an isometric isomorphism.*

**Proof** For the linearity of $T^*$ it remains only to prove the homogeneity, that is, $T^*(\lambda x) = \lambda T^*(x)$. As $\overline{\overline{\lambda}} = \lambda$, the homogeneity of $T^*$ follows from the linearity of $T'$ and the application of (7.6) to $u_{H_1}^{-1}$ and to $u_{H_2}$. The operator $T^*$ is continuous as it is the composition of continuous operators. Given $x \in H_1$ and $y \in H_2$,

$$\langle x, T^*(y) \rangle = u_{H_1}(T^*(y))(x) = [(u_{H_1} \circ T^*)(y)](x) = [(T' \circ u_{H_2})(y)](x)$$
$$= T'(u_{H_2}(y))(x) = u_{H_2}(y)(T(x)) = \langle T(x), y \rangle.$$

On the one hand,

$$\|T^*(y)\| = \|u_{H_1}^{-1} \circ T' \circ u_{H_2}(y)\| = \|u_{H_1}^{-1}(T' \circ u_{H_2}(y))\| = \|T' \circ u_{H_2}(y)\|$$
$$\leq \|T'\| \cdot \|u_{H_2}(y)\| = \|T\| \cdot \|y\|,$$

proving that $\|T^*\| \leq \|T\|$. On the other hand, proceeding analogously for the equality $T' = u_{H_1} \circ T^* \circ u_{H_2}^{-1}$ we obtain $\|T\| = \|T'\| \leq \|T^*\|$.

We leave the remaining information as an exercise. ∎

**Definition 7.4.2** *Let $H$ be a Hilbert space. An operator $T \in \mathcal{L}(H, H)$ is called self-adjoint if $T = T^*$.*

**Example 7.4.3**

(a) It is clear that the null operator and the identity operator on a Hilbert space are self-adjoint.
(b) What we proved in Example 4.3.10 shows that the backward shift operator

$$T : \ell_2 \longrightarrow \ell_2 \; , \; T((a_1, a_2, a_3, \ldots,)) = (a_2, a_3, \ldots),$$

is not self-adjoint.
(c) Let $(X, \Sigma, \mu)$ be a measure space and $g \in L_\infty(\mu)$. As $|g| \leq \|g\|_\infty$ $\mu$-almost everywhere, the multiplication operator

$$T_g : L_2(\mu) \longrightarrow L_2(\mu) \, , \, T_g(f) = fg,$$

is linear and continuous ($\|T_g\| \leq \|g\|_\infty$). We shall see under which conditions the operator $T_g$ is self-adjoint. We know that

$$u_{L_2(\mu)}: L_2(\mu) \longrightarrow L_2(\mu)' , \quad u_{L_2(\mu)}(h)(f) = \int_X f\bar{h}\, d\mu, \text{ and}$$

$$T_g': L_2(\mu)' \longrightarrow L_2(\mu)' , \quad T_g'(\varphi)(f) = \varphi(fg).$$

Observe that

$$T_g^*(f) = (u_{L_2(\mu)})^{-1}(T_g'(u_{L_2(\mu)}(f))) = (u_{L_2(\mu)})^{-1}\left(T_g'\left(h \mapsto \int_X h\bar{f}\, d\mu\right)\right)$$

$$= (u_{L_2(\mu)})^{-1}\left(h \mapsto \int_X hg\bar{f}\, d\mu\right).$$

Applying $u_{L_2(\mu)}$ it follows that

$$u_{L_2(\mu)}(T_g^*(f)) = \left(h \mapsto \int_X hg\bar{f}\, d\mu\right),$$

that is,

$$u_{L_2(\mu)}(T_g^*(f))(h) = \int_X hg\bar{f}\, d\mu.$$

Imposing the condition $T_g^*(f) = T_g(f)$ results in

$$\int_X hg\bar{f}\, d\mu = u_{L_2(\mu)}(T_g^*(f))(h) = u_{L_2(\mu)}(T_g(f))(h) = \int_X h\overline{T_g(f)}\, d\mu$$

$$= \int_X h\bar{f}\bar{g}\, d\mu$$

for every $h \in L_2(\mu)$. Thus

$$T_g^* = T_g \iff T_g^*(f) = T_g(f) \text{ for every } f \in L_2(\mu)$$

$$\iff \int_X hg\bar{f}\, d\mu = \int_X h\bar{f}\bar{g}\, d\mu \text{ for all } f, h \in L_2(\mu)$$

$$\iff g = \bar{g} \text{ in } L_2(\mu) \iff g(x) \in \mathbb{R} \ \mu\text{-almost everywhere.}$$

In words, $T_g$ is self-adjoint if and only if the function $g$ assumes real values $\mu$-almost everywhere. In the real case, $T_g$ is always self-adjoint.

In the complex case, it is easy to identify self-adjoint operators:

## 7.4 Self-adjoint Operators on Hilbert Spaces

**Proposition 7.4.4** *Let $H$ be a complex Hilbert space and let $T \in \mathcal{L}(H, H)$ be given. Then $T$ is self-adjoint if and only if $\langle T(x), x \rangle \in \mathbb{R}$ for every $x \in H$.*

*Proof* Suppose $T$ is self-adjoint. For every $x \in H$,

$$\langle T(x), x \rangle = \langle x, T^*(x) \rangle = \langle x, T(x) \rangle = \overline{\langle T(x), x \rangle}.$$

Conversely, assuming that $\langle T(x), x \rangle \in \mathbb{R}$ for every $x \in H$, we have

$$\langle T(x), x \rangle + \overline{\lambda} \langle T(x), y \rangle + \lambda \langle T(y), x \rangle + |\lambda|^2 \langle T(y), y \rangle$$
$$= \langle T(x + \lambda y), x + \lambda y \rangle = \overline{\langle T(x + \lambda y), x + \lambda y \rangle}$$
$$= \langle T(x), x \rangle + \lambda \langle y, T(x) \rangle + \overline{\lambda} \langle x, T(y) \rangle + |\lambda|^2 \langle T(y), y \rangle,$$

therefore

$$\overline{\lambda} \langle T(x), y \rangle + \lambda \langle T(y), x \rangle = \lambda \langle y, T(x) \rangle + \overline{\lambda} \langle x, T(y) \rangle,$$

for all $x, y \in H$ and $\lambda \in \mathbb{C}$. Taking $\lambda = 1$ and $\lambda = -i$ we get

$$\langle T(x), y \rangle + \langle T(y), x \rangle = \langle y, T(x) \rangle + \langle x, T(y) \rangle \quad \text{and}$$

$$\langle T(x), y \rangle - \langle T(y), x \rangle = -\langle y, T(x) \rangle + \langle x, T(y) \rangle.$$

Adding the two equations we have $\langle T(x), y \rangle = \langle x, T(y) \rangle$ for all $x, y \in H$. Item (c) of Proposition 7.4.1 reveals that $T^* = T$. ∎

The next result provides a useful characterization of the norm of a self-adjoint operator.

**Theorem 7.4.5** *Let $H$ be a Hilbert space and let $T \in \mathcal{L}(H, H)$ be a self-adjoint operator. Then*

$$\|T\| = \sup\{|\langle T(x), x \rangle| : \|x\| = 1\}.$$

*Proof* If $T = 0$, the result is immediate. Suppose $T \neq 0$. Note that, from the Cauchy–Schwarz inequality (Proposition 5.1.2),

$$|\langle Tx, x \rangle| \leq \|T(x)\| \cdot \|x\| \leq \|T\| \cdot \|x\|^2$$

for every $x \in H$, therefore

$$\sup\{|\langle T(x), x \rangle| : \|x\| = 1\} \leq \|T\|.$$

It remains to prove the reverse inequality. As $T \neq 0$, choose $x_0 \in H$ with $\|x_0\| = 1$ and $T(x_0) \neq 0$. Call

$$x := \|T(x_0)\|^{\frac{1}{2}} \cdot x_0 \text{ and } y := \|T(x_0)\|^{-\frac{1}{2}} \cdot T(x_0).$$

Then $\|x\|^2 = \|y\|^2 = \|T(x_0)\|$. It is clear that $\langle T(x), y \rangle = \|T(x_0)\|^2$ and, as $T$ is self-adjoint,

$$\langle T(y), x \rangle = \langle y, T(x) \rangle = \overline{\langle T(x), y \rangle} = \|T(x_0)\|^2,$$

proving that

$$\langle T(x), y \rangle = \langle T(y), x \rangle = \|T(x_0)\|^2.$$

Defining $u = x + y$ and $v = x - y$, and subtracting the equations

$$\langle T(u), u \rangle = \langle T(x), x \rangle + \langle T(x), y \rangle + \langle T(y), x \rangle + \langle T(y), y \rangle$$
$$\langle T(v), v \rangle = \langle T(x), x \rangle - \langle T(x), y \rangle - \langle T(y), x \rangle + \langle T(y), y \rangle,$$

we obtain

$$\langle T(u), u \rangle - \langle T(v), v \rangle = 2\langle T(x), y \rangle + 2\langle T(y), x \rangle = 4\|T(x_0)\|^2.$$

To simplify the notation, write $C = \sup\{|\langle T(z), z \rangle| : \|z\| = 1\}$ and see that

$$|\langle T(w), w \rangle| \leq C \|w\|^2 \text{ for every } w \in H.$$

Indeed, for $w = 0$ the result is immediate and, for $w \neq 0$,

$$\frac{1}{\|w\|^2} \cdot |\langle T(w), w \rangle| = \left|\left\langle T\left(\frac{w}{\|w\|}\right), \frac{w}{\|w\|} \right\rangle\right| \leq C.$$

Using the Parallelogram Law and remembering that $\langle T(w), w \rangle \in \mathbb{R}$ for every $w \in H$ (Proposition 7.4.4), we have

$$4\|Tx_0\|^2 = \langle T(u), u \rangle - \langle T(v), v \rangle \leq |\langle T(u), u \rangle| + |\langle T(v), v \rangle|$$
$$\leq C\|u\|^2 + C\|v\|^2 = C\|x+y\|^2 + C\|x-y\|^2$$
$$= 2C\left(\|x\|^2 + \|y\|^2\right) = 4C\|T(x_0)\|.$$

As $T(x_0) \neq 0$ we conclude that $\|T(x_0)\| \leq C$. This holds for every $x_0 \in H$ such that $\|x_0\| = 1$ and $T(x_0) \neq 0$. It follows that

$$\|T\| = \sup\{\|T(x_0)\| : \|x_0\| = 1\} = \sup\{\|T(x_0)\| : \|x_0\| = 1, T(x_0) \neq 0\} \leq C.$$

∎

## 7.5 Spectral Theory of Self-adjoint Operators

In this section, we will explore the spectral properties of self-adjoint operators and also of operators that are simultaneously compact and self-adjoint. We begin by showing that the eigenvalues of a self-adjoint operator are always real and that eigenvectors associated with distinct eigenvalues are orthogonal:

**Proposition 7.5.1** *Let $H$ be a Hilbert space and let $T \in \mathcal{L}(H, H)$ be a self-adjoint operator. Then:*

(a) *The eigenvalues of $T$ are real numbers.*
(b) *If $\lambda$ and $\mu$ are distinct eigenvalues of $T$ then $V_\lambda \perp V_\mu$, that is, $\langle x, y \rangle = 0$ for any $x \in V_\lambda$ and $y \in V_\mu$.*

*Proof*

(a) Given an eigenvalue $\lambda$ of $T$, let $x_0$ be a non-zero eigenvector associated with $\lambda$. From Proposition 7.4.4 we know that $\langle T(x_0), x_0 \rangle \in \mathbb{R}$, hence

$$\lambda = \frac{1}{\|x_0\|^2} \cdot \langle \lambda x_0, x_0 \rangle = \frac{1}{\|x_0\|^2} \cdot \langle T(x_0), x_0 \rangle \in \mathbb{R}.$$

(b) By (a) we know that $\lambda$ and $\mu$ are real numbers. For $x \in V_\lambda$ and $y \in V_\mu$,

$$\lambda \langle x, y \rangle = \langle \lambda x, y \rangle = \langle T(x), y \rangle = \langle x, T(y) \rangle = \langle x, \mu y \rangle = \mu \langle x, y \rangle.$$

Hence $(\lambda - \mu)\langle x, y \rangle = 0$, therefore $\langle x, y \rangle = 0$ because $\lambda \neq \mu$. ∎

Our next goal (Theorem 7.5.3) will refine the information above and prove that the entire spectrum of a self-adjoint operator is real.

**Proposition 7.5.2** *Let $H$ be a Hilbert space, let $T \in \mathcal{L}(H, H)$ be a self-adjoint operator and $\lambda \in \mathbb{K}$. Then $\lambda \in \rho(T)$ if and only if there exists $C > 0$ such that $\|T(x) - \lambda x\| \geq C \|x\|$ for every $x \in H$.*

*Proof* If $\lambda \in \rho(T)$, then $T - \lambda I : E \longrightarrow E$ is bijective, hence it is an isomorphism by the Open Mapping Theorem. Then $\|x\| \leq \|(T - \lambda I)^{-1}\| \cdot \|T(x) - \lambda x\|$ for every $x \in H$ and the result follows.

Conversely, suppose there exists $C > 0$ such that $\|T(x) - \lambda x\| \geq C \|x\|$ for every $x \in H$. We must prove that $T - \lambda I$ is bijective. Injectivity is simple: if $T(x) - \lambda x = 0$, then

$$0 \leq C\|x\| \leq \|T(x) - \lambda x\| = 0,$$

proving that $x = 0$. For surjectivity it is enough to prove that $(T - \lambda I)(H)$ is closed and dense in $H$. Suppose there exists $0 \neq x_0 \in \overline{(T - \lambda I)(H)}^\perp$. In this case, for every $x \in H$,

$$0 = \langle T(x) - \lambda x, x_0 \rangle = \langle T(x), x_0 \rangle - \lambda \langle x, x_0 \rangle.$$

As $T$ is self-adjoint, we get

$$\langle x, T(x_0) \rangle = \langle T(x), x_0 \rangle = \lambda \langle x, x_0 \rangle = \langle x, \overline{\lambda} x_0 \rangle$$

for every $x \in H$, proving that $T(x_0) = \overline{\lambda} x_0$. This means that $\overline{\lambda}$ is an eigenvalue of the self-adjoint operator $T$, hence $\lambda$ is a real number by Proposition 7.5.1(a). In this case we also get

$$0 = \|T(x_0) - \lambda x_0\| \geq C \|x_0\| > 0.$$

The contradiction above assures that $\overline{(T - \lambda I)(H)}^{\perp} = \{0\}$. From Theorem 5.2.5(a) we conclude that

$$H = \overline{(T - \lambda I)(H)} \oplus \overline{(T - \lambda I)(H)}^{\perp} = \overline{(T - \lambda I)(H)},$$

therefore $(T - \lambda I)(H)$ is dense in $H$.

Let now $y \in \overline{(T - \lambda I)(H)}$ be given and let $(y_n)_{n=1}^{\infty}$ be a sequence in $(T - \lambda I)(H)$ that converges to $y$. For each $n$ let $x_n \in H$ be such that $y_n = T(x_n) - \lambda x_n$. Therefore

$$\|x_n - x_m\| \leq \frac{1}{C} \|T(x_n - x_m) - \lambda(x_n - x_m)\| = \frac{1}{C} \|y_n - y_m\|$$

for all natural numbers $n, m$. From the convergence of the sequence $(y_n)_{n=1}^{\infty}$ we conclude that the sequence $(x_n)_{n=1}^{\infty}$ is Cauchy, therefore convergent, say $x_n \longrightarrow x \in H$. The continuity of $T - \lambda I$ reveals that

$$y_n = (T - \lambda I)(x_n) \longrightarrow (T - \lambda I)(x) \in (T - \lambda I)(H).$$

This proves that $(T - \lambda I)(H)$ is closed. ∎

**Theorem 7.5.3** *Let $H$ be a Hilbert space. The spectrum $\sigma(T)$ of a self-adjoint operator $T \in \mathcal{L}(H, H)$ is real.*

**Proof** It is enough to show that if $\lambda = a + ib$ with $a$ and $b$ real and $b \neq 0$, then $\lambda \in \rho(T)$. Let $\lambda$ be of this form. Given $0 \neq x \in H$, from Proposition 7.4.4 we know that $\langle T(x), x \rangle \in \mathbb{R}$. Subtracting the equations

$$\langle (T - \lambda I)(x), x \rangle = \langle T(x), x \rangle - \lambda \langle x, x \rangle,$$

$$\overline{\langle (T - \lambda I)(x), x \rangle} = \overline{\langle T(x), x \rangle} - \overline{\lambda \langle x, x \rangle} = \langle T(x), x \rangle - \overline{\lambda} \langle x, x \rangle,$$

## 7.5 Spectral Theory of Self-adjoint Operators

we obtain

$$-2i\,\mathrm{Im}(\langle(T-\lambda I)(x),x\rangle) = \overline{\langle(T-\lambda I)(x),x\rangle} - \langle(T-\lambda I)(x),x\rangle$$
$$= (\lambda - \overline{\lambda})\langle x,x\rangle = 2ib\,\|x\|^2.$$

Therefore,

$$|b|\cdot\|x\|^2 = |\mathrm{Im}(\langle(T-\lambda I)(x),x\rangle)| \le |\langle(T-\lambda I)(x),x\rangle| \le \|(T-\lambda I)(x)\|\cdot\|x\|.$$

Dividing by $\|x\|$ we get $|b|\cdot\|x\| \le \|T(x) - \lambda x\|$ for every $x \neq 0$. Since this inequality holds trivially for $x = 0$, from Proposition 7.5.2 we conclude that $\lambda \in \rho(T)$. ∎

We have already seen that both compact operators and self-adjoint operators have spectra with special properties. It is natural to expect that operators that are both compact and self-adjoint have an even richer spectral theory. Indeed, this is the case: we are now in search of the spectral decomposition of compact self-adjoint operators, which says that if $T \in \mathcal{L}(H,H)$ is compact self-adjoint, then $H$ admits a complete orthonormal system formed exclusively by eigenvectors of $T$. In this case, every vector in $H$ can be written as a sum (possibly infinite) of eigenvectors. To prove this result we need the

**Proposition 7.5.4** *Let $H$ be a Hilbert space and let $T \in \mathcal{L}(H,H)$ be a non-zero compact self-adjoint operator. Then $\|T\|$ or $-\|T\|$ is an eigenvalue of $T$. Furthermore, there exists an eigenvector $x \in H$ associated with this eigenvalue such that $\|x\| = 1$ and $|\langle T(x), x\rangle| = \|T\|$.*

**Proof** By Theorem 7.4.5 there is a sequence of unit vectors $(z_n)_{n=1}^\infty$ such that $|\langle T(z_n), z_n\rangle| \longrightarrow \|T\|$. The sequence $(\langle T(z_n), z_n\rangle)_{n=1}^\infty$ is then bounded, hence it admits a convergent subsequence $(\langle T(x_n), x_n\rangle)_{n=1}^\infty$. It is clear that $|\langle T(x_n), x_n\rangle| \longrightarrow \|T\|$, so $(\langle T(x_n), x_n\rangle)_{n=1}^\infty$ converges to $\|T\|$ or to $-\|T\|$. This means that defining $\lambda = \lim_n \langle T(x_n), x_n\rangle$, we have $\lambda = \|T\|$ or $\lambda = -\|T\|$. We must then prove that $\lambda$ is an eigenvalue of $T$. As $\|x_n\| = 1$ for every $n$,

$$0 \le \|T(x_n) - \lambda x_n\|^2 = \langle T(x_n) - \lambda x_n, T(x_n) - \lambda x_n\rangle$$
$$= \|T(x_n)\|^2 - \lambda\langle T(x_n), x_n\rangle - \lambda\langle x_n, T(x_n)\rangle + \lambda^2\|x_n\|^2$$
$$\le \|T\|^2 - 2\lambda\langle T(x_n), x_n\rangle + \lambda^2 \longrightarrow \lambda^2 - 2\lambda^2 + \lambda^2 = 0.$$

It follows that $T(x_n) - \lambda x_n \longrightarrow 0$. As $T$ is compact, from the sequence $(T(x_n))_{n=1}^\infty$ we can extract a convergent subsequence $(T(y_n))_{n=1}^\infty$. Therefore $T(y_n) - \lambda y_n \longrightarrow 0$, hence $(\lambda y_n)_{n=1}^\infty$ also converges, say $\lambda y_n \longrightarrow y$. Note that $y \neq 0$ since $\|y_n\| = 1$ for every $n$ and $\lambda \neq 0$. On the one hand it is clear that $T(y_n) \longrightarrow y$; on the other hand, from the continuity of $T$ we have

$$\lambda T(y_n) = T(\lambda y_n) \longrightarrow T(y).$$

We can conclude that $T(y) = \lambda y$, proving that $\lambda$ is an eigenvalue of $T$. It is easily checked that the eigenvector $\frac{y}{\lambda}$ satisfies the desired conditions. ∎

**Corollary 7.5.5** *Let $H$ be a Hilbert space and let $T \in \mathcal{L}(H, H)$ be a compact self-adjoint operator. Then:*

(a) $\sigma(T) \neq \emptyset$.
(b) *If $\sigma(T) = \{0\}$, then $T = 0$.*

*Proof*

(a) If $T = 0$, then 0 is an eigenvalue of $T$. If $T \neq 0$, then $T$ has an eigenvalue by Proposition 7.5.4.
(b) If $T$ were non-zero, $T$ would have a non-zero eigenvalue by Proposition 7.5.4. ∎

From Linear Algebra, the reader recalls that when $V$ is a finite-dimensional vector space, a linear transformation $T: V \longrightarrow V$ is diagonalizable if and only if $V$ admits a basis consisting of eigenvectors of $T$. In this sense, conveniently replacing the notion of algebraic basis, the theorem below states that compact self-adjoint operators on Hilbert spaces are diagonalizable.

**Theorem 7.5.6 (Spectral Decomposition of Compact Self-adjoint Operators)**
*Let $H$ be a Hilbert space and let $T: H \longrightarrow H$ be a compact self-adjoint operator. Then $H$ admits a complete orthonormal system formed by eigenvectors of $T$. Moreover, there exist sequences (finite or infinite) of eigenvalues $(\lambda_n)_n$ of $T$ and vectors $(v_n)_n$ such that each $v_n$ is an eigenvector associated with $\lambda_n$ and*

$$T(x) = \sum_n \lambda_n \langle x, v_n \rangle v_n$$

*for every $x \in H$.*

*Proof* From Theorem 7.3.7 we know that $T$ has at most a countable number of eigenvalues. Let $(\alpha_1, \alpha_2, \ldots)$ be an enumeration of the distinct and non-zero eigenvalues of $T$. Note that this sequence can be finite or infinite. We call $\alpha_0 = 0$ and define

$$E_0 = \ker(T) \text{ and } E_n = V_{\alpha_n} = \ker(T - \alpha_n I), \ n \geq 1.$$

As a closed subspace of a Hilbert space, $E_0$ is also a Hilbert space, so it admits a complete orthonormal system $\mathcal{B}_0$ by Theorem 5.4.6. Proposition 7.3.1(a) yields $0 < \dim(E_n) < \infty$ for every $n \geq 1$, so we can consider a finite complete orthonormal system $\mathcal{B}_n$ in each $E_n$, $n \geq 1$. From Proposition 7.5.1(b) we know that the subspaces $E_0, E_1, E_2, \ldots$ are pairwise orthogonal, so the set formed by the vectors of $\mathcal{B}_0, \mathcal{B}_1, \mathcal{B}_2, \ldots$, is orthonormal and contains only eigenvectors of $T$.

Defining $S := \mathcal{B}_0 \cup \mathcal{B}_1 \cup \mathcal{B}_2 \cup \cdots$, by Theorem 5.3.10 it is enough to prove that $[S]$ is dense in $H$. As every element of $[S]$ is a linear combination of eigenvectors of $T$, it is immediate that $T(S) \subseteq T([S]) \subseteq [S]$. Given $x \in S^\perp$ and $y \in S$, as $T(y) \in [S]$,

$$\langle T(x), y \rangle = \langle x, T(y) \rangle = 0,$$

proving that $T(S^\perp) \subseteq S^\perp$. It is clear that the restriction of a compact operator is compact and the restriction of a self-adjoint operator is self-adjoint, so $T_0 := T|_{S^\perp} : S^\perp \longrightarrow S^\perp$ is compact self-adjoint. Suppose there exists $\lambda \in \sigma(T_0)$, $\lambda \neq 0$. In this case, by Theorem 7.3.6 we know that $\lambda$ is an eigenvalue of $T_0$, hence there exists $x \in S^\perp$, $x \neq 0$, such that $T(x) = T_0(x) = \lambda x$. Being a non-zero eigenvalue of $T$, $\lambda = \alpha_n$ for some $n \geq 1$. Thus $x \in E_n \cap S^\perp \subseteq [S] \cap S^\perp$, therefore $x = 0$. This contradiction proves that $\sigma(T_0) \subseteq \{0\}$, hence $\sigma(T_0) = \{0\}$ by Corollary 7.5.5(a). From Corollary 7.5.5(b) we conclude that $T_0 = 0$. Consequently,

$$S^\perp \subseteq \ker(T) = \overline{[\mathcal{B}_0]} \subseteq \overline{[S]},$$

which, by the continuity of the inner product, gives $S^\perp = \{0\}$. As $\overline{[S]}^\perp \subseteq S^\perp$, we have $\overline{[S]}^\perp = \{0\}$. From Theorem 5.2.5(a) we know that $H = \overline{[S]} \oplus \overline{[S]}^\perp$, so $[S]$ is dense in $H$. This completes the proof that $\mathcal{B}_0 \cup \mathcal{B}_1 \cup \mathcal{B}_2 \cup \cdots$ is a complete orthonormal system for $H$ formed by eigenvectors of $T$. As $E_n$ is finite-dimensional for every $n \geq 1$, we can consider a basis (finite or countable) $(v_n)_n$ for the space generated by $E_1, E_2, \ldots$. It is clear that each $v_n$ is an eigenvector of $T$. We also consider $\lambda_n$, $n \geq 1$, eigenvalue of $T$ to which $v_n$ is associated. Given $x \in H$, by Theorem 5.3.10 we have $x = x_0 + \sum_n \langle x, v_n \rangle v_n$ where $x_0 \in E_0 = \ker(T)$. Suppose there are infinitely many $v_1, v_2, \ldots$. As $T$ is linear and continuous,

$$T(x) = T\left(x_0 + \sum_{n=1}^\infty \langle x, v_n \rangle v_n\right) = T\left(\sum_{n=1}^\infty \langle x, v_n \rangle v_n\right) = T\left(\lim_{n \to \infty} \sum_{j=1}^n \langle x, v_j \rangle v_j\right)$$

$$= \lim_{n \to \infty} T\left(\sum_{j=1}^n \langle x, v_j \rangle v_j\right) = \lim_{n \to \infty} \sum_{j=1}^n \langle x, v_j \rangle T(v_j) = \sum_{n=1}^\infty \lambda_n \langle x, v_n \rangle v_n.$$

In the case where the sequence $(v_n)_n$ is finite, the linearity of $T$ implies, *mutatis mutandis*, this same formula. ∎

## 7.6 Comments and Historical Notes

The notion of compact operator dates back to Hilbert's 1906 work on the resolution of systems of infinitely many linear equations. The definition came with Riesz in

1918 under the name of *completely continuous operator*. Since 1950, following a suggestion by E. Hille, the term *compact operator* has been used. The term *completely continuous operator* is reserved for operators that transform weakly convergent sequences into convergent sequences, that is, satisfy implication (7.1). As we saw in Proposition 7.2.8, for operators defined on reflexive spaces, in particular for operators between Hilbert spaces, these two notions coincide. From Propositions 7.2.2(b) and 7.2.5 it follows that every operator that is the limit of a sequence of finite-rank operators is compact. In 1931, T. W. Hildebrandt asked if the converse was true. This problem remained unanswered for several decades and Pietsch [51, p. 54] states that this was the most important question in the entire history of Banach space theory. A Banach space $E$ is said to have the *approximation property* if every compact operator with values in $E$ is the limit of a sequence of finite-rank operators. The question then becomes: is it true that every Banach space has the approximation property? P. Enflo solved the problem in 1972 by proving that for $1 < p < \infty$, $p \neq 2$, $\ell_p$ contains a subspace that does not have the approximation property. The study of the approximation property and its many variants has become a trend that remains very active to this day. A good reference for the subject is [36, Chapter 7]. In Exercise 7.7.17 the reader will prove that Hilbert spaces have the approximation property.

Recently, compact operators have starred in a major event in the mathematical community. One of the main questions that emerged at the end of the Gowers and Maurey theory mentioned in the Introduction of this book was the following: does there exist an infinite-dimensional Banach space $E$ such that every operator $T \in \mathcal{L}(E, E)$ can be written in the form

$$T = \lambda \cdot \mathrm{id}_E + U,$$

where $U$ is a compact operator? In other words: is there an infinite-dimensional space $E$ such that every operator from $E$ to $E$ is a compact perturbation of the identity? For obvious reasons, such a space is said to have very few operators and, according to Gowers' words on his weblog, such a space has almost no non-trivial structure. In 2009, S. Argyros and R. Haydon [4] announced the proof of the existence of such a space, a discovery published in 2011 in the famous journal Acta Mathematica.

Corollary 7.3.5 was first proven by I. Fredholm in 1903 for the integral operators of Example 7.2.4. Fredholm interpreted integral equations as limits of systems of linear equations considering integrals as limits of Riemann sums and passing to the limit in Cramer's rule for the calculation of the determinants associated with these linear systems. The terms *resolvent set* and *spectrum* were coined by Hilbert in 1906. The spectral theory of compact operators is also known as *Riesz–Schauder theory*. The aim of this theory, initiated by Riesz in 1918 and completed by Schauder in 1930, was precisely to avoid Fredholm's 'infinite determinants'.

Self-adjoint operators are sometimes called *Hermitian operators*, in honor of C. Hermite, a nineteenth-century French mathematician. Some books use the term self-adjoint operator for the real case and Hermitian operator for the complex case.

Corollary 7.5.5 holds for operators that are merely self-adjoint (not necessarily compact). For a proof see [9, Corollary 6.10].

The spectral theorem (Theorem 7.5.6) has many applications in the theory of differential equations; see, for example, [9, 8.6 and 9.8] and [13, II.6].

Unitary operators and normal operators on Hilbert spaces are introduced in Exercises 7.7.36 and 7.7.37, respectively. Normal operators enjoy a spectral decomposition similar to the decomposition of compact self-adjoint operators that we saw in Sect. 7.5. For more details see [21, Theorem 7.53].

## 7.7 Exercises

**Exercise 7.7.1** Let $E$ be a Banach space. Prove that $\sigma(T) = \sigma(T')$ for every $T \in \mathcal{L}(E, E)$.

**Exercise 7.7.2** Let $E$ be a Banach space, $T \in \mathcal{L}(E, E)$ and $\lambda \in \mathbb{K}$. Suppose that

(i) $(T - \lambda I)^{-1} \colon (T - \lambda I)(E) \longrightarrow E$ exists;
(ii) The linear operator $(T - \lambda I)^{-1} \colon (T - \lambda I)(E) \longrightarrow E$ is continuous;
(iii) $(T - \lambda I)(E)$ is dense in $E$.

Prove that $(T - \lambda I)(E) = E$, therefore $\lambda \in \rho(T)$.

**Exercise 7.7.3** Let $E$ and $F$ be normed spaces. Prove that $\mathcal{K}(E, F)$ is a vector subspace of $\mathcal{L}(E, F)$

**Exercise 7.7.4** Show that the integral operator $T$ from Example 7.2.4 is well defined and is linear.

**Exercise 7.7.5** Show that the operator $T \colon \ell_2 \longrightarrow \ell_2$ given by $T((a_j)_{j=1}^\infty) = \left(\frac{a_j}{j}\right)_{j=1}^\infty$ is compact but not of finite rank.

**Exercise 7.7.6** Decide whether the operator $T \colon \ell_2 \longrightarrow \ell_2$ given by $T((a_j)_{j=1}^\infty) = (a_1, 0, a_3, 0, \ldots)$ is compact or not.

**Exercise 7.7.7** Let $E$ be an infinite-dimensional normed space and let $T \colon E \longrightarrow E$ be a compact bijective linear operator. Show that the inverse operator $T^{-1}$ is not continuous.

**Exercise 7.7.8** Let $E$ be an infinite-dimensional normed space and let $T \colon E \longrightarrow E$ be a compact bijective linear operator. Show that $E$ is not complete.

**Exercise 7.7.9** Let $1 \leq p < \infty$ and $T \colon (C[a, b], \|\cdot\|_p) \longrightarrow (C[a, b], \|\cdot\|_\infty)$ be the integral operator defined exactly as in Example 7.2.4. Prove that $T$ is well defined, linear and compact.

**Exercise 7.7.10** Prove that $\mathcal{K}(E, \ell_1) = \mathcal{L}(E, \ell_1)$ for every reflexive space $E$.

**Exercise 7.7.11** Prove that $\mathcal{K}(c_0, E) = \mathcal{L}(c_0, E)$ for every reflexive space $E$.

**Exercise 7.7.12** Given a compact operator $T \in \mathcal{K}(E, E)$ and $n \in \mathbb{N}$, prove that there exists a compact operator $S \in \mathcal{K}(E, E)$ such that $(I - T)^n = I - S$.

**Exercise 7.7.13** Let $\widehat{E}$ and $\widehat{F}$ be the completions of the normed spaces $E$ and $F$, let $T \in \mathcal{L}(E, F)$ be a compact operator, and let $\widehat{T} \in \mathcal{L}(\widehat{E}, \widehat{F})$ be the linear bounded extension of $T$ to $\widehat{E}$. Prove that:

(a) $T(\widehat{E}) \subseteq F$ and $\widehat{T}$ is compact.
(b) If $E = F$, then $T$ and $\widehat{T}$ have the same non-zero eigenvalues.

**Exercise 7.7.14** Let $(\lambda_n)_{n=1}^\infty$ be a sequence of scalars that converges to 0. Construct a compact operator $T$ such that $\sigma(T) = \{0\} \cup \{\lambda_n : n \in \mathbb{N}\}$.

**Exercise 7.7.15** Show that the reflexivity of $E$ is an essential hypothesis in Proposition 7.2.8(b).

**Exercise 7.7.16** Let $E$ be a Banach space such that for every compact $K \subseteq E$ and every $\varepsilon > 0$ there exists a finite rank operator $T \in \mathcal{L}(E, E)$ such that $\sup_{x \in K} \|x - T(x)\| < \varepsilon$. Prove that $E$ has the approximation property (see definition in Sect. 7.6). The converse of this exercise also holds—see [44, Theorem 3.4.32].

**Exercise 7.7.17\*** Let $E$ be a normed space, let $H$ be a Hilbert space, and let $T: E \longrightarrow H$ be a compact operator. Show that there exists a sequence of finite rank operators that converges to $T$ in $\mathcal{L}(E, H)$.

**Exercise 7.7.18** Let $E$ and $F$ be Banach spaces. An operator $T \in \mathcal{L}(E, F)$ is said to be *nuclear* if there exist sequences $(\varphi_n)_{n=1}^\infty \subseteq E'$ and $(b_n)_{n=1}^\infty \subseteq F$ such that $\sum_{n=1}^\infty \|\varphi_n\| \cdot \|b_n\| < \infty$ and $T(x) = \sum_{n=1}^\infty \varphi_n(x) b_n$ for every $x \in E$. Prove that every nuclear operator is compact.

**Exercise 7.7.19\*** Let $E$ be a Banach space. Prove that

(a) If $T, S \in \mathcal{L}(E, E)$ are such that $T$ is an isomorphism and $\|S - T\| < \|T^{-1}\|^{-1}$, then $S$ is an isomorphism and

$$\|S^{-1} - T^{-1}\| \leq \frac{\|T^{-1}\|^2 \cdot \|S - T\|}{1 - \|T^{-1}\| \cdot \|S - T\|}.$$

(b) The set $\mathcal{C}$ of isomorphisms from $E$ to $E$ is an open subset of $\mathcal{L}(E, E)$ and the function $T \longrightarrow T^{-1}$ is a homeomorphism from $\mathcal{C}$ to $\mathcal{C}$.

**Exercise 7.7.20** Let $E$ be a Banach space and let $T, S \in \mathcal{L}(E, E)$ be such that $T \circ S = S \circ T$. Prove that $S \circ T$ is an isomorphism if and only if $T$ and $S$ are isomorphisms.

**Exercise 7.7.21** Let $E$ be a Banach space and let $T_1, \ldots, T_n \in \mathcal{L}(E, E)$ be such that $T_{\pi(1)} \circ \cdots \circ T_{\pi(n)} = T_1 \circ \cdots \circ T_n$ for every bijection $\pi: \{1, \ldots, n\} \longrightarrow \{1, \ldots, n\}$. Show that $T_1 \circ \cdots \circ T_n$ is an isomorphism if and only if $T_i$ is an isomorphism for every $i = 1, \ldots, n$.

## 7.7 Exercises

**Exercise 7.7.22** Let $E$ be a complex Banach space, $T \in \mathcal{L}(E, E)$ and $n \in \mathbb{N}$. Prove that $\sigma(T^n) = \{\lambda^n : \lambda \in \sigma(T)\}$. (Remember that $T^n = T \circ \overset{(n)}{\cdots} \circ T$)

**Exercise 7.7.23** Prove item (c) of Proposition 7.4.1.

**Exercise 7.7.24** Let $H_1$ and $H_2$ be Hilbert spaces and let $S, T \in \mathcal{L}(H_1, H_2)$ be given. Prove that:

(a) $(S + T)^* = S^* + T^*$.
(b) $(\lambda T)^* = \bar{\lambda} T^*$.
(c) $\|T \circ T^*\| = \|T^* \circ T\| = \|T\|^2$.
(d) Item (d) of Proposition 7.4.1 holds.

**Exercise 7.7.25** Let $H$ be a complex Hilbert space, $T \in \mathcal{L}(H, H)$ and $\lambda \in \mathbb{C}$. Prove that $\lambda \in \sigma(T)$ if and only if $\bar{\lambda} \in \sigma(T^*)$.

**Exercise 7.7.26** Let $H$ be a Hilbert space and $T \in \mathcal{L}(H, H)$. Prove that $\ker(T) = T^*(H)^\perp$, therefore $H = \ker(T) \oplus T^*(H)$.

**Exercise 7.7.27** Let $H$ be a Hilbert space and $S, T \in \mathcal{L}(H, H)$. Show that $(T \circ S)^* = S^* \circ T^*$.

**Exercise 7.7.28** Let $H_1$ and $H_2$ be Hilbert spaces, let $T \in \mathcal{L}(H_1, H_2)$ be given, and let $M$ and $N$ be closed subspaces of $H_1$ and $H_2$, respectively. Show that $T(M) \subseteq N$ if and only if $T^*(N^\perp) \subseteq M^\perp$.

**Exercise 7.7.29** Show that the operator $T: \ell_2 \longrightarrow \ell_2$ given by $T((a_j)_{j=1}^\infty) = (a_1, \frac{a_2}{2}, \frac{a_3}{3}, \ldots)$, is self-adjoint.

**Exercise 7.7.30** Let $H$ be a Hilbert space, let $T \in \mathcal{L}(H, H)$ be a self-adjoint operator and $n \in \mathbb{N}$. Prove that $T^n$ is self-adjoint and $\|T^n\| = \|T\|^n$.

**Exercise 7.7.31** Let $H$ be a Hilbert space and let $S, T \in \mathcal{L}(H, H)$ be self-adjoint operators. Show that $T \circ S$ is self-adjoint if and only if $T \circ S = S \circ T$.

**Exercise 7.7.32** Let $H$ be a Hilbert space and let $T \in \mathcal{L}(H, H)$ be a self-adjoint isomorphism. Prove that if $T$ is *positive*, that is, $\langle T(x), x \rangle \geq 0$ for every $x \in H$, then the expression

$$[x, y] := \langle T(x), y \rangle, \quad x, y \in H,$$

defines an inner product on $H$ whose induced norm is equivalent to the original norm of $H$.

**Exercise 7.7.33** Let $H$ be a complex Hilbert space and let $T: H \longrightarrow H$ be a bounded linear operator. Show that there exist unique self-adjoint operators $T_1, T_2 \in \mathcal{L}(H, H)$ such that $T = T_1 + iT_2$.

**Exercise 7.7.34** Let $H$ be a Hilbert space in which there exists an injective compact self-adjoint operator $T \in \mathcal{L}(H, H)$. Prove that $H$ is separable.

**Exercise 7.7.35** Let $H$ be a separable Hilbert space and let $T \in \mathcal{L}(H, H)$ be a compact self-adjoint operator. Prove that $\sigma(T)$ is the closure of the set of eigenvalues of $T$.

**Exercise 7.7.36** Let $H_1$ and $H_2$ be Hilbert spaces. Prove that the following statements are equivalent for an operator $T \in \mathcal{L}(H_1, H_2)$:

(a) $T$ is invertible, $T \circ T^* = \mathrm{id}_{H_2}$ and $T^* \circ T = \mathrm{id}_{H_1}$.
(b) $T$ is surjective and $\langle T(x), T(y) \rangle = \langle x, y \rangle$ for all $x, y \in H_1$.

In this case $T$ is called a *unitary operator*.

**Exercise 7.7.37** Let $H$ be a Hilbert space. Prove that the following statements are equivalent for an operator $T \in \mathcal{L}(H, H)$:

(a) $T \circ T^* = T^* \circ T$.
(b) $\langle T(x), T(y) \rangle = \langle T^*(x), T^*(y) \rangle$ for all $x, y \in H$.

In this case $T$ is called a *normal operator*.

# Chapter 8
# Topological Vector Spaces

Let $A$ be a subset of a normed space $E$. We ask the reader for a bit of patience to follow us in the proof of the following implication:

$$\text{If } A \text{ is convex, then } \overline{A} \text{ is also convex.} \tag{8.1}$$

Given $x, y \in \overline{A}$ and $0 \leq \lambda \leq 1$, we must prove that $(1-\lambda)x + \lambda y \in \overline{A}$. Consider sequences $(x_n)_{n=1}^\infty$ and $(y_n)_{n=1}^\infty$ in $A$ such that $x_n \longrightarrow x$ and $y_n \longrightarrow y$. From the convexity of $A$ we know that $(1-\lambda)x_n + \lambda y_n \in A$ for every $n \in \mathbb{N}$, and as the algebraic operations in $E$ — addition and scalar multiplication — are continuous,

$$(1-\lambda)x_n + \lambda y_n \longrightarrow (1-\lambda)x + \lambda y.$$

This is enough to conclude that $(1-\lambda)x + \lambda y \in \overline{A}$, proving that $\overline{A}$ is convex.

What is of interest in the argument above is the following: at no point did we use the norm of $E$, that is, at no point was it necessary to know that to each vector $x \in E$ is associated a real number $\|x\|$ satisfying certain conditions. The only information we used was that the algebraic operations in $E$ are continuous when we consider on $E$ the topology induced by the norm. The central point is that it was not the norm that solved the problem, but the fact that the topology induced by the norm makes the algebraic operations continuous. The more attentive reader may argue that, to argue through sequences, we also used that the topology of the norm is metrizable. This is true, but abstracting the metrizability one can replace sequences with nets so that the argument continues to work. That is, in fact the argument depends only on the continuity of the algebraic operations.

In fact, we have already gone through this situation before in this book (see, for example, the proof of Theorem 6.3.9). This means that many of the properties of normed spaces depend only on the fact that the topology induced by the norm makes the algebraic operations continuous. The theme of this chapter is precisely the study of vector spaces equipped with a topology with respect to which the operations of addition and scalar multiplication are continuous. Such spaces include normed spaces, but as we will see throughout the chapter, they also include many interesting and useful spaces that are not normed spaces. In the course of the chapter we will revisit some results previously obtained for normed spaces, for example the Hahn–Banach and Goldstine theorems, and also weak topologies.

## 8.1 Examples and First Properties

The discussion above has prepared us for the following definition:

**Definition 8.1.1** Let $E$ be a vector space over the field $\mathbb{K}$ and let $\tau$ be a topology on $E$. We say that the pair $(E, \tau)$ is a *topological vector space* if the algebraic operations

$$AD \colon E \times E \longrightarrow E, \quad AD(x, y) = x + y \text{ and}$$

$$SM \colon \mathbb{K} \times E \longrightarrow E, \quad SM(a, y) = ay,$$

are continuous when we consider on $E \times E$ and on $\mathbb{K} \times E$ the respective product topologies.

In words, in a topological vector space, the topological and algebraic structures are compatible.

**Example 8.1.2**

(a) Normed spaces are topological vector spaces.
(b) Consider in the vector space $E$ the chaotic topology $\tau_c = \{\emptyset, E\}$. It is immediate that $(E, \tau_c)$ is a topological vector space. We shall see that this space is not normed when $E \neq \{0\}$. Suppose there is a norm $\|\cdot\|$ on $E$ that induces the topology $\tau_c$. In this case, being an open set, the open ball $B(0; 1) = \{x \in E : \|x\| < 1\}$ coincides with $\emptyset$ or with $E$. Since the origin belongs to the ball, $B(0; 1) = E$. This clearly cannot occur because non-trivial normed spaces are not bounded (take $0 \neq x \in E$ and observe that the set $\{ax : a \in \mathbb{K}\}$ is unbounded).
(c) Consider on the vector space $E \neq \{0\}$ the discrete topology $\tau_d = \{A : A \subseteq E\}$. We shall prove that $(E, \tau_d)$ is not a topological vector space. Let $A = \{a\} \subseteq E$ with $a \neq 0$. Since every subset of $E$ is open, it is clear that $A$ is open. Next we show that $SM^{-1}(A)$ is not open. We will prove this by showing that $(1, a) \in SM^{-1}(A)$ but $(1, a)$ is not an interior point of $SM^{-1}(A)$. This is true because if $V \subseteq \mathbb{K} \times E$ is an open set containing $(1, a)$, there are open sets $1 \in V_1 \subseteq \mathbb{K}$ and $a \in V_2 \subseteq E$ such that $V_1 \times V_2 \subseteq V$. Thus, we can take $1 \neq \alpha \in V_1$, and in this case $(\alpha, a) \in V_1 \times V_2 \subseteq V$, while

$$SM(\alpha, a) = \alpha a \notin \{a\} = A.$$

8.1 Examples and First Properties    181

This proves that there is no open set in $\mathbb{K} \times E$ containing $(1, a)$ and contained in $SM^{-1}(A)$. Therefore $(1, a)$ is not an interior point of $SM^{-1}(A)$, and the scalar multiplication $SM$ is not continuous.

(d) Let $E$ be a normed space, let $\sigma(E, E')$ be the weak topology on $E$, and let $\sigma(E', E)$ be the weak-star topology on $E'$. The spaces $(E, \sigma(E, E'))$ and $(E', \sigma(E', E))$ are topological vector spaces. Let $(x_\lambda)_\lambda$ and $(y_\lambda)_\lambda$ be nets in $E$ such that $x_\lambda \xrightarrow{w} x$ and $y_\lambda \xrightarrow{w} y$. From the definition of the weak topology we know that every functional $\varphi \in E'$ is continuous in the weak topology, hence $\varphi(x_\lambda) \longrightarrow \varphi(x)$ and $\varphi(y_\lambda) \longrightarrow \varphi(y)$ for every $\varphi \in E'$. We have

$$\varphi(x_\lambda + y_\lambda) = \varphi(x_\lambda) + \varphi(y_\lambda) \longrightarrow \varphi(x) + \varphi(y) = \varphi(x+y)$$

for every $\varphi \in E'$. From Proposition 6.2.2(c) we conclude that $x_\lambda + y_\lambda \xrightarrow{w} x+y$. This proves that addition is continuous. A similar argument proves that scalar multiplication is also continuous, hence $(E, \sigma(E, E'))$ is a topological vector space. We leave to the reader the task of showing that $(E', \sigma(E', E))$ is also a topological vector space. As we saw in Proposition 6.7.2, these two topological vector spaces are not even metrizable, hence they are not normed.

(e) For $0 < p < 1$, define $\ell_p$ exactly as before, that is:

$$\ell_p = \left\{ (a_j)_{j=1}^\infty : a_j \in \mathbb{K} \text{ for every } j \in \mathbb{N} \text{ and } \sum_{j=1}^\infty |a_j|^p < \infty \right\}.$$

First we prove that $\ell_p$ is a vector space with the usual operations of sequences and that

$$d_p : \ell_p \times \ell_p \longrightarrow \mathbb{R}, \ d_p\left((a_j)_{j=1}^\infty, (b_j)_{j=1}^\infty\right) = \sum_{j=1}^\infty |a_j - b_j|^p,$$

is a metric on $\ell_p$. Fix $s \geq 0$ and consider the differentiable function

$$f_s : [0, \infty) \longrightarrow \mathbb{R}, \ f_s(t) = s^p + t^p - (s+t)^p.$$

The condition $0 < p < 1$ easily implies that $f_s$ has a non-negative derivative, hence it is non-decreasing on $[0, +\infty)$. As $f_s(0) = 0$, it follows that $f_s(t) \geq 0$ for every $t \geq 0$. This reveals that

$$(s+t)^p \leq s^p + t^p \text{ for all } s, t \geq 0. \tag{8.2}$$

It follows immediately that $\ell_p$ is closed under addition, and as it is obviously closed under scalar multiplication, we conclude that $\ell_p$ is a vector space. The triangle inequality for $d_p$ also follows from (8.2), and as the other axioms of metric are immediate, it follows that $d_p$ is a metric on $\ell_p$.

Next we see that the algebraic operations in $\ell_p$ are continuous with respect to the topology induced by the metric $d_p$. It is clear that the metric $d_p$ is translation invariant, that is,

$$d_p(x+z, y+z) = d_p(x, y) \text{ for all } x, y, z \in \ell_p.$$

Given then $x, y \in \ell_p$ and sequences $(x_n)_{n=1}^{\infty}$, $(y_n)_{n=1}^{\infty}$ in $\ell_p$ such that $x_n \longrightarrow x$ and $y_n \longrightarrow y$,

$$d_p(x_n + y_n, x + y) = d_p(x_n + y_n - x, x + y - x) = d_p(x_n + y_n - x, y)$$
$$= d_p(x_n + y_n - x - y_n, y - y_n) = d_p(x_n - x, y - y_n)$$
$$\leq d_p(x_n - x, 0) + d_p(0, y - y_n)$$
$$= d_p(x_n, x) + d_p(y_n, y) \longrightarrow 0 + 0 = 0.$$

This proves that $x_n + y_n \longrightarrow x + y$, therefore addition is continuous. For scalar multiplication we will use the following properties of the metric $d_p$, equally obvious:

$$d_p(ax, ay) = |a|^p \cdot d_p(x, y) \text{ for all } x, y \in \ell_p \text{ and } a \in \mathbb{K},$$

$$d_p(ax, bx) = |a - b|^p \cdot d_p(x, 0) \text{ for all } x \in \ell_p \text{ and } a, b \in \mathbb{K}.$$

Given $x \in \ell_p$, $a \in \mathbb{K}$ and sequences $(x_n)_{n=1}^{\infty}$ in $\ell_p$ and $(a_n)_{n=1}^{\infty}$ in $\mathbb{K}$ such that $x_n \longrightarrow x$ and $a_n \longrightarrow a$,

$$d_p(a_n x_n, ax) \leq d_p(a_n x_n, a_n x) + d_p(a_n x, ax)$$
$$= |a_n|^p \cdot d_p(x_n, x) + |a_n - a|^p \cdot d_p(x, 0) \longrightarrow 0 + 0 = 0,$$

because the sequence $(|a_n|^p)_{n=1}^{\infty}$ is bounded (as it is convergent). This proves that scalar multiplication is continuous, therefore $(\ell_p, d_p)$ is a metrizable topological vector space. Soon, we will see that, for $0 < p < 1$, this space is not normed.

(f) Let $X$ be a non-empty topological space. It is easy to see that the set

$$C(X, \mathbb{K}) = \left\{ f: X \longrightarrow \mathbb{K} : f \text{ is continuous and } \sup_{t \in X} |f(t)| < \infty \right\},$$

is a vector space with the pointwise operations of functions. For $f \in C(X, \mathbb{K})$ and $n \in \mathbb{N}$, consider the sets

$$U(n) = \left\{ g \in C(X, \mathbb{K}) : \sup_{t \in X} |g(t)| < \frac{1}{n} \right\} \text{ and } U(n, f) = \{f + g : g \in U(n)\}.$$

## 8.1 Examples and First Properties

We leave to the reader the task of showing that, for each $f \in C(X, \mathbb{K})$, the sets $(U(n, f))_{n=1}^{\infty}$ form a neighborhood basis for $f$, and that $C(X, \mathbb{K})$ becomes a topological vector space when equipped with the (unique) topology generated by these neighborhood bases.

(g) Let $(E, \tau)$ be a topological vector space and let $M$ be a vector subspace of $E$. As the product topology on $M \times M$ coincides with the topology induced by the product topology on $E \times E$ (and the same for $\mathbb{K} \times M$), and as the restriction of a continuous function is continuous, $M$ equipped with the topology induced by $E$ is a topological vector space.

When there is no need or danger of ambiguity, when referring to a topological vector space we will simply write $E$ instead of $(E, \tau)$.

**Definition 8.1.3** Let $E$ and $F$ be topological vector spaces. A linear homeomorphism $T: E \longrightarrow F$ is called an *isomorphism* between $E$ and $F$. In this case we say that the topological vector spaces $E$ and $F$ are *isomorphic*.

**Proposition 8.1.4** *The translations and homotheties are homeomorphisms on a topological vector space. More precisely, if $E$ is a topological vector space, $x_0 \in E$ and $a \in \mathbb{K}$, $a \neq 0$, then the functions*

$$x \in E \mapsto x + x_0 \in E \text{ and } x \in E \mapsto ax \in E,$$

*are homeomorphisms. The homothety is also an isomorphism.*

*Proof* It is clear that these functions are bijective. For continuity, we just need to see the translation as the composition of the maps

$$x \in E \mapsto (x, x_0) \in E \times \{x_0\} \text{ and } (x, x_0) \in E \times \{x_0\} \mapsto x + x_0 \in E,$$

and represent the homothety as the composition of the maps

$$x \in E \mapsto (a, x) \in \{a\} \times E \text{ and } (a, x) \in \{a\} \times E \mapsto ax \in E, a \in \mathbb{K},$$

and observe that the definition of topological vector space and the definition of the product topology ensure that all the functions involved are continuous. The inverse functions are also continuous because the inverse of the translation by $x_0$ is the translation by $-x_0$ and the inverse of the homothety by $a$ is the homothety by $\frac{1}{a}$. The linearity of the homothety is clear. ∎

**Corollary 8.1.5** *The following statements are equivalent for a subset $V$ of the topological vector space $E$:*

(a) *$V$ is a neighborhood of the origin in $E$.*
(b) *$\lambda V = \{\lambda x : x \in V\}$ is a neighborhood of the origin in $E$ for some $\lambda \in \mathbb{K}$, $\lambda \neq 0$.*
(c) *$\lambda V = \{\lambda x : x \in V\}$ is a neighborhood of the origin in $E$ for every $\lambda \in \mathbb{K}$, $\lambda \neq 0$.*

(d) $x_0 + V = \{x_0 + x : x \in V\}$ is a neighborhood of $x_0$ for some $x_0 \in E$.
(e) $x_0 + V = \{x_0 + x : x \in V\}$ is a neighborhood of $x_0$ for every $x_0 \in E$.

**Proof** It is enough to remember that homeomorphisms transform open sets into open sets. ∎

Some algebraic properties play a relevant role in the theory of topological vector spaces. One of them, convexity, is already known to us. Two others are as follows:

**Definition 8.1.6** A subset $A$ of the vector space $E$ is said to be:

(a) *absorbing* if for every $x \in E$ there exists $\lambda_0 > 0$ such that $x \in \lambda A$ for every $\lambda \in \mathbb{K}$ with $|\lambda| \geq \lambda_0$.
(b) *balanced* if $\lambda x \in A$ for every $x \in A$ and every $\lambda \in \mathbb{K}$ with $|\lambda| \leq 1$; that is, $\lambda A \subseteq A$ whenever $|\lambda| \leq 1$.

**Proposition 8.1.7** *Let $\mathcal{U}$ be the family of all neighborhoods of the origin of the topological vector space $E$. If $U \in \mathcal{U}$, then*

(a) *$U$ is absorbing.*
(b) *There exists $V \in \mathcal{U}$ such that $V + V \subseteq U$.*
(c) *There exists $V \in \mathcal{U}$, $V$ balanced, such that $V \subseteq U$.*

**Proof**

(a) Let $x_0 \in E$. The map

$$f : \mathbb{K} \longrightarrow E, \quad f(\lambda) = \lambda x_0,$$

is continuous, as it is the composition of the continuous maps

$$\lambda \in \mathbb{K} \mapsto (\lambda, x_0) \in \mathbb{K} \times \{x_0\} \text{ and } (\lambda, x_0) \in \mathbb{K} \times \{x_0\} \mapsto \lambda x_0 \in E.$$

Since $U$ is a neighborhood of the origin in $E$, there exists $\delta > 0$ such that $f(\lambda) \in U$ whenever $|\lambda| < \delta$, that is,

$$\lambda x_0 \in U \text{ whenever } |\lambda| < \delta,$$

or even,

$$x_0 \in \mu U \text{ whenever } |\mu| > 1/\delta.$$

(b) As the addition $AD \colon E \times E \longrightarrow E$ is continuous and $AD(0, 0) = 0$, there exists $V \in \mathcal{U}$ such that $AD(V \times V) \subseteq U$, that is, $V + V \subseteq U$.
(c) Let $\overline{\Delta} = \{\lambda \in \mathbb{K} : |\lambda| \leq 1\}$ be the closed disk in $\mathbb{K}$. As the scalar multiplication $SM \colon \mathbb{K} \times E \longrightarrow E$ is continuous and $SM(\lambda, 0) = 0$, for each $\lambda \in \overline{\Delta}$ there exist $A_\lambda$, an open neighborhood of $\lambda$ in $\mathbb{K}$, and $V_\lambda \in \mathcal{U}$ such that $\mu x \in U$ whenever $\mu \in A_\lambda$ and $x \in V_\lambda$. From the compactness of the disk $\overline{\Delta}$ there are $\lambda_1, \ldots, \lambda_n \in \overline{\Delta}$ such that

## 8.1 Examples and First Properties

$$\overline{\Delta} \subseteq A_{\lambda_1} \cup \cdots \cup A_{\lambda_n}.$$

Taking $W = V_{\lambda_1} \cap \cdots \cap V_{\lambda_n} \in \mathcal{U}$ we see that

$$\mu x \in U \text{ whenever } \mu \in \overline{\Delta} \text{ and } x \in W.$$

In fact, if $\mu \in \overline{\Delta}$, then $\mu \in A_{\lambda_i}$ for some $i \in \{1, \ldots, n\}$; and if $x \in W$, then $x \in V_{\lambda_i}$. Defining

$$V = \{\mu x : x \in W \text{ and } \mu \in \overline{\Delta}\},$$

it follows that $V$ is balanced and $V \subseteq U$. Moreover, $V \in \mathcal{U}$ since $W \subseteq V$. ∎

**Proposition 8.1.8** *Let $A$ be a subset of a topological vector space $E$.*
*(a) If $A$ is convex then $\overline{A}$ is also convex.*
*(b) If $A$ is balanced then $\overline{A}$ is also balanced.*

*Proof* We have already proven item (a) in the normed case by proving (8.1) at the beginning of this chapter. We also said that the general case follows from a similar argument with nets. The reader is invited to fill in the details in Exercise 8.6.10.

We prove item (b). By Proposition 8.1.4, if $\lambda \in \mathbb{K}$ is a non-zero scalar, the homothety

$$M_\lambda : E \longrightarrow E, \ M_\lambda(x) = \lambda x,$$

is an isomorphism. Thus, if $0 < |\lambda| \leq 1$, as $A$ is balanced,

$$\lambda \overline{A} = M_\lambda(\overline{A}) = \overline{M_\lambda(A)} = \overline{\lambda A} \subseteq \overline{A},$$

proving that $\overline{A}$ is balanced. ∎

**Proposition 8.1.9** *Every neighborhood of the origin in a topological vector space contains a closed and balanced neighborhood of the origin.*

*Proof* Let $V$ be a neighborhood of the origin. From items (b) and (c) of Proposition 8.1.7 there exists a balanced neighborhood $U$ of the origin such that $U + U \subseteq V$. From Proposition 8.1.8(b) we already know that $\overline{U}$ is balanced, so it is enough to show that $\overline{U} \subseteq V$. Let $x \in \overline{U}$. From Corollary 8.1.5 we know that $x + U$ is a neighborhood of $x$, so $(x + U) \cap U \neq \emptyset$, that is, there exists $y \in U$ such that $x + y \in U$. Since $U$ is balanced, $-y \in U$, therefore

$$x \in -y + U \subseteq U + U \subseteq V.$$

∎

Next, we will see how linear operators behave between topological vector spaces. Remembering that the essence of the definition of uniformly continuous functions between metric spaces is that close points have close images; this concept can be naturally generalized to the scope of topological vector spaces:

**Definition 8.1.10** A function $f: E \longrightarrow F$ between topological vector spaces is *uniformly continuous* if for every neighborhood $U$ of the origin in $F$ there is a neighborhood $V$ of the origin in $E$ such that $f(x) - f(y) \in U$ whenever $x, y \in E$ and $x - y \in V$.

**Proposition 8.1.11** *The following statements are equivalent for a linear operator* $T: E \longrightarrow F$ *between topological vector spaces.*

(a) *$T$ is continuous.*
(b) *$T$ is continuous at the origin of $E$.*
(c) *$T$ is continuous at some point of $E$.*
(d) *$T$ is uniformly continuous.*

**Proof** The implications (a) $\implies$ (b) $\implies$ (c) and (d) $\implies$ (a) are obvious. Suppose (c) and let $T$ be continuous at $x_0$. Let $U$ be a neighborhood of the origin in $F$. By Corollary 8.1.5 we know that $T(x_0) + U$ is a neighborhood of $T(x_0)$. From the continuity of $T$ at $x_0$ there is a neighborhood $V$ of $x_0$ such that $T(V) \subseteq T(x_0) + U$. Again by Corollary 8.1.5 we know that $-x_0 + V$ is a neighborhood of the origin in $E$. Given $x, y \in E$ with $x - y \in -x_0 + V$, from the linearity of $T$ we conclude that

$$T(x) - T(y) = T(x - y) \in T(-x_0 + V)$$
$$= -T(x_0) + T(V) \subseteq -T(x_0) + T(x_0) + U = U.$$

This proves the uniform continuity of $T$ and completes the proof. ∎

For linear functionals, some additional characterizations hold that bring the theory even closer to the normed case:

**Proposition 8.1.12** *The following statements are equivalent for a linear functional* $\varphi: E \longrightarrow \mathbb{K}$ *on the topological vector space $E$.*

(a) *$\varphi$ is continuous.*
(b) *$\ker(\varphi)$ is closed in $E$.*
(c) *$\varphi$ is bounded in some neighborhood of the origin of $E$, that is, there exists a neighborhood $W$ of the origin of $E$ such that $\varphi(W)$ is bounded in $\mathbb{K}$.*

**Proof** (a) $\implies$ (b) This implication is clear because the unit set $\{0\}$ is closed in $\mathbb{K}$.
(b) $\implies$ (c) If $\ker(\varphi) = E$ there is nothing to do. Assuming $\ker(\varphi) \neq E$, the set $V := E - \ker(\varphi)$ is a non-empty open set in $E$. Let $x \in V$. As $V$ is an open neighborhood of $x$, $U := -x + V$ is an open neighborhood of the origin. By Proposition 8.1.7(c) there exists a balanced neighborhood $W$ of the origin contained in $U$. Thus,

## 8.2 Locally Convex Spaces

$$W \subseteq U = -x + V,$$

so $x + W \subseteq V$, and

$$(x + W) \cap \ker(\varphi) = \emptyset. \tag{8.3}$$

The linearity of $\varphi$ ensures that $\varphi(W)$ is a balanced subset of $\mathbb{K}$. Then $\varphi(W)$ is bounded or is $\mathbb{K}$ itself (see Exercise 8.6.9). Suppose $\varphi(W) = \mathbb{K}$. In this case there exists $w \in W$ such that $\varphi(w) = -\varphi(x)$, hence $\varphi(x + w) = 0$. By (8.3) we know that this does not occur, so $\varphi(W)$ is bounded, proving (c).

(c) $\Longrightarrow$ (a) By hypothesis there exist a neighborhood $W$ of the origin in $E$ and some $K > 0$ such that $|\varphi(v)| < K$ for every $v \in W$. Given $\varepsilon > 0$, we therefore conclude that $|\varphi(x)| < \varepsilon$ whenever $x \in \frac{\varepsilon}{K} W$. As $\frac{\varepsilon}{K} W$ is a neighborhood of the origin by Corollary 8.1.5, it follows that $\varphi$ is continuous at the origin. The continuity of $\varphi$ follows from Proposition 8.1.11. ∎

## 8.2 Locally Convex Spaces

The study of normed spaces was favored by the fact, guaranteed mainly by the Hahn–Banach Theorem, that non-trivial normed spaces have a beautiful supply of continuous linear functionals. The applications of Corollaries 3.1.3 and 3.1.4 that we made throughout the text are clear proof of this. However, as we will see below, this does not repeat in the context of topological vector spaces. Keeping the notation of $E'$ for the vector space of continuous linear functionals on $E$, there exist topological vector spaces $E \neq \{0\}$ such that $E' = \{0\}$ (see Example 8.2.3(e)). One way to avoid this unfavorable situation is to discover necessary conditions for having $E' \neq \{0\}$, for example:

**Proposition 8.2.1** *Let $E$ be a topological vector space such that $E' \neq \{0\}$. Then there exists a convex neighborhood of the origin $V \neq E$.*

*Proof* Let $\varphi \in E'$, $\varphi \neq 0$, and let $x_0 \in E$ be such that $\varphi(x_0) \neq 0$. The set

$$V := |\varphi|^{-1}\left(\left(-\infty, \frac{|\varphi(x_0)|}{2}\right)\right) = \left\{x \in E : |\varphi(x)| < \frac{|\varphi(x_0)|}{2}\right\},$$

is open because the function $|\varphi|$ is continuous and, obviously, contains the origin. It is also clear that $x_0 \notin V$, so $V \neq E$. A routine calculation shows that $V$ is convex. ∎

It is clear then that the existence of non-null continuous linear functionals depends on the existence of non-trivial convex open sets. Therefore, we study topological vector spaces that contain abundant stocks of convex open sets:

**Definition 8.2.2** A *locally convex space* is a topological vector space in which the origin admits a basis of convex neighborhoods, that is, every neighborhood of the origin contains a convex open set containing the origin.

Since translations are homeomorphisms (Proposition 8.1.4) and transform convex sets into convex sets, a topological vector space $E$ is locally convex if and only if every vector in $E$ admits a basis of convex neighborhoods.

**Example 8.2.3**

(a) Normed spaces are locally convex spaces. Indeed, the open balls $(B(0; \varepsilon))_{\varepsilon>0}$ form a basis of convex neighborhoods of the origin.
(b) It is clear that if $E$ is a vector space and $\tau_c = \{\emptyset, E\}$ is the chaotic topology, then $(E, \tau_c)$ is a locally convex space.
(c) Let $E$ be a normed space. Elementary calculations show that the bases of open neighborhoods of the weak topology $\sigma(E, E')$ on $E$ and of the weak-star topology $\sigma(E', E)$ on $E'$ described in Propositions 6.2.2(b) and 6.3.2(b), respectively, are formed by convex sets. Thus, $(E, \sigma(E, E'))$ and $(E', \sigma(E', E))$, which we already know to be non-metrizable topological vector spaces by Example 8.1.2(d), are locally convex spaces.
(d) Let $0 < p < 1$. We will now prove that the metrizable topological vector space $\ell_p$ from Example 8.1.2(e) is not locally convex. Let $r > 0$. Then the open ball $B(0; r) = \{x \in \ell_p : d_p(x, 0) < r\}$ is a neighborhood of the origin. Suppose there exists a convex neighborhood of the origin $V$ contained in the ball $B(0; r)$. Then there exists $\varepsilon > 0$ such that $B(0; \varepsilon) \subseteq V$. Choose $0 < \delta < \varepsilon$ and consider the canonical unit vectors $e_1, e_2, \ldots$ of $\ell_p$. As $d_p(\delta^{1/p} e_n, 0) = \delta < \varepsilon$, it follows that $\delta^{1/p} e_n \in V$ for every $n$. By convexity,

$$\frac{\delta^{1/p}}{n} \cdot \sum_{j=1}^{n} e_j = \frac{1}{n}\delta^{1/p} e_1 + \cdots + \frac{1}{n}\delta^{1/p} e_n \in V$$

for every $n$. Since $0 < p < 1$,

$$d_p\left(\frac{\delta^{1/p}}{n} \cdot \sum_{j=1}^{n} e_j, 0\right) = \sum_{j=1}^{n} \left(\frac{\delta^{1/p}}{n}\right)^p = \sum_{j=1}^{n} \frac{\delta}{n^p} = \delta n^{1-p} \xrightarrow{n \to \infty} \infty.$$

This shows that $V$ is unbounded, which is a contradiction because $V \subseteq B(0; r)$. We then conclude that $B(0; r)$ does not contain any convex neighborhood of the origin, proving that the topological vector space $\ell_p$, despite being metrizable, is not locally convex. In particular, $\ell_p$ is not normed for $0 < p < 1$.
(e) Let $0 < p < 1$ and let $L_p[0, 1]$ be the set of (classes of) Lebesgue-measurable functions $f : [0, 1] \longrightarrow \mathbb{R}$ such that $\int_{[0,1]} |f|^p \, dm < \infty$, where $m$ is the Lebesgue measure on $[0, 1]$. In Exercise 8.6.6 the reader will check that $L_p[0, 1]$ is a vector space with the usual operations of (classes of) functions and that the expression

## 8.2 Locally Convex Spaces

$$d_p(f, g) = \int_{[0,1]} |f - g|^p \, dm$$

defines a metric on $L_p[0, 1]$ which thus becomes a topological vector space. Let $U$ be a convex neighborhood of the origin of $L_p[0, 1]$. Choose $\varepsilon > 0$ such that $B(0; \varepsilon) \subseteq U$. Given $f \in L_p[0, 1]$, $f \neq 0$, take $n \in \mathbb{N}$ such that $d_p(f, 0) < n^{1-p} \cdot \varepsilon$, which is possible because $0 < p < 1$. From the Fundamental Theorem of Calculus (see [25, Theorem 3.36]), the function

$$F: [0, 1] \longrightarrow \mathbb{R}, \quad F(t) = \int_{[0,t)} |f|^p \, dm,$$

is absolutely continuous, hence continuous. Since $F$ is non-decreasing and non-zero, $F(1) > 0$. Therefore $F(0) = 0 < \frac{n-1}{n} \cdot F(1) < F(1)$. From the Intermediate Value Theorem we can find $0 < t_{n-1} < t_n := 1$ such that $F(t_{n-1}) = \frac{n-1}{n} \cdot F(1)$. Then

$$F(t_n) - F(t_{n-1}) = F(1) - \frac{n-1}{n} \cdot F(1) = \frac{1}{n} F(1).$$

Since $F(0) = 0 < \frac{n-2}{n} \cdot F(1) < \frac{n-1}{n} \cdot F(1) = F(t_{n-1})$, again by the Intermediate Value Theorem there exists $0 < t_{n-2} < t_{n-1}$ such that $F(t_{n-2}) = \frac{n-2}{n} \cdot F(1)$. Thus,

$$F(t_{n-1}) - F(t_{n-2}) = \frac{n-1}{n} \cdot F(1) - \frac{n-2}{n} \cdot F(1) = \frac{1}{n} F(1).$$

Calling $t_0 = 0$, after $n$ steps we will have constructed a partition $0 = t_0 < t_1 < \cdots < t_n = 1$ of $[0, 1]$ such that

$$\int_{[t_{j-1}, t_j)} |f|^p \, dm = F(t_j) - F(t_{j-1}) = \frac{1}{n} F(1) = \frac{1}{n} \int_{[0,1]} |f|^p \, dm,$$

for every $j = 1, \ldots, n$. For each $j = 1, \ldots, n$, recall that $\mathcal{X}_{[t_{j-1}, t_j)}$ denotes the characteristic function of the interval $[t_{j-1}, t_j)$ and consider the function $f_j \in L_p[0, 1]$ given by

$$f_j(t) = nf(t) \mathcal{X}_{[t_{j-1}, t_j)}(t).$$

In this way,

$$d_p(f_j, 0) = n^p \int_{[t_{j-1}, t_j)} |f|^p \, dm \leq n^{p-1} \int_{[0,1]} |f|^p \, dm = n^{p-1} d_p(f, 0) < \varepsilon,$$

for each $j = 1, \ldots, n$. Thus each $f_j \in B(0; \varepsilon) \subseteq U$. As

$$\sum_{j=1}^{n} \frac{1}{n} f_j = \frac{1}{n} f_1 + \cdots + \frac{1}{n} f_n = f \mathcal{X}_{[t_0,t_1)} + \cdots + f \mathcal{X}_{[t_{n-1},t_n)} = f,$$

and $U$ is convex, it follows that $f \in U$, that is $U = L_p[0, 1]$. We have proved that the only convex neighborhood of the origin is $L_p[0, 1]$ itself. It follows that $L_p[0, 1]$ is not locally convex, in particular it is not normed. Furthermore, by Proposition 8.2.1 we conclude that $L_p[0, 1]' = \{0\}$.

(f) Let $E$ be a locally convex space and let $M$ be a vector subspace of $E$. Then $M$ equipped with the topology induced by $E$ is also a locally convex space.

For locally convex spaces, the following stronger version of Proposition 8.1.7 holds, which says that in every locally convex space the origin has a neighborhood basis formed by balanced, convex and closed sets:

**Proposition 8.2.4** *Let $\mathcal{U}$ be the family of all neighborhoods of the origin of a locally convex space $E$. If $U \in \mathcal{U}$, then*

(a) *$U$ is absorbing.*
(b) *There exists $V \in \mathcal{U}$ such that $V + V \subseteq U$.*
(c) *There exists $V_0 \in \mathcal{U}$, $V_0$ balanced, convex and closed, such that $V_0 \subseteq U$.*

*Proof* Conditions (a) and (b) are already known. We shall prove (c). Let $U$ be a neighborhood of the origin. Contained in $U$ there is a closed neighborhood $U_1$ of the origin; and contained in $U_1$ we can consider a convex neighborhood $U_2$ of the origin, because $E$ is a locally convex space. By Proposition 8.1.7(c) we know that $U_2$ contains a balanced neighborhood $W$ of the origin. Since $W$ is balanced, we have

$$sW \subseteq W \text{ and } s^{-1}W \subseteq W$$

for every scalar $s$ with $|s| = 1$. Therefore

$$W = s^{-1}(sW) \subseteq s^{-1}W \text{ and } W = s(s^{-1}W) \subseteq sW,$$

and we conclude that

$$sW = W = s^{-1}W$$

for every scalar $s$ with modulus 1. In particular, for each such $s$,

$$s^{-1}W = W \subseteq U_2,$$

therefore $W \subseteq sU_2$. Define $V = \bigcap_{|s|=1} sU_2$. Thus $W \subseteq V \subseteq U_2$ and $V$ is a neighborhood of the origin contained in $U_2$. In addition, each $sU_2$ is convex, and as the intersection of convex sets is still convex, it follows that the neighborhood $V$ is convex.

## 8.2 Locally Convex Spaces

We will now show that the neighborhood $V$ is balanced. Let $t$ be a scalar with $|t| \leq 1$. If $v \in V$, from the definition of $V$ it follows that $v \in sU_2$ for every scalar $s$ with $|s| = 1$. Therefore, for any such scalar $s$,

$$tv \in tsU_2 = s_1 |t| U_2,$$

for $s_1 = se^{i\theta}$, where $t = |t| e^{i\theta}$. Now we show that

$$s_1 |t| U_2 \subseteq s_1 U_2.$$

Indeed, if $u \in U_2$, then

$$s_1 |t| u = |t|(s_1 u) + (1 - |t|)0 \in s_1 U_2,$$

because $s_1 U_2$ is convex and contains the origin, as $U_2$ is convex and contains the origin. Therefore $tv \in s_1 U_2$. As $s$ is any scalar of modulus 1, $s_1$ is any of all these scalars as well, hence

$$tv \in \bigcap_{|s_1|=1} s_1 U_2 = V.$$

Then we find a convex balanced neighborhood $V$ of the origin contained in $U_2$. From Proposition 8.1.8 it follows that $\overline{V}$ is also a convex balanced neighborhood of the origin. As $U_1$ is closed and $V \subseteq U_2 \subseteq U_1 \subseteq U$, it follows that $V_0 := \overline{V}$ is a convex, closed and balanced neighborhood of the origin and $V_0 \subseteq U_1 \subseteq U$. ∎

Convex balanced sets play a central role in the theory of locally convex spaces. Next we look at some stability properties of such sets, the proofs of which we leave to the reader.

**Proposition 8.2.5**

(a) *Linear combinations and arbitrary intersections of convex balanced sets are convex balanced.*
(b) *Let $E$ and $F$ be vector spaces, let $T: E \longrightarrow F$ be linear, and let $A \subseteq E$ and $B \subseteq F$ be convex balanced sets. Then the sets $T(A) \subseteq F$ and $T^{-1}(B) \subseteq E$ are convex balanced.*

As homeomorphisms map neighborhood bases to neighborhood bases, the following result now follows immediately:

**Proposition 8.2.6** *A topological vector space isomorphic to a locally convex space is also locally convex.*

The next result generalizes Proposition 8.2.1 and reinforces the need for convex neighborhoods for the existence of linear and continuous operators.

**Proposition 8.2.7** *Let $E$ be a topological vector space without proper convex neighborhoods of the origin and let $F$ be a Hausdorff locally convex space. Then the only continuous linear operator from $E$ to $F$ is the null operator.*

*Proof* Let $\mathcal{U}$ be a basis of convex balanced neighborhoods of the origin of $F$. The Hausdorff property ensures that $\{0\} = \bigcap_{A \in \mathcal{U}} A$. Let $T \colon E \longrightarrow F$ be linear and continuous. From Proposition 8.2.5 we know that, for each $A \in \mathcal{U}$, $T^{-1}(A)$ is convex balanced. The continuity of $T$ implies that $T^{-1}(A)$ is a neighborhood of the origin in $E$. By hypothesis it follows that $T^{-1}(A) = E$ for every $A \in \mathcal{U}$. Thus,

$$\ker(T) = T^{-1}(\{0\}) = T^{-1}\left(\bigcap_{A \in \mathcal{U}} A\right) = \bigcap_{A \in \mathcal{U}} T^{-1}(A) = E,$$

proving that $T$ is the null operator. ∎

Combining the proposition above with Example 8.2.3(e) we obtain:

**Corollary 8.2.8** *For $0 < p < 1$, the only continuous linear operator from $L_p[0, 1]$ to any Hausdorff locally convex space is the null operator.*

## 8.3 Seminorms and Topologies

In this section, we will learn a widely used method to introduce locally convex topologies in vector spaces. Moreover, we will see that every locally convex space can be generated by this method. In a normed space we usually consider the topology generated by the norm. We will see that families of seminorms also generate topologies, not always normed but always locally convex.

**Definition 8.3.1** A *seminorm* on the vector space $E$ is a function $p \colon E \longrightarrow \mathbb{R}$ that satisfies the following conditions:

(a) $p(x) \geq 0$ for every $x \in E$.
(b) $p(ax) = |a| \, p(x)$ for all $a \in \mathbb{K}$ and $x \in E$.
(c) $p(x + y) \leq p(x) + p(y)$ for any $x, y \in E$.

From conditions (b) and (c) we conclude that $p(0) = 0$ and that

$$|p(x) - p(y)| \leq p(x - y) \text{ for all } x, y \in E. \tag{8.4}$$

**Example 8.3.2** It is clear that every norm is a seminorm. Let $E$ be a vector space and let $\varphi$ be a linear functional on $E$. Then the correspondence $x \in E \mapsto |\varphi(x)|$ is a seminorm on $E$. More generally, given a vector space $E$, a normed space $F$ and $T \colon E \longrightarrow F$ linear, the correspondence $x \in E \mapsto \|T(x)\|$ is a seminorm on $E$.

## 8.3 Seminorms and Topologies

We will now introduce the topology determined on a vector space by a family of seminorms. Let $\mathcal{P}$ be a family of seminorms on the vector space $E$. Given $x_0 \in E$, $\varepsilon > 0$, $n \in \mathbb{N}$ and $p_1, \ldots, p_n \in \mathcal{P}$, we define

$$V(x_0, p_1, \ldots, p_n; \varepsilon) = \{x \in E : p_i(x - x_0) < \varepsilon, \ i = 1, \ldots, n\}.$$

For each $x \in E$, we call $\mathcal{V}_x$ the collection of all subsets of $E$ of the form $V(x, p_1, \ldots, p_n; \varepsilon)$, with $n \in \mathbb{N}$, $p_1, \ldots, p_n \in \mathcal{P}$ and $\varepsilon > 0$.

**Theorem 8.3.3** *Let $\mathcal{P}$ be a family of seminorms on the vector space $E$. Then:*

(a) *There exists a topology $\tau_\mathcal{P}$ on $E$ such that $\mathcal{V}_x$ is a neighborhood basis for each $x \in E$, that is,*

$$\tau_\mathcal{P} = \{G \subseteq E : \text{for each } x \in G \text{ there exists } U \in \mathcal{V}_x \text{ such that } U \subseteq G\}.$$

(b) *$(E, \tau_\mathcal{P})$ is a locally convex space.*
(c) *Each seminorm $p \in \mathcal{P}$ is $\tau_\mathcal{P}$-continuous.*
(d) *$(E, \tau_\mathcal{P})$ is a Hausdorff space if and only if for each $0 \neq x \in E$ there exists a seminorm $p \in \mathcal{P}$ such that $p(x) \neq 0$.*

The topology $\tau_\mathcal{P}$ will be called the *topology determined (or generated) by the family of seminorms* $\mathcal{P}$. Despite the terminology, the reader should be aware that $\tau_\mathcal{P}$ *is not* the smallest topology on $E$ that makes the seminorms of the family $\mathcal{P}$ continuous in the sense of Sect. 6.1 (see Exercise 8.6.24). In Exercise 8.6.27 the reader will understand why we do not use here the construction of Sect. 6.1.

**Proof** The reader will have no difficulty in checking that the sets $\{\mathcal{V}_x\}_{x \in E}$ satisfy conditions (a)–(c) of Theorem B.15. This proves item (a).

To prove the continuity of the addition $AD: E \times E \longrightarrow E$, let $(x_0, y_0) \in E \times E$ be given and let $W$ be a neighborhood of $x_0 + y_0$ in $E$. Without loss of generality, consider $W$ as a basic neighborhood of $x_0 + y_0$. In this case, there exist $n \in \mathbb{N}$, $p_1, \ldots, p_n \in \mathcal{P}$ and $\varepsilon > 0$ such that $W = V(x_0+y_0, p_1, \ldots, p_n; \varepsilon)$. Defining $U := V(0, p_1, \ldots, p_n; \varepsilon)$, we have $W = x_0 + y_0 + U$. Then $V := V(0, p_1, \ldots, p_n; \varepsilon/2)$ is a neighborhood of the origin, therefore $(x_0 + V) \times (y_0 + V)$ is a neighborhood of $(x_0, y_0)$. As $V + V \subseteq U$, we have

$$AD((x_0 + V) \times (y_0 + V)) = x_0 + V + y_0 + V = x_0 + y_0 + V + V$$
$$\subseteq x_0 + y_0 + U = W,$$

proving the continuity of the addition $AD$ at $(x_0, y_0)$.

We now move on to checking the continuity of the scalar multiplication operation $SM: \mathbb{K} \times E \longrightarrow E$. To do so, consider $(a_0, x_0) \in \mathbb{K} \times E$ and $W$ as a basic neighborhood of $a_0 x_0$ in $E$. As before, there is a basic neighborhood of the origin, which we will denote by $W_0$, such that $a_0 x_0 + W_0 = W$. Consider $n \in \mathbb{N}$, $p_1, \ldots, p_n \in \mathcal{P}$ and $\varepsilon > 0$ such that $W_0 = V(0, p_1, \ldots, p_n; \varepsilon)$. For each $j = 1, \ldots, n$, define

$$\eta_j = \frac{\varepsilon}{2(p_j(x_0)+1)}, \quad \eta = \min\{\eta_j, j = 1, \ldots, n\},$$

$$\delta = \frac{\varepsilon}{2(|a_0|+\eta)}, \quad A = \{a \in \mathbb{K} : |a| < \eta\} \text{ and } V = V(0, p_1, \ldots, p_n; \delta).$$

Thus, $(a_0 + A) \times (x_0 + V)$ is a neighborhood of $(a_0, x_0)$. As $a_0 x_0 + W_0 = W$, to ensure the continuity of $SM$ it is enough to prove that

$$SM((a_0 + A) \times (x_0 + V)) \subseteq a_0 x_0 + W_0. \tag{8.5}$$

Indeed this occurs because, given $j \in \{1, \ldots, n\}$, $a \in a_0 + A$ and $x \in x_0 + V$,

$$p_j(ax - a_0 x_0) = p_j(ax - ax_0 + ax_0 - a_0 x_0) \leq p_j(ax - ax_0) + p_j(ax_0 - a_0 x_0)$$

$$= |a| p_j(x - x_0) + |a - a_0| p_j(x_0) \leq |a| \delta + \eta p_j(x_0)$$

$$\leq |a| \frac{\varepsilon}{2(|a_0|+\eta)} + \frac{\varepsilon}{2(p_j(x_0)+1)} p_j(x_0)$$

$$\leq (|a_0|+\eta) \frac{\varepsilon}{2(|a_0|+\eta)} + \frac{\varepsilon}{2(p_j(x_0)+1)} p_j(x_0) < \varepsilon,$$

hence $(ax - a_0 x_0) \in W_0$, which implies $ax \in a_0 x_0 + W_0$ and proves (8.5). The convexity of sets of the form $V(x, p_1, \ldots, p_n; \varepsilon)$ is immediate, so (b) is proved.

To prove (c), let $p \in \mathcal{P}$, $x_0 \in E$ and $\varepsilon > 0$ be given. Then $V(x_0, p; \varepsilon)$ is a neighborhood of $x_0$. For every $x \in V(x_0, p; \varepsilon)$, from (8.4) it follows that

$$|p(x) - p(x_0)| \leq p(x - x_0) < \varepsilon,$$

proving that $p$ is continuous at $x_0$.

Finally, we prove (d). Suppose that $E$ is a Hausdorff space. Then, given $x \in E$, $x \neq 0$, there is a basic neighborhood $V = V(0, p_1, \ldots, p_n; \varepsilon)$ of the origin that does not contain $x$. It follows that

$$p_j(x) = p_j(x - 0) \geq \varepsilon > 0$$

for some $j \in \{1, \ldots, n\}$. Conversely, suppose that for each $x \in E$ with $x \neq 0$, there exists $p \in \mathcal{P}$ such that $p(x) \neq 0$. Given $x, y \in E$, $x \neq y$, choose $p \in \mathcal{P}$ such that $p(x - y) > 0$. Then $V(x, p; p(x - y)/2)$ is a neighborhood of $x$ and $V(y, p; p(x - y)/2)$ is a neighborhood of $y$. From the triangle inequality it follows that $V(x, p; p(x - y)/2) \cap V(y, p; p(x - y)/2) = \emptyset$, consequently $(E, \tau_\mathcal{P})$ is a Hausdorff space. ∎

### Example 8.3.4

(a) Considering Theorem 8.3.3(a) and Propositions 6.2.2(b) and 6.3.2(b), it is clear that the weak topology $\sigma(E, E')$ on the normed space $E$ is generated

## 8.3 Seminorms and Topologies

by the family of seminorms $(x \in E \mapsto |\varphi(x)|)_{\varphi \in E'}$, and that the weak-star topology $\sigma(E', E)$ on $E'$ is generated by the family of seminorms $(\varphi \in E' \mapsto |\varphi(x)|)_{x \in E}$.

(b) Let $X$ be any set and let $E$ be a vector space (with the usual operations of functions) formed by functions $f : X \longrightarrow \mathbb{K}$. For example, we can consider the vector space of all functions from $X$ to $\mathbb{K}$. For each $t \in X$, the function

$$p_t : E \longrightarrow \mathbb{R}, \quad p_t(f) = |f(t)|,$$

is a seminorm on $E$. Taking $\mathcal{P} = \{p_t\}_{t \in X}$ we know that $(E, \tau_\mathcal{P})$ is a locally convex space. We shall check that the convergence of sequences in the topology $\tau_\mathcal{P}$ coincides with the pointwise convergence of sequences of functions (the same can be done for nets). That is, given $(f_n)_{n=1}^\infty$ and $f$ in $E$,

$$f_n \xrightarrow{\tau_\mathcal{P}} f \iff f_n(t) \longrightarrow f(t) \text{ for every } t \in X. \tag{8.6}$$

Suppose $f_n \xrightarrow{\tau_\mathcal{P}} f$ and take $t \in X$. Given $\varepsilon > 0$, $V(f, p_t; \varepsilon)$ is an open set in $E$ containing $f$. Therefore, there exists $n_0 \in \mathbb{N}$ such that $f_n \in V(f, p_t; \varepsilon)$ for every $n \geq n_0$. Then

$$|f_n(t) - f(t)| = p_t(f_n - f) < \varepsilon \text{ for every } n \geq n_0,$$

proving that $f_n(t) \longrightarrow f(t)$. Conversely, suppose that $f_n(t) \longrightarrow f(t)$ for every $t \in X$. Given a neighborhood $U$ of $f$, consider $k \in \mathbb{N}$, $t_1, \ldots, t_k \in X$ and $\varepsilon > 0$ such that $V(f, p_{t_1}, \ldots, p_{t_k}; \varepsilon) \subseteq U$. For each $j = 1, \ldots, k$, there exists $n_j \in \mathbb{N}$ such that $|f_n(t_j) - f(t_j)| < \varepsilon$ whenever $n \geq n_j$. Taking $n_0 = \max\{n_1, \ldots, n_k\}$ it is immediate that $f_n \in V(f, p_{t_1}, \ldots, p_{t_k}; \varepsilon) \subseteq U$ for every $n \geq n_0$. This shows that $f_n \xrightarrow{\tau_\mathcal{P}} f$. In view of (8.6), $\tau_\mathcal{P}$ is called the *topology of pointwise convergence*.

The goal now is to show that every locally convex space has its topology generated by a family of seminorms. It is necessary to establish the very close relationship between seminorms and the Minkowski functional, which we have already studied in Sect. 3.4. This is the purpose of the next two results.

**Proposition 8.3.5** *Let $A$ be a convex, balanced and absorbing subset of the vector space $E$. Then the Minkowski functional of $A$, defined by*

$$p_A : E \longrightarrow \mathbb{R}, \quad p_A(x) = \inf\{\lambda > 0 : x \in \lambda A\},$$

*is a seminorm on $E$. Furthermore,*

$$\{x \in E : p_A(x) < 1\} \subseteq A \subseteq \{x \in E : p_A(x) \leq 1\}. \tag{8.7}$$

**Proof** The function $p_A$ is well defined because $A$ is absorbing. It is clear that $p_A(x) \geq 0$. We first show that $p(ax) = |a| \, p(x)$ for all $a \in \mathbb{K}$ and $x \in E$. If $a = 0$, the desired equality follows trivially. Suppose $a \neq 0$ and write $a = |a|e^{i\theta}$. Let $\lambda > 0$ be such that $x \in \lambda A$. Therefore

$$ax \in a\lambda A = |a|\lambda e^{i\theta} A = |a|\lambda A,$$

because $A$ is balanced. Thus, $p_A(ax) \leq |a| \lambda$ whenever $x \in \lambda A$. It follows that

$$p_A(ax) \leq |a| \cdot \inf\{\lambda > 0 : x \in \lambda A\} = |a| \, p_A(x).$$

For the reverse inequality, let $\lambda > 0$ be such that $ax \in \lambda A$. Since $a \neq 0$,

$$x \in \frac{\lambda}{a} A = \frac{\lambda}{|a|} e^{-i\theta} A = \frac{\lambda}{|a|} A,$$

because $A$ is balanced, therefore $p_A(x) \leq \frac{\lambda}{|a|}$. Then $|a|p_A(x) \leq \lambda$ for every $\lambda > 0$ such that $ax \in \lambda A$. It follows that $|a|p_A(x) \leq p_A(ax)$.

We now prove the triangle inequality $p_A(x+y) \leq p_A(x) + p_A(y)$ for any $x, y \in E$. Let $\lambda, \mu > 0$ be such that $x \in \lambda A$ and $y \in \mu A$. Therefore $x + y \in \lambda A + \mu A$. Since $A$ is convex, $\frac{\lambda}{\lambda + \mu} \leq 1$, $\frac{\mu}{\lambda + \mu} \leq 1$ and $\frac{\lambda}{\lambda + \mu} + \frac{\mu}{\lambda + \mu} = 1$, we have

$$x + y \in (\lambda + \mu)\left(\frac{\lambda A + \mu A}{\lambda + \mu}\right) = (\lambda + \mu)\left(\frac{\lambda A}{\lambda + \mu} + \frac{\mu A}{\lambda + \mu}\right) \subseteq (\lambda + \mu)A.$$

Therefore $p_A(x+y) \leq \lambda + \mu$, from which follows the triangle inequality.

It remains to prove (8.7). The second inclusion is obvious; we prove the first. If $p_A(x) < 1$, then there exists $\lambda > 0$ such that $p_A(x) \leq \lambda < 1$ and $x \in \lambda A$. Then $x \in A$ because $A$ is balanced. Hence $\{x \in E : p_A(x) < 1\} \subseteq A$. ∎

It is an interesting exercise for the reader to compare the proof above with the proof of Proposition 3.4.5. The converse also holds:

**Proposition 8.3.6** *Let $p: E \longrightarrow \mathbb{R}$ be a seminorm on a vector space $E$. Then the set $A := \{x \in E : p(x) \leq 1\}$ is convex, balanced, absorbing and $p_A(x) = p(x)$ for every $x \in E$.*

**Proof** Let $a, b > 0$, $a + b = 1$, and $x, y \in A$ be given. From

$$p(ax + by) \leq ap(x) + bp(y) \leq a + b = 1,$$

we conclude that $ax + by \in A$, hence $A$ is convex.

If $|\lambda| \leq 1$ and $x \in A$, then $p(\lambda x) = |\lambda| \, p(x) \leq 1$, proving that $A$ is balanced. From the definition of $A$ and the properties of the seminorm, it easily follows that if $x \in \lambda_0 A$ and $|\lambda_1| \geq |\lambda_0|$, then $x \in \lambda_1 A$. To show that $A$ is absorbing, let $x \in E$.

## 8.3 Seminorms and Topologies

We first consider the case where $x \in A$. Since $x \in A = 1A$, it follows that $x \in \lambda A$ whenever $|\lambda| \geq 1$, which solves the problem in this case. Now we consider the case where $x \notin A$. In this case $p(x) > 1$, therefore $x = p(x) \frac{x}{p(x)} \in p(x)A$. Consequently, $x \in \lambda A$ whenever $|\lambda| \geq p(x)$.

We will now prove that $p_A(x) = p(x)$ for every $x$ in $E$. From Proposition 8.3.5 we know that

$$\{x \in E : p_A(x) < 1\} \subseteq A \subseteq \{x \in E : p_A(x) \leq 1\}.$$

Let $x \in E$. Suppose $p(x) = 0$. The definition of $A$ provides $\frac{x}{\lambda} \in A$ for every $\lambda > 0$, hence $x \in \lambda A$ for every $\lambda > 0$. It follows that $p_A(x) = 0 = p(x)$. Suppose now $p(x) \neq 0$. In this case $\frac{rx}{p(x)} \notin A$ regardless of the $r > 1$. By the first inclusion above we conclude that $p_A(\frac{rx}{p(x)}) \geq 1$, therefore $r p_A(x) = p_A(rx) \geq p(x)$ for every $r > 1$. Let $r \longrightarrow 1^+$ to obtain $p_A(x) \geq p(x)$. On the other hand, as $\frac{x}{p(x)} \in A$, we have $p_A(\frac{x}{p(x)}) \leq 1$ by the second inclusion above. Therefore, $p_A(x) \leq p(x)$. ∎

Next we will see that the topologies generated by families of seminorms exhaust all locally convex topologies. Recall that every locally convex space admits a neighborhood basis of the origin consisting of convex balanced sets (Proposition 8.2.4).

**Theorem 8.3.7** *Let $\mathcal{B}_0$ be a basis of convex balanced neighborhoods of the origin in the locally convex space $E$. Then the topology of $E$ is generated by the family of seminorms $\mathcal{P} = \{p_V\}_{V \in \mathcal{B}_0}$.*

**Proof** Each neighborhood $V \in \mathcal{B}_0$ is absorbing by Proposition 8.1.7, so $p_V$ is a seminorm on $E$ and

$$\{x \in E : p_V(x) < 1\} \subseteq V \subseteq \{x \in E : p_V(x) \leq 1\}$$

by Proposition 8.3.5. It follows that, for all $V \in \mathcal{B}_0$ and $\varepsilon > 0$,

$$\{x \in E : p_V(x) < \varepsilon\} \subseteq \varepsilon V \subseteq \{x \in E : p_V(x) \leq \varepsilon\} \subseteq \{x \in E : p_V(x) < 2\varepsilon\}.$$

It is then enough to show that the sets of the form $\{x \in E : p_V(x) < \varepsilon\}$, with $V \in \mathcal{B}_0$ and $\varepsilon > 0$, form a neighborhood basis of the origin in the topology $\tau_\mathcal{P}$. It is clear that each set of this form is an open neighborhood of the origin in the topology $\tau_\mathcal{P}$. Let $0 \in U \in \tau_\mathcal{P}$. From Theorem 8.3.3 there exist $\varepsilon > 0$, $n \in \mathbb{N}$ and $V_1, \ldots, V_n \in \mathcal{B}_0$ such that $V(0, p_{V_1}, \ldots, p_{V_n}; \varepsilon) \subseteq U$. As $\mathcal{B}_0$ is a neighborhood basis of the origin, there exists $V \in \mathcal{B}_0$ such that $V \subseteq V_1 \cap \cdots \cap V_n$. From the definition of the Minkowski functional it follows that $p_{V_j} \leq p_V$ for each $j = 1, \ldots, n$, so

$$\{x \in E : p_V(x) < \varepsilon\} \subseteq V(0, p_{V_1}, \ldots, p_{V_n}; \varepsilon) \subseteq U,$$

which completes the proof. ∎

Combine Theorems 8.3.3 and 8.3.7 to obtain the

**Corollary 8.3.8** *A topological vector space is a locally convex space if and only if its topology is generated by a family of seminorms.*

## 8.4 Revisiting Hahn–Banach and Goldstine

To give the reader an idea of which important parts of the theory of normed spaces can be extended to locally convex spaces, we will prove in this section some locally convex versions of the Hahn–Banach Theorem. And, to settle a debt with the reader in Sect. 6.4, we will prove the complex case of the Goldstine Theorem.

**Theorem 8.4.1 (Hahn–Banach Theorem for Locally Convex Spaces)** *Let $E$ be a locally convex space and let $M$ be a vector subspace of $E$. Then every bounded linear functional $\varphi_0 \in M'$ can be extended to $E'$ preserving linearity and continuity, that is, there exists $\varphi \in E'$ such that $\varphi(x) = \varphi_0(x)$ for every $x \in M$.*

*Proof* The continuity of $\varphi_0$ implies that the set

$$U = \{x \in M : |\varphi_0(x)| \leq 1\}$$

is a closed neighborhood of the origin in $M$. There then exists an open neighborhood $W$ of the origin in $E$ such that

$$U \supseteq \operatorname{int}(U) = M \cap W,$$

where $\operatorname{int}(U)$ denotes the interior of the set $U$. As $E$ is locally convex, there exists a convex balanced neighborhood $V$ of the origin such that $V \subseteq W$. In particular, $M \cap V \subseteq U$. Since every neighborhood of the origin is absorbing (Proposition 8.2.4), it follows that $V$ is absorbing. From Proposition 8.3.5 we know that the Minkowski functional $p_V$ is a seminorm on $E$ and

$$\{x \in E : p_V(x) < 1\} \subseteq V \subseteq \{x \in E : p_V(x) \leq 1\}.$$

Next we check that

$$|\varphi_0(x)| \leq p_V(x) \text{ for every } x \in M. \tag{8.8}$$

Let $x \in M$ and $\varepsilon > 0$. As $p_V\left(\frac{x}{p_V(x)+\varepsilon}\right) < 1$, it follows that $\frac{x}{p_V(x)+\varepsilon} \in V$ and consequently

$$\frac{x}{p_V(x)+\varepsilon} \in V \cap M \subseteq U.$$

## 8.4 Revisiting Hahn–Banach and Goldstine

From this and from the definition of $U$,

$$\left|\varphi_0\left(\frac{x}{p_V(x)+\varepsilon}\right)\right| \leq 1,$$

and consequently $|\varphi_0(x)| \leq p_V(x) + \varepsilon$. Let $\varepsilon \longrightarrow 0^+$ to obtain (8.8). As the seminorm $p_V$ satisfies conditions (3.2) and (3.3), by Theorem 3.1.2 there exists a linear functional $\varphi \colon E \longrightarrow \mathbb{K}$ that extends $\varphi_0$ to $E$ and

$$|\varphi(x)| \leq p_V(x) = \inf\{\lambda > 0 : x \in \lambda V\}$$

for every $x \in E$. It follows that, for any $\delta > 0$, $|\varphi(x)| \leq \delta$ for every $x \in \delta V$. But $\delta V$ is a neighborhood of the origin, therefore the continuity of $\varphi$ follows from Proposition 8.1.12. ∎

Local convexity came into play in Proposition 8.2.1 as a necessary condition for the existence of non-zero continuous linear functionals. Next we will see that, when accompanied by the Hausdorff property, local convexity is also a sufficient condition:

**Corollary 8.4.2** *Let $E$ be a locally convex Hausdorff space. For each $0 \neq x \in E$ there exists $\varphi \in E'$ such that $\varphi(x) = 1$. In particular, $E' \neq \{0\}$ if $E \neq \{0\}$.*

*Proof* Given $0 \neq x \in E$, call $M$ the 1-dimensional subspace of $E$ generated by $x$ and consider the functional

$$\varphi_0 \colon M \longrightarrow \mathbb{K}, \quad \varphi_0(\lambda x) = \lambda.$$

The linearity of $\varphi_0$ is obvious. As $E$ is a Hausdorff space, the set $\{0\} = \ker(\varphi_0)$ is closed in $E$, so it is also closed in $M$. From Proposition 8.1.12 it follows that the linear functional $\varphi_0$ is continuous in $M$. The result follows from Theorem 8.4.1. ∎

We will now work towards proving a geometric version of the Hahn-Banach Theorem for locally convex spaces. We need some preliminary results.

**Proposition 8.4.3** *The following statements are equivalent for a seminorm $p$ on a topological vector space $E$:*

(a) *$p$ is continuous at the origin.*
(b) *$p$ is continuous.*
(c) *$p$ is bounded in some neighborhood of the origin.*

*Proof* (a) $\Longrightarrow$ (b) Let $x_0 \in E$ and let $(x_\alpha)_\alpha$ be a net in $E$ converging to $x_0$. From Proposition 8.1.4 we know that the translation

$$x \in E \mapsto x - x_0 \in E$$

is a homeomorphism, therefore $x_\alpha - x_0 \longrightarrow 0$. From the continuity of $p$ at the origin it follows that $p(x_\alpha - x_0) \longrightarrow p(0) = 0$. From inequality (8.4) we have

$$0 \leq |p(x_\alpha) - p(x_0)| \leq p(x_\alpha - x_0) \longrightarrow 0,$$

therefore $p(x_\alpha) \longrightarrow p(x_0)$. This proves the continuity of $p$ at $x_0$.
(a) $\Longrightarrow$ (c) Since $p(0) = 0$, by hypothesis there exists a neighborhood $U$ of the origin such that $p(U) \subseteq \{t \in \mathbb{K} : |t| < 1\}$. Therefore, $p$ is bounded in $U$.
(c) $\Longrightarrow$ (a) By assumption there exist a neighborhood $V$ of the origin and a constant $C > 0$ such that $|p(x)| < C$ for every $x \in V$. Given $\varepsilon > 0$, $\frac{\varepsilon}{C}V$ is a neighborhood of the origin and

$$|p(x) - p(0)| = |p(x)| < \varepsilon$$

for every $x \in \frac{\varepsilon}{C}V$. This proves the continuity of $p$ at the origin. ∎

**Corollary 8.4.4** *Let $A$ be a convex balanced neighborhood of the origin of a topological vector space. Then the Minkowski functional $p_A$ is continuous.*

**Proof** The functional $p_A$ is well defined because $A$ is absorbing by Proposition 8.1.7. As $p_A(x) \leq 1$ for every $x \in A$, $p_A$ is bounded in the neighborhood of the origin $A$, therefore continuous by Proposition 8.4.3. ∎

**Proposition 8.4.5** *If $V$ is a convex balanced neighborhood of the origin of the topological vector space $E$, then*

$$\overline{V} = \{x \in E : p_V(x) \leq 1\}.$$

*In particular, if $V$ is a convex, balanced, and closed neighborhood of the origin, then*

$$V = \{x \in E : p_V(x) \leq 1\}.$$

**Proof** The continuity of $p_V$ ensures that the set $\{x \in E : p_V(x) \leq 1\}$ is closed. As $V$ is absorbing by being a neighborhood of the origin, from Proposition 8.3.5 we know that

$$\{x \in E : p_V(x) < 1\} \subseteq V \subseteq \{x \in E : p_V(x) \leq 1\},$$

therefore $\overline{V} \subseteq \{x \in E : p_V(x) \leq 1\}$. On the other hand, let $x_0 \in E$ be such that $p_V(x_0) \leq 1$ and let $W$ be a neighborhood of $x_0$. It is easy to see that the linear operator

$$T : \mathbb{K} \longrightarrow E, \quad T(a) = ax_0,$$

## 8.4 Revisiting Hahn–Banach and Goldstine

is continuous. As $T(1) = x_0$, there exists $\varepsilon > 0$ such that $ax_0 = T(a) \in W$ whenever $|a - 1| < \varepsilon$. By Proposition 8.3.5 we know that $p_V$ is a seminorm, therefore

$$p_V(ax_0) = |a| p_V(x_0) \leq |a|.$$

Thus, $p_V(ax_0) < 1$ — hence $ax_0 \in V$ — whenever $|a| < 1$. Taking then $a \in \mathbb{K}$ such that $|a - 1| < \varepsilon$ and $|a| < 1$, we have $ax_0 \in V \cap W$. In particular, $V \cap W \neq \emptyset$, therefore $x_0 \in \overline{V}$. ∎

In addition to extending the geometric forms of the Hahn-Banach Theorem to the context of locally convex spaces, the next result will be particularly useful in proving the Goldstine Theorem.

**Theorem 8.4.6 (Geometric Form of the Hahn-Banach Theorem for Locally Convex Spaces)** *Let $A$ be a convex, balanced, and closed subset of a locally convex space $E$, with $0 \in A$. For each $b \in E$, $b \notin A$ and $c > 0$, there exists a continuous linear functional $\varphi \in E'$ such that*

$$\varphi(b) = c > \sup\{|\varphi(x)| : x \in A\}.$$

*Proof* First we prove that there exists a convex, balanced, and closed neighborhood of the origin $U$ such that

$$(b + U) \cap A = \emptyset. \tag{8.9}$$

Indeed, as $A$ is closed and $b \notin A$, $E - A$ is open and contains $b \neq 0$. Thus, $-b + (E - A)$ is an open neighborhood of the origin (Corollary 8.1.5). By Proposition 8.2.4 there exists a convex, balanced, and closed neighborhood of the origin $U$ such that

$$0 \in U \subseteq (E - A) - b.$$

Therefore $b + U \subseteq E - A$, proving (8.9). Now we see that we can conclude that

$$\left(b + \frac{1}{2}U\right) \cap \left(A + \frac{1}{2}U\right) = \emptyset. \tag{8.10}$$

To do so, suppose that there exists $z \in \left(b + \frac{1}{2}U\right) \cap \left(A + \frac{1}{2}U\right)$. In this case consider $x, y \in U$ and $a \in A$ such that

$$b + \frac{1}{2}x = z = a + \frac{1}{2}y,$$

and we obtain

$$b + \left(\frac{1}{2}x + \frac{1}{2}(-y)\right) = a \in A.$$

As $U$ is balanced, $-y \in U$; and as $U$ is convex, $\frac{1}{2}x + \frac{1}{2}(-y) \in U$. Therefore

$$b + \left(\frac{1}{2}x + \frac{1}{2}(-y)\right) \in (b + U) \cap A,$$

which contradicts (8.9) and, therefore, proves (8.10). In particular, $b \notin \left(b + \frac{1}{2}U\right) \cap \left(A + \frac{1}{2}U\right)$, from which it follows that

$$b \notin V := \overline{A + \frac{1}{2}U},$$

because $\left(b + \frac{1}{2}U\right)$ is a neighborhood of $b$. As $A$ and $U$ are convex balanced, $A + \frac{1}{2}U$ is a convex balanced neighborhood of the origin. Therefore $V = \overline{A + \frac{1}{2}U}$ is a convex, balanced and closed neighborhood of the origin by Proposition 8.1.8. From Proposition 8.4.5,

$$V = \{x \in E : p_V(x) \leq 1\},$$

so $p_V(b) > 1$ because $b \notin V$. Consider the linear functional

$$\varphi_0 : [b] \longrightarrow \mathbb{K}, \quad \varphi_0(\lambda b) = \lambda c,$$

and the seminorm

$$q : E \longrightarrow \mathbb{K}, \quad q(x) = \frac{c p_V(x)}{p_V(b)}.$$

Note that

$$|\varphi_0(\lambda b)| = |c\lambda| = q(\lambda b)$$

for every scalar $\lambda$. Theorem 3.1.2 then guarantees the existence of a linear functional $\varphi : E \longrightarrow \mathbb{K}$ such that $\varphi(\lambda b) = \varphi_0(\lambda b)$ for every scalar $\lambda$ and $|\varphi(x)| \leq q(x)$ for every $x \in E$. The seminorm $q$ is continuous, because the Minkowski functional $p_V$ is continuous by Corollary 8.4.4. It follows then that $\varphi$ is continuous. Observe that $A \subseteq V$ because $0 \in U$. Then, for every $x \in A$, $p_V(x) \leq 1$, hence

$$|\varphi(x)| \leq q(x) \leq \frac{c}{p_V(b)} < c = \varphi(b).$$

Thus, $\sup\{|\varphi(x)| : x \in A\} \le \frac{c}{pv(b)} < \varphi(b)$. ∎

We finish the chapter with a proof of Goldstine's Theorem that works in both real and complex cases. Recall that, given a normed space $E$, $J_E \colon E \longrightarrow E''$ denotes the canonical embedding given by $J_E(x)(\varphi) = \varphi(x)$ for all $x \in E$ and $\varphi \in E'$.

**Theorem 8.4.7 (Goldstine's Theorem)** *Let $E$ be a Banach space. Then $J_E(B_E)$ is dense in $B_{E''}$ in the weak-star topology $\sigma(E'', E')$ of $E''$.*

*Proof* The inclusion $\overline{J_E(B_E)}^{\sigma(E'',E')} \subseteq B_{E''}$ was proved at the beginning of the proof of Theorem 6.4.4, so it is enough to prove the reverse inclusion. By Proposition 8.2.5(b), the set $J_E(B_E)$ is convex balanced. Therefore, the set $\overline{J_E(B_E)}^{\sigma(E'',E')}$ is convex balanced by Proposition 8.1.8 and, obviously, it is closed in the weak-star topology and contains the origin. Let $f \in E'' - \overline{J_E(B_E)}^{\sigma(E'',E')}$. As the weak-star topology is locally convex, by Theorem 8.4.6 there exists $\Phi \in (E'', \sigma(E'', E'))'$ such that $\Phi(f) \in \mathbb{R}$ and

$$\Phi(f) > \sup\left\{|\Phi(g)| : g \in \overline{J_E(B_E)}^{\sigma(E'',E')}\right\}.$$

*A fortiori*,

$$\Phi(f) > \sup\{|\Phi(g)| : g \in J_E(B_E)\}. \tag{8.11}$$

But, by Corollary 6.3.7, $(E'', \sigma(E'', E'))' = J_{E'}(E')$; therefore there exists $\varphi \in E'$ such that $J_{E'}(\varphi) = \Phi$. The inequality (8.11) can then be rewritten in the form

$$f(\varphi) > \sup\{|g(\varphi)| : g \in J_E(B_E)\}.$$

As for each $g \in J_E(B_E)$ there exists a unique $x \in B_E$ such that $g = J_E(x)$. Then,

with $f(\varphi) > \sup\{|\varphi(x)| : x \in B_E\} = \|\varphi\|$

and $f\left(\frac{\varphi}{\|\varphi\|}\right) > 1$, from which it follows that $f \notin B_{E''}$. ∎

## 8.5 Comments and Historical Notes

The topic of topological vector spaces is very vast and there are many books specifically dealing with the subject. Therefore, the material seen in this chapter should be understood as a brief introduction to the topic. For more in-depth studies we suggest [35, 48, 59].

The concept of topological vector space was formalized by A. N. Kolmogorov in 1934. However, in 1933 S. Mazur and W. Orlicz already considered vector spaces equipped with metrizable topologies. The first to generalize to vector spaces equipped with topologies with bases of convex neighborhoods was J. von Neumann in 1935. In the same year, A. N. Tychonoff coined the term *locally convex space*.

Balanced convex sets of topological vector spaces are usually called *absolutely convex*. The reason will become clear in Exercise 8.6.11. Proposition 8.2.4 then says that in every locally convex space the origin admits a neighborhood basis formed by absolutely convex closed sets.

In Example 8.2.3(e) we proved that $L_p[0, 1]' = \{0\}$ for $0 < p < 1$, a fact established by M. M. Day in 1940. For the reader's knowledge, we inform that the same does not happen with the spaces $\ell_p$. In fact, for $0 < p < 1$, $(\ell_p)'$ is normed, and, in a certain sense, quite large, since it is isometrically isomorphic to the Banach space $\ell_\infty$ by the same duality relation that establishes the formula $(\ell_1)' = \ell_\infty$ (see [35, Example 6.10B]).

The spaces $\ell_p$ and $L_p[a, b]$, $0 < p < 1$, are particular cases of an important class of metrizable topological vector spaces. Given $0 < p < 1$, a *p-norm* on the vector space $E$ is a function $\|\cdot\| \colon E \longrightarrow \mathbb{R}$ such that

(a) $\|x\| \geq 0$ for every $x \in E$ and $\|x\| = 0 \Longleftrightarrow x = 0$;
(b) $\|ax\| = |a| \cdot \|x\|$ for all $a \in \mathbb{K}$ and $x \in E$;
(c) $\|x + y\|^p \leq \|x\|^p + \|y\|^p$ for all $x, y \in E$.

In this case, the expression

$$d_p \colon E \times E \longrightarrow \mathbb{R}, \ d_p(x, y) = \|x - y\|^p,$$

defines a metric on $E$ and the topology induced by this metric makes $E$ a topological vector space. The pair $(E, \|\cdot\|)$ is called a *p-normed space* and, when it is complete in the $d_p$ metric, it is called a *p-Banach space*. As we have seen in the text, these spaces are generally not locally convex. The article [38] is a good read on the topic.

Other versions, equivalent statements and applications of the Hahn–Banach Theorem in locally convex spaces can be found in [48, Chapter 7]. For vector versions, along the lines of what we did in Sect. 3.2, Chapter 10 of the same reference is a good research source.

There are versions of the Banach–Steinhaus, the Closed Graph and the Open Mapping theorems for continuous linear operators between topological vector spaces satisfying certain conditions, many of them related to local convexity and metrizability. A good reference for the subject is [48, Section 11.9 and Chapter 14].

Theorem 8.4.7 was proven by H. Goldstine in 1938.

## 8.6 Exercises

**Exercise 8.6.1** Let $E$ be a normed space. Prove that $E'$ equipped with the weak-star topology is a topological vector space.

## 8.6 Exercises

**Exercise 8.6.2** It is said that $(G, \cdot, \tau)$ is a *topological group* if $(G, \cdot)$ is a group and $\tau$ is a topology on $G$ that makes the functions

$$(x, y) \in G \times G \mapsto x \cdot y \in G \text{ and } x \in G \mapsto x^{-1} \in G,$$

continuous. Let $(E, \tau)$ be a topological vector space. Prove that $(E, +, \tau)$ is an abelian (commutative) topological group.

**Exercise 8.6.3** Complete the details of Example 8.1.2(f).

**Exercise 8.6.4** Let $B$ be a subset of a topological vector space. Prove that $\bigcap_{t>1} (tB) \subseteq \overline{B}$.

**Exercise 8.6.5** Let $E$ and $F$ be real topological vector spaces with $F$ being Hausdorff. Prove that if $T: E \longrightarrow F$ is a continuous map and $T(x + y) = T(x) + T(y)$ for any $x, y \in E$, then $T$ is linear.

**Exercise 8.6.6\*** Let $0 < p < 1$ and let $(X, \Sigma, \mu)$ be a measure space. Define $L_p(X, \Sigma, \mu)$ as the set of measurable functions $f: X \longrightarrow \mathbb{R}$ such that $\int_X |f|^p \, d\mu < \infty$, with the usual convention of identifying functions that are equal $\mu$-almost everywhere. Prove that:

(a) $L_p(X, \Sigma, \mu)$ is a vector space with the pointwise operations of (classes of) functions.
(b) The expression

$$d_p(f, g) = \int_X |f - g|^p \, d\mu$$

defines a metric on $L_p(X, \Sigma, \mu)$.
(c) The metric space $(L_p(X, \Sigma, \mu), d_p)$ is complete.
(d) The topology induced by the metric $d_p$ makes $L_p(X, \Sigma, \mu)$ a topological vector space.

**Exercise 8.6.7** Show that the metric space $(\ell_p, d_p)$ from Example 8.1.2(e) is complete.

**Exercise 8.6.8** Call $s$ the vector space of all sequences of scalars with the usual operations of sequences. Prove that the expression

$$d\left((a_n)_{n=1}^\infty, (b_n)_{n=1}^\infty\right) = \sum_{n=1}^\infty 2^{-n} \frac{|a_n - b_n|}{1 + |a_n + b_n|},$$

defines a metric that makes $s$ a complete metrizable topological vector space.

**Exercise 8.6.9** Prove that if $A \subseteq \mathbb{K}$ is balanced, then $A$ is bounded or $A = \mathbb{K}$.

**Exercise 8.6.10** Let $A$ be a subset of a topological vector space $E$. Prove that:

(a) If $A$ is convex then $\overline{A}$ is also convex.
(b) If $A$ is a vector subspace of $E$ then $\overline{A}$ is also a vector subspace of $E$.

**Exercise 8.6.11** A subset $A$ of the vector space $E$ is said to be *absolutely convex* if $ax + by \in A$ for any $x, y \in A$ and scalars $a, b$ with $|a| + |b| \leq 1$. Show that $A$ is absolutely convex if and only if $A$ is convex balanced.

**Exercise 8.6.12** Consider the set $B = \{(x, y) \in \mathbb{R}^2 : y > x^2\}$ and call $X$ the topological space $(\mathbb{R}^2, \tau)$ where $\tau = \{\mathbb{R}^2, \emptyset, B\}$.

(a) Find a subspace $A$ of $X$ such that $\overline{A}$ is not convex.
(b) Does item (a) contradict Proposition 8.1.8(a)?

**Exercise 8.6.13** Let $A$ be a balanced subset of a vector space. Prove that the convex hull of $A$ (see Exercise 1.8.17) is also a balanced set.

**Exercise 8.6.14** Exhibit a convex set $A \subseteq \mathbb{R}^2$ whose balanced hull

$$\text{eq}(A) = \{\lambda x : |\lambda| \leq 1 \text{ and } x \in A\}$$

is not convex.

**Exercise 8.6.15** Consider $\mathbb{C}$ as a vector space over itself. Show that the set $\{z \in \mathbb{C} : |\text{Im } z| < 1\}$ is convex, absorbing, but not balanced.

**Exercise 8.6.16 (Product Space)** Let $E_1, \ldots, E_n$ be topological vector spaces. Prove that:

(a) $E_1 \times \cdots \times E_n$ equipped with the product topology is a topological vector space.
(b) $E_1 \times \cdots \times E_n$ is a locally convex space if and only if $E_1, \ldots, E_n$ are locally convex spaces.

**Exercise 8.6.17** Prove Proposition 8.2.5.

**Exercise 8.6.18** Prove inequality (8.4).

**Exercise 8.6.19** Let $\varphi$ be a linear functional on the vector space $E \neq \{0\}$. Prove that the seminorm $x \in E \mapsto |\varphi(x)|$ is a norm if and only if $E$ is 1-dimensional.

**Exercise 8.6.20** Let $p_1, \ldots, p_n$ be seminorms on the vector space $E$, $x_0 \in E$ and $\varepsilon > 0$. Show that $V(x_0, p_1, \ldots, p_n; \varepsilon) = x_0 + V(0, p_1, \ldots, p_n; \varepsilon)$.

**Exercise 8.6.21** Let $p_1$ and $p_2$ be seminorms on the vector space $E$. Show that

$$p(x) := p_1(x) + p_2(x) \quad \text{and} \quad q(x) := \max\{p_1(x), p_2(x)\}$$

are seminorms on $E$.

**Exercise 8.6.22** Prove item (a) of Theorem 8.3.3.

## 8.6 Exercises

**Exercise 8.6.23** Let $\mathcal{P}$ be a family of seminorms on a vector space $E$. Show that a net $(x_\alpha)_\alpha$ converges to zero in $(E, \tau_\mathcal{P})$ if and only if $p(x_\alpha) \longrightarrow 0$ for every seminorm $p \in \mathcal{P}$.

**Exercise 8.6.24** Under the same conditions of the previous exercise, show that it can occur $p(x_\alpha) \longrightarrow p(x)$ for every seminorm $p \in \mathcal{P}$ without the net $(x_\alpha)_\alpha$ being convergent to $x$ in the topology $\tau_\mathcal{P}$. Conclude that $\tau_\mathcal{P}$ does not always coincide with the smallest topology on $E$ that makes the seminorms of $\mathcal{P}$ continuous.

**Exercise 8.6.25** If $E$ is a normed space and $B$ denotes the closed ball of radius $r$ centered at the origin, show that $p_B(x) = \frac{\|x\|}{r}$ for every $x \in E$.

**Exercise 8.6.26** Check if the topology on $\mathbb{R}$ generated by the unitary family $\mathcal{P} = \{p\}$, composed by the seminorm $p(x) = |x|$, is Hausdorff.

**Exercise 8.6.27** Call $\tau$ the smallest topology on $\mathbb{R}$ that makes the seminorm $p(x) = |x|$ continuous. Show that $(\mathbb{R}, \tau)$ is not a topological vector space.

**Exercise 8.6.28** Let $E$ and $F$ be locally convex topological vector spaces and let $T: E \longrightarrow F$ be a linear operator. Show that $T$ is continuous if and only if for each continuous seminorm $q: F \longrightarrow \mathbb{R}$, $q \circ T$ is a continuous seminorm on $E$.

**Exercise 8.6.29** Let $p$ and $q$ be seminorms on the topological vector space $E$ such that $q$ is continuous and $p(x) \leq q(x)$ for every $x \in E$. Prove that $p$ is continuous.

**Exercise 8.6.30** Prove that a locally convex space $E$ is Hausdorff if and only if its dual $E'$ separates points of $E$, that is, for all $x, y \in E$, $x \neq y$, there exists $\varphi \in E'$ such that $\varphi(x) \neq \varphi(y)$.

**Exercise 8.6.31\*** In $C[0, 1]$, call $\tau_1$ the topology induced by the metric

$$d(f, g) = \int_0^1 \frac{|f(x) - g(x)|}{1 + |f(x) - g(x)|} dx,$$

and $\tau_2$ the topology generated by the family of seminorms $\{p_x\}_{x \in [0,1]}$ defined by $p_x(f) = |f(x)|$. Show that the identity $(C[0, 1], \tau_2) \longrightarrow (C[0, 1], \tau_1)$:

(a) Is sequentially continuous, that is, $f_n \xrightarrow{\tau_1} f$ whenever $f_n \xrightarrow{\tau_2} f$.
(b) Is not continuous.

**Exercise 8.6.32\*** Let $(X, \Sigma, \mu)$ be a measure space and call $L_0(X, \Sigma, \mu)$ the vector space of (classes of) measurable functions $f: X \longrightarrow \mathbb{R}$.

(a) Prove that the expression

$$d(f, g) = \inf\{\{1\} \cup \{\varepsilon > 0 : \mu\{x \in X : |f(x) - g(x)| > \varepsilon\} < \varepsilon\}$$

defines a metric on $L_0(X, \Sigma, \mu)$.
(b) Prove that a sequence in $L_0(X, \Sigma, \mu)$ is Cauchy with respect to the metric $d$ if and only if it is Cauchy in measure.

(c) Prove that convergence in the metric space $(L_0(X, \Sigma, \mu), d)$ coincides with convergence in measure.
(d) Prove that the metric space $(L_0(X, \Sigma, \mu), d)$ is complete.

The topology on $L_0(X, \Sigma, \mu)$ induced by the metric $d$ is called the *topology of convergence in measure*. From now on, we will consider $L_0(X, \Sigma, \mu)$ equipped with this topology.

(e) Prove that the addition $AD: L_0(X, \Sigma, \mu) \times L_0(X, \Sigma, \mu) \longrightarrow L_0(X, \Sigma, \mu)$ is continuous.
(f) In the case of the Lebesgue measure in $\mathbb{R}$ we write $L_0(\mathbb{R})$. Find a function $f \in L_0(\mathbb{R})$ such that the sequence $(n^{-1}f)_{n=1}^{\infty}$ does not converge in measure to zero. Conclude that, in this case, scalar multiplication is not continuous, therefore $L_0(\mathbb{R})$ is not a topological vector space.
(g) Prove that if $\mu(X) < \infty$, then scalar multiplication in $L_0(X, \Sigma, \mu)$ is continuous, so $L_0(X, \Sigma, \mu)$ is a topological vector space.
(h) In the case of the Lebesgue measure on the interval $[0, 1]$, we write $L_0[0, 1]$. Prove that the only convex neighborhood of the origin in $L_0[0, 1]$ is $L_0[0, 1]$ itself, and conclude that $L_0[0, 1]$ is not locally convex and has a trivial dual.

# Chapter 9
# Introduction to Nonlinear Analysis

*"... the world is nonlinear"*. In this chapter, we distance ourselves a bit from the rigid, yet very useful linear structure of the previous chapters, to venture into a nonlinear world. The aim of this chapter is to offer a small sample of the theory of nonlinear operators between Banach spaces, introducing some basic tools in the study of nonlinear problems, usually found in mathematical models.

From Sect. 9.1 to Sect. 9.5 we will explore properties related to the continuity of nonlinear operators. In Sects. 9.6 and 9.7 we will extend the techniques of Differential and Integral Calculus to functions between Banach spaces.

Throughout this chapter, the scalar field will always be the field of real numbers, although some of the results have natural versions for spaces over the field of complex numbers.

## 9.1 Continuity in the Norm Topology

Throughout the first chapters of this book, we have already worked with some examples of nonlinear maps. Certainly, the first example is the norm of a normed vector space itself; another illustrative example is the inner product on Hilbert spaces $H$, as a map from $H \times H$ to $\mathbb{R}$.

The first relevant notion in the study of maps between Banach spaces is continuity with respect to natural topologies. The definition of continuity in the general context of topological spaces can be found in Appendix B (Definition B.22); and, for functions between normed spaces, the usual characterization using $\varepsilon$ and $\delta$ is stated at the beginning of Chap. 2. In this section, we discuss continuity with respect to the strong topology, that is, the topology generated by the norm.

According to what we did in Example 1.1.2, we say that an arbitrary function $\Phi \colon M_1 \longrightarrow M_2$ between metric spaces is *bounded* if the set $\Phi(A)$ is bounded in $M_2$ whenever $A$ is bounded in $M_1$. From Theorem 2.1.1 we know that the notions of boundedness and continuity are equivalent for linear operators between normed spaces. However, in general, boundedness does not imply continuity. On the other hand, any continuous function defined on a finite dimensional normed vector space is bounded.

**Example 9.1.1** In Sect. 2.1 we defined the notion of Lipschitz function between metric spaces. For the particular case of normed spaces, it is clear that a function $\Phi\colon E \longrightarrow F$ will be Lipschitz if there exists a constant $K > 0$ such that $\|\Phi(x) - \Phi(y)\| \leq K\|x - y\|$ for all $x, y \in E$. In this case, the number $\inf\{K : \|\Phi(x) - \Phi(y)\| \leq K\|x - y\|\}$ is called the *Lipschitz norm of* $\Phi$ and denoted by $\|\Phi\|_{\text{Lip}}$. It is easily checked that every Lipschitz function is uniformly continuous and bounded.

**Example 9.1.2** Consider the map

$$m\colon L_p[0,1] \longrightarrow L_p[0,1]\,,\ m(f)(x) := |f(x)|.$$

By the triangle inequality we see that $\|m(f) - m(g)\|_p \leq \|f - g\|_p$, that is, $m$ is Lipschitz and $\|m\|_{\text{Lip}} = 1$.

The map studied in the example above suggests an important class of nonlinear operators acting between $L_p$-spaces, known in the literature as *substitution operators* or *Nemytskii operators*, which we will study next.

**Definition 9.1.3** Let $(X, \Sigma, \mu)$ be a measure space. We say that a map $\varphi\colon X \times \mathbb{R} \longrightarrow \mathbb{R}$ *satisfies the Carathéodory condition* if:

(a) The function

$$\varphi(\cdot, r)\colon X \longrightarrow \mathbb{R},\ x \in X \mapsto \varphi(x, r),$$

is measurable for every fixed $r \in \mathbb{R}$,
(b) The function

$$\varphi(x, \cdot)\colon \mathbb{R} \longrightarrow \mathbb{R},\ r \in \mathbb{R} \mapsto \varphi(x, r),$$

is continuous for almost every $x \in X$, that is, there exists $A \in \Sigma$ such that $\mu(A) = 0$ and $\varphi(x, \cdot)$ is continuous for every $x \notin A$.

From now on we will write $L_p(X)$ instead of $L_p(X, \Sigma, \mu)$.

**Theorem 9.1.4** *Let $(X, \Sigma, \mu)$ be a measure space, let $\varphi\colon X \times \mathbb{R} \longrightarrow \mathbb{R}$ be a map satisfying the Carathéodory condition, let $1 \leq p, q < \infty$ be given, and let $k \in L_q(X)$ be a positive function. Suppose there exists a constant $C > 0$ such that*

$$|\varphi(x, t)| \leq C|t|^{p/q} + k(x)\ \text{for almost every } x \in X \text{ and every } t \in \mathbb{R}.$$

*Then the substitution operator*

## 9.1 Continuity in the Norm Topology

$$\Phi\colon L_p(X) \longrightarrow L_q(X)\,, \quad \Phi(f)(x) := \varphi(x, f(x)),$$

*is well defined, bounded and continuous.*

**Proof** We leave it to the reader to prove that $\Phi(f)$ is measurable whenever $f$ is measurable. Note that

$$\|\Phi(f)\|_q \leq C \cdot \|f\|_p^{p/q} + \|k\|_q,$$

thus $\Phi$ is indeed well defined as a map from $L_p(X)$ to $L_q(X)$ and is bounded.

To show the continuity of $\Phi$, consider a sequence $(f_n)_{n=1}^\infty$ such that $f_n \longrightarrow f$ in $L_p(X)$. Suppose that $(\Phi(f_n))_{n=1}^\infty$ does not converge to $\Phi(f)$ in $L_q(X)$. In this case there exists a neighborhood $V$ of $\Phi(f)$ in $L_q(X)$ and a subsequence $(f_{n_j})_{j=1}^\infty$ of $(f_n)_{n=1}^\infty$ such that $\Phi(f_{n_j}) \notin V$ for every $j$. Since $f_{n_j} \longrightarrow f$ in $L_p(X)$, we know from Exercise 9.9.5 that there exists a subsequence $(f_{n_{j_k}})_{k=1}^\infty$ of $(f_{n_j})_{j=1}^\infty$ such that $f_{n_{j_k}} \longrightarrow f$ $\mu$-almost everywhere, and that there exists a function $g \in L_p(X)$ such that $|f_{n_{j_k}}| \leq g$ $\mu$-almost everywhere. Thus,

$$|\Phi(f_{n_{j_k}})(x)| \leq C|g(x)|^{p/q} + k(x) \in L_q(X),$$

for every $k$. It follows from the Dominated Convergence Theorem that $\Phi(f_{n_{j_k}}) \longrightarrow \Phi(f)$ in $L_q(X)$. This contradicts the fact that $\Phi(f_{n_j}) \notin V$ for every $j$, therefore $\Phi(f_n) \longrightarrow \Phi(f)$ in $L_q(X)$, proving the continuity of $\Phi$. ∎

Next example gives another family of continuous maps acting on Banach spaces.

**Example 9.1.5** Let $\varphi\colon \mathbb{R} \longrightarrow \mathbb{R}$ be a continuous function and let $E := C[0,1]$ be endowed with the supremum norm. Let us see that the composition map

$$\Phi\colon E \longrightarrow E\,, \quad \Phi(f)(x) := \varphi(f(x)),$$

is continuous. To verify this, we have to prove that $(\Phi(f_n))_{n=1}^\infty$ converges uniformly to $\Phi(f)$ in $[0,1]$ provided $f_n \longrightarrow f$ uniformly in $[0,1]$. Easily one checks that $(\Phi(f_n))_{n=1}^\infty$ converges pointwisely to $\Phi(f)$. Hence, it suffices to show that $(\Phi(f_n(x)))_{n=1}^\infty$ is a Cauchy sequence (Why?). By uniform convergence, there exists a compact set $K \subseteq \mathbb{R}$ such that $\bigcup_{n=1}^\infty f_n([0,1]) \subseteq K$. Hence, $\varphi$ restricted to $K$ is uniformly continuous. Given $\varepsilon > 0$, there exists a $\delta > 0$ such that $|\varphi(y_1) - \varphi(y_2)| < \varepsilon$ whenever $|y_1 - y_2| < \delta$. Since $f_n \longrightarrow f$ uniformly, $(f_n)_{n=1}^\infty$ is a Cauchy sequence. There exists $n_0 \in \mathbb{N}$ such that if $n, m \geq n_0$, then $|f_n(x) - f_m(x)| < \delta$ for every $x \in [0,1]$. Thus, for all $n, m \geq n_0$,

$$\|\Phi(f_n) - \Phi(f_m)\|_\infty = \sup_{x \in [0,1]} |\varphi(f_n(x)) - \varphi(f_m(x))| < \varepsilon.$$

## 9.2 Continuity in the Weak Topology

Considering the relevance of compactness as a topological invariant and considering the lack of local compactness in the norm topology on infinite dimensional spaces, weak topologies become vital in the study of problems modeled in normed spaces. Thus, understanding the behavior of nonlinear operators between normed spaces with respect to weak convergence is fundamental in nonlinear analysis. The aim of this section is to present to the reader some results on weak sequential continuity of nonlinear operators, one of the most relevant research topics in modern analysis.

**Definition 9.2.1** Let $E$, $F$ be normed spaces and let $\Phi \colon E \longrightarrow F$ be an arbitrary map. We say that:

(a) $\Phi$ is *sequentially weakly continuous* if $\Phi(x_n) \xrightarrow{w} \Phi(x)$ in $F$ whenever $x_n \xrightarrow{w} x$ in $E$.
(b) $\Phi$ is *demicontinuous* if $\Phi(x_n) \xrightarrow{w} \Phi(x)$ in $F$ whenever $x_n \longrightarrow x$ in $E$.
(c) $\Phi$ is *completely continuous* if $\Phi(x_n) \longrightarrow \Phi(x)$ in $F$ whenever $x_n \xrightarrow{w} x$ in $E$.

Completely continuous maps are obviously continuous and sequentially weakly continuous. It is also clear that every continuous or sequentially weakly continuous map is demicontinuous. The relation between continuity and weak sequential continuity is a bit more delicate. For linear operators, Proposition 6.2.9 ensures that weak sequential continuity is equivalent to continuity but, in general, this equivalence does not hold. We will illustrate this phenomenon in the next example, for which we recall the following classic notation: given any set $X$ and an arbitrary function $f \colon X \longrightarrow \mathbb{R}$, we define

$$f^+ \colon X \longrightarrow \mathbb{R}, \quad f^+(x) := \max\{0, f(x)\}.$$

**Example 9.2.2** It is immediate that the function

$$\varphi \colon [0, 2\pi] \times \mathbb{R} \longrightarrow \mathbb{R}, \quad \varphi(x, t) = \max\{0, t\},$$

satisfies the conditions of Theorem 9.1.4, therefore its associated substitution operator

$$p \colon L_2[0, \pi] \longrightarrow L_2[0, \pi], \quad p(f)(x) := f^+(x),$$

is continuous. In fact, it is simple to show that $p$ is Lipschitz. By Exercise 9.9.6, the sequence of functions $f_n(x) = \sin(nx)$, $n \in \mathbb{N}$, converges weakly to 0. By change of variables,

$$\int_0^\pi \sin^+(nx)\,dx = \frac{1}{n}\int_0^{n\pi} \sin^+(y)\,dy.$$

## 9.2 Continuity in the Weak Topology

Considering $n = 2s - 1$ we conclude that

$$\liminf_{n\to\infty} \int_0^{\frac{\pi}{2}} p(\sin(nx))dx \geq \lim_{s\to\infty} \frac{1}{(2s-1)} \sum_{k=1}^{s} \int_{2k\pi}^{(2k+1)\pi} \sin(y)dy$$

$$= \lim_{s\to\infty} \frac{2s}{(2s-1)} = 1.$$

Therefore, the substitution operator $p(f) := f^+$ is not sequentially weakly continuous.

**Definition 9.2.3** A Banach space $E$ is *compactly embedded* in the Banach space $F$ if $E$ is a subspace of $F$ and the inclusion $i: (E, \|\cdot\|_E) \longrightarrow (F, \|\cdot\|_F)$ is a compact operator.

The definition above is only of interest when the norm of $E$ is *not* equivalent to the norm induced by $F$, otherwise $E$ is finite dimensional (Exercise 9.9.12), and in this case the inclusion is automatically compact.

The main theorem of this section, although it has an elementary statement, is very useful in a number of applications. To prove it we need the following elementary topological lemma:

**Lemma 9.2.4** *Let $X$ be a topological space, let $(x_n)_{n=1}^{\infty}$ be a sequence in $X$ and $x \in X$. If $(x_n)_{n=1}^{\infty}$ does not converge to $x$ and every subsequence of $(x_n)_{n=1}^{\infty}$ has a convergent subsequence, then there exist a subsequence $(x_{n_j})_{j=1}^{\infty}$ and $x \neq y \in X$ such that $x_{n_j} \longrightarrow y$.*

**Theorem 9.2.5** *Let $F$ be a reflexive Banach space, let $\Phi: F \longrightarrow F$ be a demicontinuous operator, and let $E$ be a reflexive Banach space compactly embedded in $F$. If $\Phi(E) \subseteq E$ and the map $\Phi: E \longrightarrow E$ is bounded, then $\Phi: E \longrightarrow E$ is sequentially weakly continuous.*

**Proof** Let $x_n \xrightarrow{w} x$ in $E$. As weakly convergent sequences are bounded (Proposition 6.2.5) and $\Phi$ is a bounded operator on $E$, the sequence $(\Phi(x_n))_{n=1}^{\infty}$ is bounded in $E$, therefore all its subsequences are bounded on $E$. As $E$ is reflexive, Theorem 6.5.4 guarantees that every subsequence of $(\Phi(x_n))_{n=1}^{\infty}$ admits a weakly convergent subsequence in $E$. Suppose that $(\Phi(x_n))_{n=1}^{\infty}$ does not converge weakly to $\Phi(x)$ in $E$. By Lemma 9.2.4 there exists a subsequence $(\Phi(x_{n_j}))_{j=1}^{\infty}$ and there exists $y \in E$ with $\Phi(x) \neq y$ such that $\Phi(x_{n_j}) \xrightarrow{w} y$ in $E$. As the inclusion of $E$ in $F$ is compact, $E$ is reflexive and $x_{n_j} \xrightarrow{w} x$ in $E$, it follows from Proposition 7.2.8(b) that $x_{n_j} \longrightarrow x$ in $F$. By the demicontinuity of $\Phi$ in $F$ we conclude that

$$\Phi(x_{n_j}) \xrightarrow{w} \Phi(x) \text{ in } F. \tag{9.1}$$

Recalling again that the inclusion from $E$ to $F$ is compact and that $E$ is reflexive, the condition $\Phi(x_{n_j}) \xrightarrow{w} y$ in $E$ implies, using Proposition 7.2.8(b) once again, that

$$\Phi(x_{n_j}) \longrightarrow y \text{ in } F. \tag{9.2}$$

From (9.1) and (9.2) it follows that $\Phi(x) = y$. This contradiction proves that $\Phi(x_n) \xrightarrow{w} \Phi(x)$ in $E$ and completes the proof. ∎

Now we present an example of a completely continuous, hence sequentially weakly continuous, operator.

**Example 9.2.6** Let $T: L_2[0, 1] \longrightarrow C[0, 1]$ be a continuous linear operator. For example, choose a function $g \in C[0, 1]$ and define

$$T: L_2[0, 1] \longrightarrow C[0, 1], \quad T(f)(t) := \int_0^1 f(s)g(t-s)\,ds.$$

Initially, we note that $T: L_2[0, 1] \longrightarrow L_2[0, 1]$ is a compact operator. As $L_2[0, 1]$ is reflexive, by Proposition 7.2.8(b) it is enough to prove that $T$ is completely continuous. Indeed, given a sequence $f_n \xrightarrow{w} f$ in $L_2[0, 1]$, by Proposition 6.2.9 it follows that $T(f_n) \xrightarrow{w} T(f)$ in $C[0, 1]$. It is clear that, for each $t \in [0, 1]$, the correspondence

$$g \in C[0, 1] \mapsto g(t),$$

defines a continuous linear functional on $C[0, 1]$. Therefore $T(f_n)(t) \longrightarrow T(f)(t)$ for every $t \in [0, 1]$. By Proposition 6.2.5(a) there exists $K_1 > 0$ such that $\|T(f_n)\|_\infty \leq K_1$ for every $n$, therefore it follows from Lebesgue's Dominated Convergence Theorem that $T(f_n) \longrightarrow T(f)$ in $L_2[0, 1]$.

Next, consider any continuous function $\varphi: \mathbb{R} \longrightarrow \mathbb{R}$ and define the nonlinear operator

$$\Phi: L_2[0, 1] \longrightarrow L_2[0, 1], \quad \Phi(f)(x) := \varphi(T(f)(x)).$$

We will see that the operator $\Phi$ is completely continuous, therefore sequentially weakly continuous. Indeed, if $f_n \xrightarrow{w} f$ in $L_2[0, 1]$, then, as we saw, $T(f_n) \longrightarrow T(f)$ in $L_2[0, 1]$ and $|T(f_n(x))| \leq K_1$ for all $x \in [0, 1]$ and $n \in \mathbb{N}$. As continuous functions are bounded on bounded sets, there exists $K_2 > 0$ such that $|\varphi(T(f_n)(x))| \leq K_2$ for all $x \in [0, 1]$ and $n \in \mathbb{N}$. From Lebesgue's Dominated Convergence Theorem it follows that $\Phi(f_n) \longrightarrow \Phi(f)$ in $L_2[0, 1]$.

## 9.3 Minimization Problems

Several mathematical models present themselves as minimization problems of a certain function $\varphi \colon E \longrightarrow \mathbb{R}$ defined on a Banach space $E$. As we know from the first calculus course, the existence of minima and/or maxima is intrinsically related to compactness properties. In arbitrary infinite dimensional Banach spaces, we do not even have the compactness of the closed unit ball in the weak topology guaranteed. We know, by Kakutani's Theorem (Theorem 6.4.5), that such compactness is only guaranteed in reflexive spaces. On the other hand, the weak continuity of nonlinear operators is difficult to be checked. These facts make minimization problems in infinite dimensional spaces quite interesting.

The aim of this section is to present and apply a minimization technique for nonlinear operators defined on Banach spaces.

**Definition 9.3.1** Let $V$ be a vector space. A function $\varphi \colon V \longrightarrow \mathbb{R}$ is *convex* if

$$\varphi(tx + (1-t)y) \leq t\varphi(x) + (1-t)\varphi(y),$$

for all $t \in [0, 1]$ and $x, y \in V$. If the inequality is strict for $t \in (0, 1)$, the function is said to be *uniformly convex*.

**Theorem 9.3.2** *Let $E$ be a reflexive Banach space and let $\varphi \colon E \longrightarrow \mathbb{R}$ be a continuous and convex function. Suppose that $\varphi$ is coercive, that is, $\lim\limits_{\|x\| \to \infty} \varphi(x) = \infty$. Then, for every convex closed subset $C$ of $E$ there exists $x_0 \in C$ such that*

$$\varphi(x_0) = \min_{y \in C} \varphi(y).$$

*Proof* Initially we observe that from the continuity and convexity of $\varphi$ it follows that, for each $\lambda \in \mathbb{R}$, the set $\{y \in E : \varphi(y) \leq \lambda\}$ is closed and convex. Therefore, by Mazur's Theorem (Theorem 6.2.11), the set $\{y \in E : \varphi(y) \leq \lambda\}$ is weakly closed, thus the set $\{y \in E : \varphi(y) > \lambda\}$ is open in the weak topology for any $\lambda \in \mathbb{R}$. Suppose $x_n \xrightarrow{w} x_0$ in $E$. Given $\varepsilon > 0$,

$$x_n \in \{y \in E : \varphi(y) > \varphi(x_0) - \varepsilon\}$$

for a sufficiently large $n$. We can conclude that

$$\liminf_{n \to \infty} \varphi(x_n) \geq \varphi(x_0) \tag{9.3}$$

whenever $x_n \xrightarrow{w} x_0$ in $E$.

Now we return to the proof of the theorem. Given $C \subseteq E$ convex and closed, we can consider a sequence $(x_n)_{n=1}^{\infty}$ in $C$ such that $\varphi(x_n) \longrightarrow \inf\limits_{y \in C} \varphi(y) < +\infty$. From the coercivity of $\varphi$ we conclude that $(x_n)_{n=1}^{\infty}$ is a bounded sequence. Since

$E$ is reflexive, by Theorem 6.5.4 there exist a subsequence $\left(x_{n_j}\right)_{j=1}^{\infty}$ and a vector $x_0 \in E$ such that $x_{n_j} \xrightarrow{w} x_0$. Applying Mazur's Theorem once more, now for the convex closed set $C$, we know that $C$ is weakly closed, hence $x_0 \in C$. And as $\varphi(x_{n_j}) \longrightarrow \inf_{y \in C} \varphi(y)$, from (9.3) it follows that

$$\varphi(x_0) \leq \liminf_{j \to \infty} \varphi(x_{n_j}) = \lim_{j \to \infty} \varphi(x_{n_j}) = \inf_{y \in C} \varphi(y) \leq \varphi(x_0).$$

We conclude that $\varphi(x_0) = \inf_{y \in C} \varphi(y) = \min_{y \in C} \varphi(y)$. ∎

As an application, we present a new proof—which is actually a generalization—of the orthogonal projection theorem in Hilbert spaces (Theorem 5.2.2).

**Proposition 9.3.3** *Let $C$ be a convex closed subset of the Hilbert space $H$. For every vector $x_0 \in H$ there exists a unique vector $y_0 \in C$ such that*

$$\mathrm{dist}(x_0, C) = \|x_0 - y_0\|.$$

*Furthermore, the vector $y_0$ enjoys the following geometric characterization: $y_0$ is the only vector in $H$ such that*

$$\langle x_0 - y_0, x - y_0 \rangle \leq 0 \tag{9.4}$$

*for every $x \in C$.*

**Proof** Given $x_0 \in H$, consider the function

$$\varphi \colon H \longrightarrow \mathbb{R}, \quad \varphi(x) = \|x_0 - x\|^2.$$

Note that $\varphi$ is continuous, convex and, by the triangle inequality, coercive. As $C$ is convex and closed, Theorem 9.3.2 guarantees the existence of a minimum $y_0$ for $\varphi$ in $C$, which clearly satisfies the desired condition. Uniqueness follows from the uniform convexity of $\varphi$. To check inequality (9.4), given $x \in C$, we define the function

$$\psi_x \colon [0, 1] \longrightarrow \mathbb{R}, \quad \psi_x(t) = \|x_0 - (tx + (1-t)y_0)\|^2.$$

As $t = 0$ is a minimum of $\psi_x$, we have

$$0 \leq \left.\frac{d}{dt}\psi_x(t)\right|_{t=0} = -\langle x_0 - y_0, x - y_0 \rangle.$$
∎

Note that, in the Proposition above, if $C$ is a closed subspace of $H$, then (9.4) is equivalent to the condition $(x_0 - y_0) \perp C$. This shows that Proposition 9.3.3 generalizes Theorem 5.2.2.

## 9.4 Fixed Point Theorems

Fixed point theorems are a classic topic in Nonlinear Analysis and, as we will see in this section and in the next one, they are very useful in applications.

**Definition 9.4.1** Let $(M_1, d_1)$ and $(M_2, d_2)$ be metric spaces. A map $\Phi \colon (M_1, d_1) \longrightarrow (M_2, d_2)$ is a *contraction* if $\|\Phi\|_{\text{Lip}} < 1$, that is, if there exists a constant $\theta < 1$ such that

$$d_2(\Phi(x), \Phi(y)) \leq \theta d_1(x, y) \text{ for all } x, y \in M_1.$$

Being Lipschitz, contractions are uniformly continuous, hence continuous.

Classic examples of contractions are differentiable functions $f \colon U \subseteq \mathbb{R}^n \longrightarrow \mathbb{R}^m$ with $\|Df(x)\| \leq \theta < 1$ for every $x \in U$. This remark is a consequence of the Mean Value Inequality for differentiable functions.

Next, we present the famous Banach's fixed point theorem, whose importance can hardly be overstated.

**Theorem 9.4.2 (Banach's Fixed Point Theorem)** *If $M$ is a complete metric space and $\Phi \colon M \longrightarrow M$ is a contraction, then there is a single point $x_0 \in M$ such that $\Phi(x_0) = x_0$.*

**Proof** Let $\theta < 1$ be the constant of the contraction $\Phi$. Consider the non-negative number

$$\iota := \inf\{d(\Phi(x), x) : x \in M\}.$$

Initially, we prove that $\iota = 0$. In fact, otherwise we would have $\theta^{-1}\iota > \iota$, then there would exist $x \in M$ such that $d(\Phi(x), x) < \theta^{-1}\iota$. We would then have

$$d(\Phi(\Phi(x)), \Phi(x)) \leq \theta d(\Phi(x), x) < \iota,$$

which is incompatible with the fact that $\iota$ is the infimum.

As $\inf\{d(\Phi(x), x) : x \in M\} = 0$, we can consider a sequence $(x_n)_{n=1}^\infty$ in $M$ such that $d(\Phi(x_n), x_n) \longrightarrow 0$. By the triangle inequality we can estimate

$$d(x_n, x_m) \leq d(\Phi(x_n), x_n) + d(\Phi(x_m), x_m) + d(\Phi(x_n), \Phi(x_m))$$
$$\leq d(\Phi(x_n), x_n) + d(\Phi(x_m), x_m) + \theta d(x_n, x_m),$$

for all $n, m \in \mathbb{N}$. Therefore,

$$d(x_n, x_m) \leq \frac{1}{1-\theta}[d(\Phi(x_n), x_n) + d(\Phi(x_m), x_m)] \longrightarrow 0,$$

when $n, m \longrightarrow \infty$. This proves that $(x_n)_{n=1}^\infty$ is a Cauchy sequence and, as $M$ is complete, $(x_n)_{n=1}^\infty$ converges to a certain $x_0 \in M$. From the continuity of $\Phi$ we have $\Phi(x_n) \longrightarrow \Phi(x_0)$, and from the continuity of the metric it follows that

$$d(\Phi(x_0), x_0) = \lim_n d(\Phi(x_n), x_n) = 0,$$

proving that $x_0$ is a fixed point for $\Phi$.

The uniqueness follows from the following remark: if $x_0$ and $x_1$ are fixed points, then $d(x_0, x_1) = d(\Phi(x_0), \Phi(x_1)) \leq \theta d(x_0, x_1)$. ∎

Fixed point theorems are very effective in solving nonlinear problems. From a computational, or numerical, point of view, it is important to discover algorithms to find fixed points and also estimate the speed of convergence of such algorithms. In general these questions are difficult to be satisfactorily answered; however, for contractions the Banach fixed point theorem guarantees the following immediate consequence (recall that $\Phi^n = \Phi \circ \overset{(n)}{\cdots} \circ \Phi$):

**Corollary 9.4.3** *Let $E$ be a Banach space, let $\Phi: E \longrightarrow E$ be a contraction with constant $\theta$, and let $x_0$ be its only fixed point. Then*

$$\|\Phi^n(x) - x_0\| \leq \theta^n \|x - x_0\|$$

*for all $x \in E$ and $n \in \mathbb{N}$. In particular, for any $x \in E$, the sequence $(\Phi^n(x))_{n=1}^\infty$ converges to the only fixed point of $\Phi$.*

Recall that Theorem 7.1.4 guarantees that the spectrum $\sigma(T)$ of a continuous linear operator $T: E \longrightarrow E$ is contained in the disk $\{\lambda \in \mathbb{K} : |\lambda| \leq \|T\|\}$. As a first application of Theorem 9.4.2, we will extend Theorem 7.1.4 to the nonlinear setting.

**Example 9.4.4 (Spectral Theory of Nonlinear Operators)** Let $E$ be a Banach space and let $\Phi: E \longrightarrow E$ be a Lipschitz function. Keeping the notation from Chap. 7, the identity operator on $E$ will be denoted by $I$. We will see that, for any $\lambda \in \mathbb{R}$ such that $|\lambda| > \|\Phi\|_{\text{Lip}}$, the map $\Phi - \lambda I$ is a Lipschitz bijection with Lipschitz inverse. In fact, to prove bijectivity, given $y \in E$ we need to show that the equation

$$\Phi(x) - \lambda x = y,$$

has a unique solution. Rewriting the equation above, we have to show that the operator

$$\varphi : E \longrightarrow E, \ \varphi(x) = \frac{1}{\lambda}(\Phi(x) - y),$$

has a unique fixed point. As $|\lambda| > \|\Phi\|_{\text{Lip}}$, it immediately follows that $\varphi$ is a contraction, and the bijectivity of $\Phi - \lambda I$ follows from Banach's fixed point theorem. It is clear that $\Phi - \lambda I$ is Lipschitz. To check that $(\Phi - \lambda I)^{-1}$ is Lipschitz, suppose that $\Phi(x_i) - \lambda x_i = y_i$, $i = 1, 2$. Subtracting the equations, the triangle inequality gives

## 9.4 Fixed Point Theorems

$$|\lambda| \cdot \|x_1 - x_2\| = \|(y_1 - y_2) + (\Phi(x_2) - \Phi(x_1))\|$$
$$\leq \|y_1 - y_2\| + \|\Phi(x_2) - \Phi(x_1)\|$$
$$\leq \|y_1 - y_2\| + \|\Phi\|_{\text{Lip}} \cdot \|x_1 - x_2\|.$$

Finally,

$$\|(\Phi - \lambda \text{Id})^{-1}(y_1) - (\Phi - \lambda \text{Id})^{-1}(y_2)\| = \|x_1 - x_2\| \leq \frac{1}{|\lambda| - \|\Phi\|_{\text{Lip}}} \cdot \|y_1 - y_2\|.$$

Next, we present a famous application of Banach's fixed point theorem in applied analysis. We will assume that a norm $\|\cdot\|$ on $\mathbb{R}^n$ has been chosen and fixed.

**Theorem 9.4.5 (Cauchy–Picard Theorem)** *Let $\varepsilon > 0$, $n \in \mathbb{N}$, $r > 0$, $z_0 \in \mathbb{R}^n$, $t_0 \in \mathbb{R}$ be given and let $f : [t_0 - \varepsilon, t_0 + \varepsilon] \times B[z_0; r] \longrightarrow \mathbb{R}^n$ be a continuous map that is Lipschitz in the second variable, that is,*

$$\|f(t, x_1) - f(t, x_2)\| \leq L \cdot \|x_1 - x_2\|,$$

*for some $L > 0$ and all $x_1, x_2 \in B[z_0; r]$ and $t \in [t_0 - \varepsilon, t_0 + \varepsilon]$. Then the ordinary differential equation*

$$\begin{cases} x'(t) = f(t, x(t)) \\ x(t_0) = z_0 \end{cases}$$

*has a unique local solution $x \colon [t_0 - \varepsilon', t_0 + \varepsilon'] \longrightarrow B[z_0; r]$, for some $0 < \varepsilon' \leq \varepsilon$.*

***Proof*** Call

$$K = \sup\{\|f(s, z)\| : (s, z) \in [t_0 - \varepsilon, t_0 + \varepsilon] \times B[z_0; r]\}$$

and

$$\varepsilon' = \min\left\{\varepsilon, \frac{r}{K}, \frac{1}{2L}\right\}.$$

Now consider the set

$$M = \{y : [t_0 - \varepsilon', t_0 + \varepsilon'] \longrightarrow B[z_0; r] : y \text{ is continuous}\},$$

equipped with the metric $d(y_1, y_2) = \|y_1 - y_2\|_\infty$. In Exercise 9.9.16 the reader will prove that the metric space $(M, d)$ is complete. Consider

$$\Phi : M \longrightarrow M, \quad \Phi(y)(t) = z_0 + \int_{t_0}^{t} f(s, y(s)) ds.$$

It is clear that $\Phi(y)$ is continuous for every $y \in M$ and, for $t \in [t_0 - \varepsilon', t_0 + \varepsilon']$, we have

$$\|\Phi(y)(t) - z_0\| \leq \int_{t_0}^{t} \|f(s, y(s))\| \, ds \leq \varepsilon' K \leq r.$$

Therefore, we can assure that $\Phi$ applies $M$ to $M$. We will now show that $\Phi$ is a contraction. To do so, for every $t \in [t_0 - \varepsilon', t_0 + \varepsilon']$, we estimate

$$\|\Phi(y_1)(t) - \Phi(y_2)(t)\| \leq \int_{t_0}^{t} \|f(s, y_1(s)) - f(s, y_2(s))\| \, ds$$

$$\leq L\varepsilon' \|y_1 - y_2\|_\infty \leq \frac{1}{2} \|y_1 - y_2\|_\infty.$$

It follows from Banach's fixed point theorem that $\Phi$ has a unique fixed point $x \in M$, which is precisely the solution of the required differential equation. ∎

## 9.5 Compact Nonlinear Operators

As we have already seen in previous chapters, many of the results that hold in Euclidean spaces do not hold in infinite dimensional environments. The *villain* is almost always the lack of local compactness. In the linear case, we saw in Chap. 7 that much can be done in the presence of local compactness. In this section we will see that also in the nonlinear case local compactness allows us to go further.

**Definition 9.5.1** Let $E$ and $F$ be Banach spaces and let $U \subseteq E$ be given. A map $\varphi \colon U \longrightarrow F$ is *compact* if:

(a) $\varphi$ is continuous with respect to the norm topologies.
(b) For every bounded subset $V \subseteq U$, $\overline{\varphi(V)}$ is compact in $F$.

We begin with some initial examples.

**Example 9.5.2** Let $\varphi \colon E \longrightarrow F$ be a map between Banach spaces.

(a) If $E$ and $F$ are finite dimensional, then $\varphi$ is compact if and only if it is continuous.
(b) If $F$ is finite dimensional, then $\varphi$ is compact if and only if it is continuous and bounded.
(c) If $E$ is reflexive, then the notion of compact map is the same as completely continuous map, as in Definition 9.2.1. In particular, the map from Example 9.2.6 is compact.
(d) Compact linear operators were studied in Sects. 7.2 and 7.3 and agree with Definition 9.5.1. For example, the integration operator,

## 9.5 Compact Nonlinear Operators

$$\mathcal{I}\colon C[0,1] \longrightarrow C[0,1], \quad \mathcal{I}(f)(t) = \int_0^t f(s)\,ds,$$

is compact.

Heuristically, an operator acting between function spaces is compact if it has some smoothing effect, that is, if the operator improves the regularity of the functions in the domain. For example, the integration operator mentioned above applies merely continuous functions to class $C^1$ functions. The operator from Example 9.2.6 associates functions from $L_2[0,1]$ to continuous functions. In general, such maps are obtained as inverses of differential operators.

**Example 9.5.3** Let $a, b, c\colon \mathbb{R} \longrightarrow \mathbb{R}$ be continuous functions with $a(x) > \gamma > 0$ for every $x \in \mathbb{R}$. Define the linear differential operator

$$L(f) := a(x)f'' + b(x)f' + c(x)f.$$

It follows from the Cauchy–Picard Theorem 9.4.5 (see Exercise 9.9.27) that, for any continuous function $g\colon [0,1] \longrightarrow \mathbb{R}$, the Initial Value Problem

$$L(f) = g, \quad f(0) = a_0, \quad f'(0) = v_0, \tag{9.5}$$

has a unique $C^2$ class solution $u\colon [0,1] \longrightarrow \mathbb{R}$. The operator $\Phi$ defined by

$$\Phi\colon C[0,1] \longrightarrow C[0,1], \quad \Phi(g) = u,$$

where $u$ is the unique solution of equation (9.5), is compact (Exercise 9.9.28).

Recall that a linear operator $T\colon E \longrightarrow F$ has *finite rank* if the range of $T$, $R(T)$, is finite dimensional. It is natural then to extend this concept to arbitrary functions in the following way:

**Definition 9.5.4** Let $E$ and $F$ be normed spaces and $U \subseteq E$. We say that an arbitrary map $\varphi\colon U \longrightarrow F$ has *finite rank* if the subspace $[\varphi(U)]$ of $F$ generated by the range of $\varphi$ is finite dimensional.

In Sect. 7.6 we discussed the importance of the problem of approximating compact linear operators by finite rank linear operators in the development of Functional Analysis. We present below a result on the approximation of compact nonlinear operators by finite rank operators (nonlinear, of course) that will be very useful in the proof of Schauder's fixed point theorem.

**Theorem 9.5.5** *Let $E$ and $F$ be Banach spaces and let $U$ be a bounded subset of $E$. A map $\varphi\colon U \longrightarrow F$ is compact if and only if for every $\varepsilon > 0$ there exists a bounded continuous function of finite rank $\varphi_\varepsilon\colon U \longrightarrow F$ such that*

$$\|\varphi(x) - \varphi_\varepsilon(x)\| < \varepsilon$$

*for every $x \in U$.*

**Proof** Assume $\varphi: U \longrightarrow F$ is compact. By definition, $\overline{\varphi(U)}$ is a compact subset of $F$. Thus, given $\varepsilon > 0$, there exist $k \in \mathbb{N}$ and vectors $y_1, y_2, \ldots, y_k \in F$ such that

$$\varphi(U) \subseteq \bigcup_{i=1}^{k} B(y_i; \varepsilon). \tag{9.6}$$

Let $G_\varepsilon = [y_1, y_2, \ldots, y_k]$. We proceed to construct a partition of unity. For each $i = 1, 2, \ldots, k$, consider the map

$$\lambda_i : U \longrightarrow \mathbb{R}, \ \lambda_i(x) = \max\{\varepsilon - \|\varphi(x) - y_i\|, 0\},$$

and, normalizing,

$$\sigma_i : U \longrightarrow \mathbb{R}, \ \sigma_i(x) = \frac{\lambda_i(x)}{\sum_{j=1}^{k} \lambda_j(x)}.$$

Note that $\sum_{j=1}^{k} \lambda_j(x) > 0$ for each $x \in U$, because, due to (9.6), $\varphi(x) \in B(y_j; \varepsilon)$ for some $j \in \{1, 2, \ldots, k\}$. The functions $\sigma_i$ are continuous, $0 \leq \sigma_i(x) \leq 1$ and, for every $x \in U$, $\sum_{i=1}^{k} \sigma_i(x) = 1$. Finally, we define the approximation

$$\varphi_\varepsilon : U \longrightarrow G_\varepsilon \subseteq F, \ \varphi_\varepsilon(x) = \sum_{i=1}^{k} \sigma_i(x) y_i. \tag{9.7}$$

It is easily checked that, for each $x \in U$,

$$\|\varphi(x) - \varphi_\varepsilon(x)\| \leq \sum_{i=1}^{k} \sigma_i(x) \cdot \|\varphi(x) - y_i\| \leq \varepsilon.$$

The converse is clear since, as the reader will check in Exercise 9.9.17, if $(\varphi_n : U \longrightarrow F)_{n=1}^{\infty}$ is a sequence of compact operators and $\varphi_n \longrightarrow \varphi$ uniformly, then $\varphi$ is compact. ∎

As we have seen in the previous section, fixed point theorems are very important in mathematical analysis, as they provide powerful tools for showing the existence of solutions to certain problems without necessarily constructing them; a task that is generally very difficult or even impossible to do. It is likely that the first fixed point theorem the reader encountered is the following immediate consequence of the Intermediate Value Theorem: every continuous function $f: [a, b] \longrightarrow [a, b]$ has a fixed point. One of the most famous fixed point theorems is an important generalization of this property:

## 9.5 Compact Nonlinear Operators

**Theorem 9.5.6 (Brouwer's Fixed Point Theorem)** *Let $C$ be a non-empty, closed, bounded and convex subset of $\mathbb{R}^n$. Any continuous map $f: C \longrightarrow C$ has at least one fixed point.*

In general, the proof of this theorem relies on advanced tools of Vector Calculus or Algebraic Topology (see, for example, [28, Theorem 18.9]). Next, we discuss the validity of Brouwer's fixed point theorem in infinite dimensional environments. We start by showing that just with continuity it is not possible to go very far:

**Example 9.5.7** Denoting by $B$ the unit closed ball of $\ell_2$, consider the continuous map

$$f: B \longrightarrow B, \ f\left(a = (a_j)_{j=1}^{\infty}\right) = \left(\left(1 - \|a\|_2^2\right)^{1/2}, a_1, a_2, \ldots\right).$$

Note that

$$\|f(a)\|_2^2 = 1 - \|a\|_2^2 + \sum_{j=1}^{\infty} a_j^2 = 1.$$

We check next that $f$ does not have a fixed point. Indeed, if there was $b \in B$ such that $f(b) = b$, we would have $\|b\|_2 = \|f(b)\|_2 = 1$, then

$$(b_1, b_2, \ldots,) = b = f(b) = \left(\left(1 - \|b\|_2^2\right)^{1/2}, b_1, b_2, \ldots\right) = (0, b_1, b_2, \ldots),$$

which is a contradiction. We conclude that $f$, even being continuous, does not have a fixed point.

The example above shows that Brouwer's fixed point theorem is yet another result confined to Euclidean spaces. However, we will show next that such a result can be extended to Banach spaces under the correct assumption: the compactness of the map.

**Theorem 9.5.8 (Schauder's Fixed Point Theorem)** *Let $E$ be a Banach space, let $C \subseteq E$ be a closed, convex and bounded set and let $f: C \longrightarrow C$ be a compact map. Then $f$ has at least one fixed point.*

**Proof** By Theorem 9.5.5 there exists a sequence of continuous, bounded and of finite rank operators $(f_n)_{n=1}^{\infty}$ that uniformly approximate $f$. Using the notation from the proof of Theorem 9.5.5 and calling $C_n$ the convex hull of the set $\{y_1, y_2, \ldots, y_n\}$, from the definition of the approximation in (9.7), it is true that $f_n(C) \subseteq C_n \subseteq C$, the last inclusion being a consequence of the convexity of $C$. Calling $G_n$ the subspace generated by $y_1, y_2, \ldots, y_n$, restricting $f_n$ to the convex and closed set $C_n \subseteq G_n$, and remembering that $f_n(C_n) \subseteq C_n$, by Brouwer's fixed point theorem, there exists $x_n \in C_n$ such that $f_n(x_n) = x_n$. As $C$ is bounded and $f$ is compact, there exist a subsequence $(x_{n_j})_{j=1}^{\infty}$ and a vector $y \in E$ such that $f(x_{n_j}) \longrightarrow y$. But $C$ is closed, so $y \in C$. Combining the uniform convergence of $(f_{n_j})_{j=1}^{\infty}$ to $f$ in $C$ with the convergence $f(x_{n_j}) \longrightarrow y$, from the inequality

$$\|x_{n_j} - y\| \leq \|x_{n_j} - f(x_{n_j})\| + \|f(x_{n_j}) - y\| = \|f_{n_j}(x_{n_j}) - f(x_{n_j})\| + \|f(x_{n_j}) - y\|,$$

we can conclude that $x_{n_j} \longrightarrow y$. From the continuity of $f$ it follows that $f(x_{n_j}) \longrightarrow f(y)$, therefore $y = f(y)$. ∎

As an application of Theorem 9.5.8, we come back to the theory of local existence of solutions to the Initial Value Problem.

$$\begin{cases} x'(t) = f(t, x(t)) \\ x(t_0) = z_0 \end{cases} \quad \text{(IVP)}$$

The Cauchy-Picard Theorem guarantees existence and uniqueness of the solution when $f(t, x)$ is Lipschitz in the variable $x$. In some applications, it is not possible to guarantee such regularity of $f$ and only continuity is assured.

**Example 9.5.9** Consider the initial value problem

$$x'(t) = \sqrt{x(t)}, \quad x(0) = 0.$$

As the functions $x(t) = 0$ and $x(t) = \frac{1}{4}t^2$ are solutions, the problem has a solution but uniqueness is no longer guaranteed. Note that $f(x) = \sqrt{x}$ is continuous but not Lipschitz near the origin.

The example above raises the question of what can be said regarding the solvability of the ordinary differential equation (IVP) when $f$ is merely a continuous function. The answer is given by the Peano Theorem, whose proof is a simple application of Schauder's fixed point theorem.

**Theorem 9.5.10 (Peano's Theorem)** *Let $\varepsilon > 0$, $y_0 \in \mathbb{R}^n$, $t_0 \in \mathbb{R}$ be given and let $f: [t_0 - \varepsilon, t_0 + \varepsilon] \times B[z_0; r] \longrightarrow \mathbb{R}^n$ be a continuous map. Then, the initial value problem*

$$\begin{cases} x'(t) = f(t, x(t)) \\ x(t_0) = z_0 \end{cases}$$

*has at least one local solution $x: [t_0 - \varepsilon', t_0 + \varepsilon'] \longrightarrow B[z_0; r]$, for some $0 < \varepsilon' \leq \varepsilon$.*

**Proof** Choose a norm $\|\cdot\|$ on $\mathbb{R}^n$ and, as in the proof of Theorem 9.4.5, define

$$K = \sup\{\|f(s, z)\| : (s, z) \in [t_0 - \varepsilon, t_0 + \varepsilon] \times B[z_0; r]\}.$$

Consider $\varepsilon' = \frac{r}{K}\varepsilon$ and the Banach space

$$C\left([t_0 - \varepsilon', t_0 + \varepsilon'], \mathbb{R}^n\right) := \{y : [t_0 - \varepsilon', t_0 + \varepsilon'] \longrightarrow \mathbb{R}^n : y \text{ is continuous}\},$$

equipped with the maximum norm. Calling the range of the function $y$ by $R(y)$, it is easy to see that the subset

$$M := \{y \in C\left([t_0 - \varepsilon', t_0 + \varepsilon'], \mathbb{R}^n\right) : R(y) \subseteq B[z_0; r]\}$$
$$\subseteq C\left([t_0 - \varepsilon', t_0 + \varepsilon'], \mathbb{R}^n\right)$$

is closed, convex and bounded. The choice of $\varepsilon'$ ensures that the map

$$\Phi : M \longrightarrow M, \quad \Phi(y)(t) = z_0 + \int_{t_0}^{t} f(s, y(s)) ds,$$

is well defined in the sense of applying $M$ to $M$. The continuity of $\Phi$ is immediate. Note that solutions of the initial value problem in the statement are exactly the fixed points of $\Phi$. By the Schauder fixed point theorem, it is enough to show that the operator $\Phi$ is compact. Let $V \subseteq M$, $y \in V$ and $t_1, t_2 \in [t_0 - \varepsilon', t_0 + \varepsilon']$ be given. We can estimate

$$\|\Phi(y)(t_1) - \Phi(y)(t_2)\| \leq \int_{t_1}^{t_2} \|f(s, y(s))\| ds \leq K \cdot |t_1 - t_2|.$$

It follows that the set $\Phi(V)$ is bounded and equicontinuous. The Ascoli Theorem (Theorem B.7) ensures that $\Phi(V)$ is relatively compact in $M$, proving that the operator $\Phi$ is compact. ∎

## 9.6 Elements of Differential Calculus in Banach Spaces

In this section we discuss the notion of differentiability of functions $f : U \subseteq E \longrightarrow F$, where $U$ is open and $E$ and $F$ are Banach spaces. According to infinitesimal calculus in finite dimension, we start our analysis by locally approximating nonlinear functions $f$ by affine functions:

$$f(x) \sim f(x_0) + f'(x_0)(x - x_0),$$

where $f'(x_0)$ is a continuous linear operator from $E$ to $F$. This goal leads to the definition of Fréchet-differentiable functions:

**Definition 9.6.1** Let $E$ and $F$ be Banach spaces and let $U$ be an open set in $E$. We say that a function $f : U \longrightarrow F$ is *Fréchet-differentiable* at $x_0 \in U$ if there exists a continuous linear operator $A : E \longrightarrow F$ such that

$$f(x_0 + h) = f(x_0) + A(h) + R(x_0, h),$$

for every $h$ such that $x_0 + h$ belongs to an open ball centered at $x_0$ and contained in $U$, where $R(x_0, h) = o(\|h\|)$, that is:

$$\lim_{h \to 0} \frac{\|R(x_0, h)\|}{\|h\|} = 0.$$

In this case $A$ is called the *Fréchet derivative of $f$ at $x_0$* and denoted by $A = Df(x_0)$.

As usual, we say that $f$ is *Fréchet-differentiable* if $f$ is Fréchet-differentiable at all vectors of $U$.

In the definition above we used the classical notation $o(\|h\|)$. We open a small digression to discuss some of the advantages of this notation, known as 'little-o'. Let $E$ and $F$ be Banach spaces. Given two functions $f \colon U \subseteq E \longrightarrow F$ and $g \colon U \subseteq E \longrightarrow [0, \infty)$, where $U$ is an open set containing the origin, we say that $f = o(g)$ if

$$\lim_{x \to 0} \frac{\|f(x)\|}{g(x)} = 0.$$

The equality $f = o(g)$ is called *asymptotic equality*. The idea is that, when $x \longrightarrow 0$, the number $\|f(x)\|$ tends to zero more quickly than $g(x)$. With this notation, saying that $f(x) \longrightarrow 0$ when $x \longrightarrow 0$ is equivalent to writing $f = o(1)$. In our context we are privileging the origin, but obviously asymptotic equalities can be defined for $x \longrightarrow x_0$ with $x_0 \neq 0$.

Observe that the definition of Fréchet-differentiability only assumes the existence of the continuous linear operator $A$ such that

$$\lim_{h \to 0} \frac{\|f(x_0 + h) - f(x_0) - A(h)\|}{\|h\|} = 0.$$

If $f$ is Fréchet-differentiable, then the Fréchet derivative of $f$ is unique:

**Proposition 9.6.2** *Let $E$ and $F$ be Banach spaces and let $U \subseteq E$ be open. If the function $f \colon U \longrightarrow F$ is Fréchet-differentiable at $x_0 \in E$, then the Fréchet derivative of $f$ at $x_0$ is unique and $f$ is continuous at $x_0$.*

**Proof** Suppose that $A, B \in \mathcal{L}(E, F)$ are such that

$$f(x_0 + h) - (f(x_0) + A(h)) = o(\|h\|) \text{ and } f(x_0 + h) - (f(x_0) + B(h)) = o(\|h\|).$$

Subtracting the asymptotic equalities above, we conclude that

$$\|(A - B)(h)\| = o(\|h\|). \tag{9.8}$$

Now, given $\varepsilon > 0$ arbitrary, there exists $h_\varepsilon$ with $\|h_\varepsilon\| = 1$ such that, for any $\delta > 0$,

## 9.6 Elements of Differential Calculus in Banach Spaces

$$\|A - B\| \leq \|(A - B)(h_\varepsilon)\| + \varepsilon = \frac{\|(A - B)(\delta h_\varepsilon)\|}{\|\delta h_\varepsilon\|} + \varepsilon.$$

In view of (9.8), letting $\delta \longrightarrow 0$ and subsequently letting $\varepsilon$ tend to zero, we conclude the uniqueness of the derivative of $f$ at $x_0$. To check the continuity of $f$ at $x_0$, it is enough to estimate, via triangle inequality,

$$\|f(x_0 + h) - f(x_0)\| \leq \|Df(x_0)\| \cdot \|h\| + \|R(x_0, h)\| \longrightarrow 0,$$

when $h \longrightarrow 0$ in $E$. ∎

We start off by analyzing some simple examples of Fréchet-differentiable functions.

**Example 9.6.3**

(a) Every constant function in an open set of a Banach space is Fréchet-differentiable and has a null derivative. For the converse of this property, see Corollary 9.6.9.
(b) Every continuous linear operator $A \colon E \longrightarrow F$ is Fréchet-differentiable and $DA(x_0) = A$ for every $x_0 \in E$.
(c) Let $H$ be a Hilbert space and consider

$$f \colon H \longrightarrow \mathbb{R}, \ f(x) = \|x\|^2.$$

Then $f$ is Fréchet-differentiable and $Df(x)(h) = 2\langle x, h \rangle$ for all $x, h \in H$. In fact,

$$\|x - h\|^2 - \|x\|^2 - 2\langle x, h \rangle = \|h\|^2 = o(\|h\|).$$

More generally, if $A \colon H \longrightarrow H$ is a continuous linear operator, then the function

$$f_A \colon H \longrightarrow \mathbb{R}, \ f(x) = \langle A(x), x \rangle,$$

is Fréchet-differentiable and

$$Df_A(x)(h) = \langle (A + A^*)x, h \rangle,$$

for all $x, h \in H$. Indeed, from

$$\langle A(x + h), x + h \rangle = \langle A(x), x \rangle + \langle A(x), h \rangle + \langle A(h), x \rangle + \langle A(h), h \rangle,$$

it follows that

$$f_A(x + h) - f_A(x) - \langle (A + A^*)x, h \rangle = \langle A(h), h \rangle = o(\|h\|).$$

Next we discuss Fréchet differentiability of Nemytskii operators.

**Example 9.6.4** Let $\Omega$ be a bounded set in $\mathbb{R}^n$, $p \geq 1$ and let $\varphi: \mathbb{R} \longrightarrow \mathbb{R}$ be differentiable. Assume that

$$|\varphi'(t)| \leq C|t|^{m-1} \text{ for every } t \in \mathbb{R}, \tag{9.9}$$

for some constant $C > 0$ and $1 \leq m \leq p$. Integrating the estimate above, we get $|\varphi(t)| \leq C'|t|^m + b$ for every $t \in \mathbb{R}$, for certain constants $b$ and $C'$. Thus, according to Theorem 9.1.4, the Nemytskii operator defined by $\Phi(f)(x) := \varphi(f(x))$ is a continuous map from $L_p(\Omega)$ to $L_{p/m}(\Omega)$.

Let us investigate its differentiability. Given $f, g \in L_p(\Omega)$, it seems natural to expect that $D\Phi(f)(g) = \varphi'(f(x)) \cdot g(x)$. Initially, it is assuring to note that, in view of (9.9), $\varphi'(f(x)) \in L_{\frac{p}{m-1}}$, then, by Hölder's inequality,

$$\int_\Omega |\varphi'(f(x))g(x)|^{p/m} dx \leq \left( \int_\Omega |g(x)|^p dx \right)^{1/m} \cdot \left( \int_\Omega |\varphi'(f(x))|^{\frac{p}{m-1}} dx \right)^{\frac{m-1}{m}}.$$

Hence, indeed the function $x \in \Omega \mapsto \varphi'(f(x)) \cdot g(x)$ is an element of $L_{p/m}$. To confirm $\varphi'(f(x))$ is the Fréchet derivative of $\Phi$ at $f$, we need to estimate the $L_{p/m}$-norm of

$$e(g) := \Phi(f+g) - \Phi(f) - \varphi'(f(x)) \cdot g(x),$$

in order to show that it is $o(\|g\|_{L_p(\Omega)})$. Fixed $x \in \Omega$, in view of the Fundamental Theorem of Calculus, we can write

$$\Phi(f+g)(x) - \Phi(f)(x) = \int_0^1 \frac{d}{dt} \varphi(f(x) + tg(x)) dt = \int_0^1 \varphi'(f(x) + tg(x)) \cdot g(x) dt.$$

Hence, since $\varphi'(f(x)) \cdot g(x) = \int_0^1 \left[ \varphi'(f(x)) \cdot g(x) \right] dt$, we can write

$$\int_\Omega |e(g)(x)|^{p/m} dx$$

$$= \int_\Omega \left| \int_0^1 \left( \varphi'(f(x) + tg(x)) \cdot g(x) - \varphi'(f(x)) \cdot g(x) \right) dt \right|^{p/m} dx.$$

Using Fubini's Theorem and Hölder's inequality, we estimate $\int_\Omega |e(g)(x)|^{p/m} dx$ from above by

$$\left( \int_0^1 \int_\Omega |\varphi'(f(x) + tg(x)) - \varphi'(f(x))|^{p/m-1} dt \right)^{\frac{m-1}{m}} \cdot \left( \int_\Omega |g(x)|^p dx \right)^{1/m}.$$

## 9.6 Elements of Differential Calculus in Banach Spaces

Finally, in view of the continuity of the Nemytskii operator $h \mapsto \varphi'(h)$ from $L_p(\Omega)$ into $L_{p/m-1}(\Omega)$, we conclude

$$\lim_{\|g\|_{L_p(\Omega)} \to 0} \frac{\|e(g)\|_{L_{p/m}(\Omega)}}{\|g\|_{L_p(\Omega)}} = 0,$$

as desired.

It is worth noting that a norm defined on a Banach space is never differentiable at the origin. Indeed, if the map $\varphi: E \longrightarrow \mathbb{R}$ given by $\varphi(x) = \|x\|$ were Fréchet differentiable at the origin, we would have, for every $x \in E$,

$$|t|\varphi(x) = \varphi(tx) = \varphi(0) + \varphi'(0)(tx) + o(t) = t\varphi'(0)(x) + o(t).$$

Dividing the equality above by $t \neq 0$ and letting $t \to 0^{\pm}$, we obtain

$$\pm \varphi(x) = \varphi'(0)(x),$$

leading to the conclusion that $\varphi(x) = 0$, which is a contradiction.

Next example explores the Fréchet differentiability of the $L_p(\Omega)$-norm away from 0.

**Example 9.6.5** For $1 < p < \infty$, consider the map

$$F: L_p(\Omega) \longrightarrow \mathbb{R}, \quad F(f) := \int_\Omega |f(x)|^p dx.$$

Using the differentiability of the real function $\psi_p: \mathbb{R} \longrightarrow \mathbb{R}$ given by $\psi^p(t) = |t|^p$ and the mean value theorem for functions of one variable, for $f, g \in L_p(\Omega)$ we can write

$$F(f - g) - F(f) = \int_\Omega |f(x) + g(x)|^p - |f(x)|^p dx$$

$$= \int_\Omega p|\xi(x)|^{p-2} \xi(x) \cdot g(x) dx, \tag{9.10}$$

where $\xi(x)$ is a measurable function (Why?) satisfying

$$|\xi(x) - f(x)| \leq |g(x)| \text{ for every } x \in \Omega. \tag{9.11}$$

Fixed $f, g \in L_p(\Omega)$, define

$$e(g) := F(f - g) - F(f) - \int_\Omega p|f(x)|^{p-2} f(x) \cdot g(x) dx.$$

In view of (9.10), using Hölder's inequality we can estimate

$$|e(g)| \le p \int_\Omega \left| |\xi(x)|^{p-2}\xi(x) - |f(x)|^{p-2}f(x) \right| \cdot |g(x)| dx$$

$$\le p \left( \int_\Omega \left| |\xi(x)|^{p-2}\xi(x) - |f(x)|^{p-2}f(x) \right|^{p/p-1} dx \right)^{1-1/p} \cdot \|g\|_{L_p(\Omega)}.$$

By (9.11) and the Dominated Convergence Theorem,

$$\int_\Omega \left| |\xi(x)|^{p-2}\xi(x) - |f(x)|^{p-2}f(x) \right|^{p/p-1} dx \longrightarrow 0 \text{ as } g \longrightarrow 0 \text{ in } L_p(\Omega).$$

Hence, $e(g) = o\left(\|g\|_{L_p(\Omega)}\right)$, which gives the differentiability of $F$ at $f$. Furthermore,

$$DF(f)(g) = p \int_\Omega |f(x)|^{p-2} f(x) \cdot g(x) dx.$$

The reader will check in Exercise 9.9.32 that, for $p = 1$ and $p = \infty$, the $L_p$-norm fails to be Fréchet differentiable. The case $p = 2$ corresponds to the computation carried out in Example 9.6.3(c).

The next results show that the Fréchet derivative behaves like the usual derivative from the theory of finite dimensional calculus.

**Proposition 9.6.6** *Let $E$ and $F$ be Banach spaces and let $U \subseteq E$ be open. If the functions $f, g: U \longrightarrow F$ are Fréchet-differentiable at $x_0 \in E$ and $\lambda \in \mathbb{R}$, then the function $f + \lambda g$ is Fréchet-differentiable at $x_0$ and $D(f + \lambda g)(x_0) = Df(x_0) + \lambda Dg(x_0)$.*

**Proof** We simply write

$$(f + \lambda g)(x_0 + h) - (f + \lambda g)(x_0) - (Df(x_0) + \lambda Dg(x_0))(h)$$
$$= o(\|h\|) + \lambda o(\|h\|) = o(\|h\|). \blacksquare$$

Next, we will prove the chain rule for the composition of Fréchet-differentiable functions. Before that, we note the following elementary fact about asymptotic analysis. Let $E$ and $F$ be Banach spaces, let $U \subseteq E$ be an open set, let $A: E \longrightarrow F$ be a continuous linear operator and let $R: U \longrightarrow F$ be a map of order $R(h) = o(\|A(h) + o(\|h\|)\|)$. Then $R(h) = o(\|h\|)$. That is,

$$R(h) = o(\|A(h) + o(\|h\|)\|) \text{ implies } R(h) = o(\|h\|). \tag{9.12}$$

Indeed,

$$\frac{\|R(h)\|}{\|h\|} = \frac{\|A(h) + o(\|h\|)\|}{\|h\|} \cdot \frac{\|R(h)\|}{\|A(h) + o(\|h\|)\|}$$

$$\leq \left(\|A\| + \frac{o(\|h\|)}{\|h\|}\right) \cdot \frac{\|R(h)\|}{\|A(h) + o(\|h\|)\|} \longrightarrow 0,$$

when $\|h\| \longrightarrow 0$.

**Theorem 9.6.7 (Chain Rule)** *Let $E, F, G$ be Banach spaces, let $f: U \subseteq E \longrightarrow F$ and $g: V \subseteq F \longrightarrow G$ be given with $U$ and $V$ open and $f(U) \subseteq V$. If $f$ is Fréchet-differentiable at $x_0 \in U$ and $g$ is Fréchet-differentiable at $f(x_0)$, then the map $g \circ f: U \longrightarrow G$ is Fréchet-differentiable at $x_0$ and*

$$D(g \circ f)(x_0) = Dg(f(x_0)) \circ Df(x_0).$$

*Proof* By the differentiability of $f$ at $x_0$, we know that

$$f(x_0 + h) = f(x_0) + Df(x_0)(h) + o(\|h\|).$$

Thus, composing with $g$, using the differentiability of $g$ at $f(x_0)$ and recalling (9.12), we get

$$g \circ f(x_0 + h) = g\left(f(x_0) + Df(x_0)(h) + o(\|h\|)\right)$$
$$= g \circ f(x_0) + Dg(f(x_0))\left[Df(x_0)(h) + o(\|h\|)\right] + o(\|Df(x_0)(h) + o(\|h\|)\|)$$
$$= g \circ f(x_0) + [Dg(f(x_0)) \circ Df(x_0)](h) + o(\|h\|).$$

∎

One of the main goals of nonlinear analysis is to infer properties of a Fréchet-differentiable map $f: U \subseteq E \longrightarrow F$ through quantitative and qualitative inspection of its derivative $Df$. For example, one of the primary interpretations of the derivative of a real function $g: (a, b) \longrightarrow \mathbb{R}$ refers to the rate of change of the function $g$. Therefore, controlling the maximum of $|g'(x)|$ is to estimate how much the function $g$ can expand. Indeed, this result is quite well known for real functions and we generalize it below for the context of functions between Banach spaces.

**Theorem 9.6.8 (Mean Value Inequality)** *Let $E$ and $F$ be Banach spaces, let $U \subseteq E$ be open and let $f: U \longrightarrow F$ be a Fréchet-differentiable map. Let $x_1, x_2 \in U$ be such that the segment $\ell := \{tx_1 + (1-t)x_2 : 0 \leq t \leq 1\}$ is contained in $U$. Then*

$$\|f(x_1) - f(x_2)\| \leq \sup_{x \in \ell} \|Df(x)\| \cdot \|x_1 - x_2\|.$$

*Proof* Fixed $\varphi \in F'$ with $\|\varphi\| = 1$, the real function

$$f_\varphi : [0, 1] \longrightarrow \mathbb{R}, \quad f_\varphi(t) = \varphi(f(tx_1 + (1-t)x_2)),$$

is differentiable. By the Chain Rule (Theorem 9.6.7),

$$\frac{d}{dt} f_\varphi(t) = \varphi(Df(tx_1 + (1-t)x_2) \cdot (x_1 - x_2)),$$

for every $t \in [0, 1]$. By the Mean Value Theorem for real functions, we know there exists $c \in [0, 1]$ such that

$$f_\varphi(1) - f_\varphi(0) = \frac{d}{dt} f_\varphi(c).$$

Therefore we can estimate

$$|\varphi(f(x_1) - f(x_2))| \leq \sup_{x \in \ell} \|Df(x)\| \cdot \|x_1 - x_2\|.$$

To complete the proof, we just need to select, via the Hahn-Banach Theorem, a functional $\varphi \in F'$ such that $|\varphi(f(x_1) - f(x_2))| = \|f(x_1) - f(x_2)\|$. ∎

**Corollary 9.6.9** *Let $E$ and $F$ be Banach spaces and let $U \subseteq E$ be open and connected. If $f : U \longrightarrow F$ is a Fréchet-differentiable map such that $Df(x) = 0$ for every $x \in U$, then $f$ is constant.*

**Proof** Let $x_0 \in U$ and consider $V := \{x \in U : f(x) = f(x_0)\}$. It is plain that $V$ is non-empty and, by continuity (Proposition 9.6.2), $V$ is closed. Given $y_0 \in V$, choose $r > 0$ such that $B(y_0; r) \subseteq U$. By Theorem 9.6.8 we have $f(z) = f(y_0)$ for every $z \in B(y_0; r)$. This proves that $B(y_0; r) \subseteq V$, therefore $V$ is open. By connectivity it follows that $V = U$. ∎

## 9.7 A Quick Foray into Vector Integration

In this section $X$ is a non-empty set, $\Sigma$ is a $\sigma$-algebra of subsets of $X$ and $\mu$ is a finite measure on $\Sigma$. Recall that the elements of $\Sigma$ are called measurable sets and that, for each measurable set $A \in \Sigma$, the characteristic function of $A$ is denoted by $\chi_A$.

The aim of this section is to provide a brief outline of the ideas and concepts of vector integration theory. Our first task, therefore, is to define the notion of $\mu$-measurable functions. For convenience, we recall the definition of simple measurable functions.

**Definition 9.7.1** A function $f : X \longrightarrow E$ is a *simple measurable function* if there exist measurable sets $A_1, \ldots, A_k \in \Sigma$ such that $A_i \cap A_j = \emptyset$ whenever $i \neq j$, and vectors $b_1, \ldots, b_k \in E$ such that

## 9.7 A Quick Foray into Vector Integration

$$f = \sum_{i=1}^{k} \chi_{A_i} b_i.$$

A function $g: X \longrightarrow E$ is $\mu$-measurable if there exists a sequence $(f_n)_{n=1}^{\infty}$ of simple measurable functions such that $f_n \longrightarrow g$ $\mu$-almost everywhere.

**Remark 9.7.2** In this section we will also suppose, in the definition above, that $\mu(A_j) < \infty$ for $j = 1, \ldots, k$. Since we are only considering $\sigma$-finite measures, this additional requirement does not restrict the results.

Initially, it is worth noting that the range of a simple measurable function is contained in a finite dimensional subspace. Thus, if $g$ is $\mu$-measurable, then there exists a measurable set $\mathcal{O}$ such that $\mu(\mathcal{O}) = 0$ and

$$g(X - \mathcal{O}) \subseteq F \subseteq E,$$

where $F$ is a closed separable subspace. To check this fact, it suffices to observe that since $g$ is $\mu$-measurable, then there exist simple measurable functions

$$f_n = \sum_{i=1}^{k_n} \chi_{A_i^n} b_i^n,$$

and a measurable set $\mathcal{O}$ such that $\mu(\mathcal{O}) = 0$ and $f_n(x) \longrightarrow g(x)$ for each $x \in X - \mathcal{O}$. Therefore

$$g(x) \in \overline{\left[b_i^n : n \in \mathbb{N}, \ i = 1, \ldots, k_n\right]}.$$

We conclude, therefore, that the theory of vector integration can be naturally restricted to functions $g: X \longrightarrow E$, where $E$ is a separable Banach space. Therefore, in this section $E$ will always denote a separable Banach space.

So, from now on $\mu$ is a finite measure on a $\sigma$-algebra $\Sigma$ of subsets of the set $X$ and $E$ is a separable Banach space.

The following result follows from Definition 9.7.1:

**Proposition 9.7.3** *Let $(f_n: X \longrightarrow E)_{n=1}^{\infty}$ be a sequence of $\mu$-measurable functions and let $f: X \longrightarrow E$ be given. If $f_n \longrightarrow f$ $\mu$-almost everywhere, then $f$ is $\mu$-measurable.*

Using the definition, it may be not easy to check whether a given function is measurable or not. The next result remedies this difficulty in part.

**Theorem 9.7.4 (Pettis' Measurability Theorem)** *The following statements are equivalent for a function $f: X \longrightarrow E$:*

(a) *$f$ is $\mu$-measurable.*
(b) *$\varphi \circ f: X \longrightarrow \mathbb{R}$ is $\mu$-measurable for every $\varphi \in E'$.*

(c) $\|f\|: X \longrightarrow \mathbb{R}$ is measurable.

**Proof** (a) $\Longrightarrow$ (b) This implication follows immediately from Definition 9.7.1.
(b) $\Longrightarrow$ (c) Since $E$ is separable, by Exercise 3.6.11 there exist linear functionals $\varphi_1, \varphi_2, \ldots$ in $E'$ such that $\|\varphi_i\| = 1$ for every $i$ and $\|y\| = \sup_i \varphi_i(y)$ for each $y \in E$. Given $\lambda \in \mathbb{R}$, by hypothesis, the sets $A_i := \{x \in X : \varphi_i \circ f(x) \leq \lambda\}$, $i \in \mathbb{N}$, are measurable. Therefore, the set

$$\{x \in X : \|f(x)\| \leq \lambda\} = \{x \in X : \sup_i \varphi_i(f(x)) \leq \lambda\}$$

$$= \{x \in X : \varphi_i(f(x)) \leq \lambda \text{ for every } i \in \mathbb{N}\} = \bigcap_{i=1}^{\infty} A_i$$

is also measurable. This proves that the function $\|f\|: X \longrightarrow \mathbb{R}$ is measurable.
(c) $\Longrightarrow$ (a) Let $(y_i)_{i=1}^{\infty}$ be a dense sequence in $E$. By hypothesis, for each $i \in \mathbb{N}$, the function

$$\tau_i : X \longrightarrow \mathbb{R}, \quad \tau_i(x) = \|f(x) - y_i\|,$$

is measurable. Therefore, for any fixed $n \in \mathbb{N}$, the set

$$X_i^n := \left\{ x \in X : \tau_i(x) \leq \frac{1}{n} \right\} = \left\{ x \in X : f(x) \in B\left[y_i, \frac{1}{n}\right] \right\}$$

is measurable. Moreover, the denseness of the sequence $(y_i)_{i=1}^{\infty}$ ensures that $X = \bigcup_{i=1}^{\infty} X_i^n$. Next, define the sequence $(A_j^n)_{j=1}^{\infty}$ of pairwise disjoint measurable sets given by

$$A_1^n = X_1^n, \quad A_j^n = X_j^n - \bigcup_{i=1}^{j-1} X_i^n, \quad j = 2, 3, \ldots,$$

and the $\mu$-measurable function

$$f_n : X \longrightarrow E, \quad f_n(x) = \sum_{i=1}^{\infty} \chi_{A_i^n} y_i.$$

It is plain that $\bigcup_{i=1}^{\infty} A_i^n = \bigcup_{i=1}^{\infty} X_i^n = X$. Thus, given $x \in X$, we can consider $j \in \mathbb{N}$ such that $x \in A_j^n$. In this case $x \in X_j^n$ and $f_n(x) = y_j$. It follows that $f_n(x) = y_j$ and $f(x) \in B\left[y_j, 1/n\right]$, therefore

$$\|f(x) - f_n(x)\| = \|f(x) - y_j\| \leq \frac{1}{n}.$$

## 9.7 A Quick Foray into Vector Integration

We conclude that $f_n(x) \longrightarrow f(x)$ for each $x \in X$, hence $f$ is $\mu$-measurable by Proposition 9.7.3. ∎

**Corollary 9.7.5** *Let $M$ be a metric space and let $\mu$ be a measure on the Borel sets of $M$. If $f : M \longrightarrow E$ is continuous, then $f$ is measurable.*

We now move on to the definition of the integral of a $\mu$-measurable function $f : X \longrightarrow E$. Just like in classical integration theory, we start by defining the integral of a simple measurable function in the natural way:

$$\int_X \left( \sum_{i=1}^k \chi_{A_i} b_i \right) d\mu := \sum_{i=1}^k \mu(A_i) b_i.$$

This definition of the integral of simple functions is consistent in the sense that it does not depend on the representation of the function. However, it would not be reasonable to extend it pointwisely to arbitrary measurable functions, as the expression on the right side of the equality above may be not summable when $k \longrightarrow \infty$. A quick reflection leads to the following definition of integral:

**Definition 9.7.6** A $\mu$-measurable function $f : X \longrightarrow E$ is *Bochner integrable* if there exists a sequence of simple measurable functions $f_n : X \longrightarrow E, n \in \mathbb{N}$, such that $f_n \longrightarrow f$ $\mu$-almost everywhere and

$$\lim_{n \to \infty} \int_X \| f - f_n \| \, d\mu = 0. \tag{9.13}$$

In this case, for each measurable set $A \in \Sigma$, we define the *Bochner integral of the function $f$ over $A$* by:

$$\int_A f \, d\mu = \lim_{n \to \infty} \int_X \chi_A f_n \, d\mu. \tag{9.14}$$

For the definition above to be consistent, it is necessary to check that the limit in (9.14) exists and that it does not depend on the sequence of simple functions that approximate $f$ in the sense (9.13). Now, if $(f_n)_{n=1}^\infty$ is a sequence of simple functions satisfying (9.13) and $A$ is a measurable set, then, by the triangle inequality,

$$\left\| \int_X \chi_A f_m \, d\mu - \int_X \chi_A f_n \, d\mu \right\| \leq \int_X \| f - f_m \| \, d\mu + \int_X \| f - f_n \| \, d\mu.$$

This is enough to conclude that the sequence $\left( \int_X \chi_A f_n \, d\mu \right)_{n=1}^\infty$ is Cauchy in $E$, therefore convergent. We will check next that the limit in (9.14) is independent of the sequence $(f_n)_{n=1}^\infty$. Indeed if $(f_n)_{n=1}^\infty$ and $(g_n)_{n=1}^\infty$ are sequences of simple functions, both satisfying (9.13), then

$$\left\| \int_X \chi_A f_n \, d\mu - \int_X \chi_A g_n \, d\mu \right\| \leq \int_X \| f - f_n \| \, d\mu + \int_X \| f - g_n \| \, d\mu \longrightarrow 0$$

when $n \longrightarrow \infty$.

Another simple remark is that if $f: X \longrightarrow E$ is Bochner integrable, then the real function $\|f\|: X \longrightarrow \mathbb{R}$ is Lebesgue-integrable: indeed, if $(f_n)_{n=1}^\infty$ is a sequence of simple functions satisfying (9.13), then

$$\left| \int_X \|f_m\| - \|f_n\| \, d\mu \right| \leq \int_X \left| \|f_m\| - \|f_n\| \right| d\mu \leq \int_X \|f_m - f_n\| \, d\mu.$$

Therefore the sequence $(\|f_n\|)_{n=1}^\infty$ is Cauchy in the Banach space $L_1(X)$. As $\|f_n\| \longrightarrow \|f\| \mu$-almost everywhere, we conclude that $\|f\| \in L_1(X)$.

The converse of this remark is true, which is of great help in checking the Bochner integrability of a given $\mu$-measurable function $f: X \longrightarrow E$:

**Theorem 9.7.7** *A $\mu$-measurable function $f: X \longrightarrow E$ is Bochner integrable if and only if the real function $\|f\|: X \longrightarrow \mathbb{R}$ is Lebesgue-integrable.*

**Proof** Assume that $\|f\| \in L_1(X)$. Let $(f_n)_{n=1}^\infty$ be a sequence of simple functions converging $\mu$-almost everywhere to $f$. For each $n \in \mathbb{N}$, consider the measurable set

$$A_n := \left\{ x \in X : \|f_n(x)\| \leq \|f(x)\|(1 + 2^{-n}) \right\}$$

and the simple function

$$g_n: X \longrightarrow E, \quad g_n(x) = f_n(x) \chi_{A_n}(x).$$

Since $f_n \longrightarrow f \mu$-almost everywhere, it is easy to observe that $g_n \longrightarrow f \mu$-almost everywhere and

$$\|g_n(x)\| \leq \|f(x)\|(1 + 2^{-n}),$$

for all $n \in \mathbb{N}$ and $x \in X$. We conclude the proof by applying the Dominated Convergence Theorem to the sequence of functions $(\|g_n - f\|)_{n=1}^\infty$. ∎

As a bonus from the proof of Theorem 9.7.7 we conclude that if $f: X \longrightarrow E$ is Bochner integrable, then there exists a sequence $(g_n)_{n=1}^\infty$ of simple functions such that $g_n \longrightarrow f \mu$-almost everywhere, $\|g_n - f\| \longrightarrow 0$ in $L_1(X)$ and $\|g_n\| \longrightarrow \|f\|$ in $L_1(X)$.

The next result, which can be understood as a triangle inequality for the Bochner integral, is very useful in practical situations.

**Corollary 9.7.8** *Let $f: X \longrightarrow E$ be a Bochner integrable function and let $A$ be a measurable set. Then*

$$\left\| \int_A f \, d\mu \right\| \leq \int_A \|f\| \, d\mu.$$

**Proof** From the remark above we can consider a sequence $(f_n)_{n=1}^\infty$ of simple functions such that $\|f_n\| \longrightarrow \|f\|$ in $L_1(X)$. The triangle inequality ensures that the inequality we want to prove is true for simple functions, so

$$\left\| \int_A f_n \, d\mu \right\| \leq \int_A \|f_n\| \, d\mu,$$

for every $n \in \mathbb{N}$. The proof is completed by letting $n \longrightarrow \infty$. ∎

We finish this brief introduction to vector integration showing a kind of commutative property between the Bochner integral and continuous linear operators:

**Proposition 9.7.9** *Let $f: X \longrightarrow E$ be a Bochner integrable function and let $T: E \longrightarrow F$ be a continuous linear operator between Banach spaces. Then the function $T \circ f: X \longrightarrow F$ is Bochner integrable and*

$$\int_X T \circ f \, d\mu = T \left( \int_X f \, d\mu \right).$$

**Proof** The Bochner integrability of $T \circ f$ is a consequence of the estimate

$$\|T(f(x))\| \leq \|T\| \cdot \|f(x)\|,$$

for each $x \in X$, in combination with Theorem 9.7.7. Now let $(f_n: X \longrightarrow E)_{n=1}^\infty$ be a sequence of simple functions such that $f_n \longrightarrow f$ $\mu$-almost everywhere and $\|f_n - f\| \longrightarrow 0$ in $L_1(X)$. Then $(T \circ f_n)_{n=1}^\infty$ is a sequence of simple functions converging $\mu$-almost everywhere to $T \circ f$ and

$$\int_X \|T \circ f_n - T \circ f\| \, d\mu \leq \|T\| \cdot \int_X \|f_n - f\| \, d\mu \longrightarrow 0,$$

when $n \longrightarrow \infty$. Finally, from the definition of Bochner integration it follows that

$$\int_X T \circ f \, d\mu = \lim_{n \to \infty} \int_X T \circ f_n \, d\mu = T \left( \lim_{n \to \infty} \int_X f_n \, d\mu \right) = T \left( \int_X f \, d\mu \right).$$

∎

## 9.8 Comments and Historical Notes

The converse of Theorem 9.1.4 was proved by M. M. Vainberg in the 1960s: if a substitution operator $\Phi(f) := \varphi(x, f(x))$ applies $L_p(X)$ to $L_q(X)$, then there exists a function $k \in L_q(X)$ such that

$$\varphi(x, t) \leq C|t|^{p/q} + k(x),$$

for almost every $x \in X$ and each $t \in \mathbb{R}$.

Theorem 9.2.5 was first proved in [62] with the purpose of studying partial differential equations involving the Hardy–Littlewood maximal operator. The classic examples of pairs $E, F$ of reflexive Banach spaces for which the inclusion $E \hookrightarrow F$ is a compact operator are the Sobolev spaces $E = W^{1,p}(X)$ (see, for example, [9, 8.2]) and $F = L_p(X)$, where $X$ is a bounded open set. In this scenario, it is proven in [62] that the maximal Hardy–Littlewood operator is sequentially weakly continuous from $W^{1,p}(X)$ to $W^{1,p}(X)$. Another didactic example of the applicability of Theorem 9.2.5 is the following:

**Example 9.8.1** Let $1 < p < \infty$ and let $\varphi \colon \mathbb{R} \longrightarrow \mathbb{R}$ be a Lipschitz function with $\|\varphi\|_{Lip(\mathbb{R})} \leq C$. Given a bounded open set $X$ of $\mathbb{R}^n$, consider the operator

$$N_\varphi \colon W^{1,p}(X) \longrightarrow W^{1,p}(X), \ N_\varphi(f) = \varphi(f(x)).$$

Then $N_\varphi$ is sequentially weakly continuous. In fact, applying Theorem 9.2.5 for $F = L_p(X)$ and $E = W^{1,p}(X)$, it is enough to check that the operator $N_\varphi \colon W^{1,p}(X) \longrightarrow W^{1,p}(X)$ is bounded. But this is simple because

$$\|N_\varphi(f)\|_{W^{1,p}(X)} = \|N_\varphi(f)\|_{L_p(X)} + \|DN_\varphi(f)\|_{L_p(X)}.$$

Now, $\|N_\varphi(f)\|_{L_p(X)} \leq C \|f\|_{L_p(X)}$ and

$$\|DN_\varphi(f)\|_{L_p(X)} = |D\varphi(f(x)) \cdot Df(x)| \leq C |Df(x)|.$$

Brouwer's fixed point theorem is generally stated for the closed unit ball of $\mathbb{R}^n$ instead of a compact and convex set. It was obtained for $n = 3$ by L. E. J. Brouwer in 1909. A similar result for differentiable functions was previously obtained in 1904 by P. Bohl. The general case was proved by J. Hadamard in 1910 and Brouwer in 1912. Theorem 9.5.6 is sometimes called the strong version of Brouwer's fixed point theorem.

Theorem 9.5.8 was proven by J. Schauder in 1930 following the lines of an earlier result by G. D. Birkhoff and O. D. Kellogg for continuous functions defined on compact convex subsets of $L_2[0, 1]$. For more details on fixed point theorems, we refer to [30, 60].

We proved Theorem 9.4.5 (Cauchy–Picard) as an application of Banach's fixed point theorem (Theorem 9.4.2). However, chronologically the facts occurred in reverse order. Theorem 9.4.5 was proven by E. Picard, in a series of articles started in 1890, as a refinement of techniques previously applied by Cauchy and by Liouville in the first half of the nineteenth century. Picard's technique became known as the *method of successive approximations*. Banach's fixed point theorem, proven by Banach in his thesis, defended in 1920 and published in 1922, is a remarkable abstraction of Picard's method and very useful in solving equations of several types.

Theorem 9.5.10, published by G. Peano in 1890—after an incorrect proof published in 1886—was also proven using the method of successive approximations. The continuity of $f$ in Theorem 9.5.10 is not an essential hypothesis: there are discontinuous functions $f$—called *absolutely continuous functions*—for which Theorem 9.5.10 holds. This result, called Carathéodory's Existence Theorem, can be proved in the same way as Theorem 9.5.10 using the fact that the Fundamental Theorem of Calculus also holds for absolutely continuous functions (see [56, Theorem 7.18]).

Let $E$ and $F$ be normed spaces. A map $f: U \subseteq E \longrightarrow F$ is *non-expansive* (or *weak contraction*) if $\|f(x) - f(y)\| \leq \|x - y\|$ for all $x, y \in U$. In Exercises 9.9.23 to 9.9.26 we discuss the existence of fixed points of non-expansive maps. The following example shows that even non-expansive maps defined on convex weakly compact sets may not have fixed points:

**Example 9.8.2 (D. Alspach, 1981)** The set

$$K := \left\{ f \in L_1[0, 1] : 0 \leq f \leq 2, \int_0^1 f \, dt = 1 \right\}$$

is convex and weakly compact in $L_1[0, 1]$ and the map

$$T: K \longrightarrow K, \quad T(f)(t) := \begin{cases} \min\{2f(2t), 2\} & \text{if } 0 \leq t \leq \frac{1}{2} \\ \max\{0, 2f(2t-1) - 2\} & \text{if } \frac{1}{2} \leq t \leq 1 \end{cases}$$

is an isometry (in particular non-expansive) that has no fixed points.

Recall that from the Open Mapping Theorem it follows that if $E$ and $F$ are Banach spaces, then a continuous linear operator $T: E \longrightarrow F$ is an isomorphism precisely when, for every $y \in F$, the equation $T(x) = y$ has a unique solution in $E$. This is a qualitative property of the operator $T$; it is natural, therefore, to question whether the fact that the derivative $Df(x_0)$ of a differentiable function $f$ being bijective guarantees that $f$ is bijective in a neighborhood of $x_0$. The answer is given by one of the main results of differential calculus: the Inverse Function Theorem, which can be transported to functions between Banach spaces as follows. Let $U \subseteq E$ be open and let $f: E \longrightarrow F$ be a Fréchet differentiable function. We say that $f$ is *of class* $C^1$ in $U$, and in this case we write $f \in C^1(U, F)$, if the derivative map $Df: U \longrightarrow \mathcal{L}(E, F)$ is continuous. If $O_1$ and $O_2$ are open in $E$ and $F$, respectively, we say that a function $f: O_1 \longrightarrow O_2$ is a *diffeomorphism of class* $C^1$ if $f$ is a bijection of class $C^1$ and $f^{-1}: O_2 \longrightarrow O_1$ is also of class $C^1$.

**Theorem 9.8.3 (Inverse Function Theorem)** *Let $E$ and $F$ be Banach spaces and let $f \in C^1(U, F)$ be given, where $U$ is an open subset of $E$. If $Df(x_0)$ is an isomorphism, then there exist neighborhoods $O_1$ of $x_0$ and $O_2$ of $f(x_0)$ such that $f: O_1 \longrightarrow O_2$ is a diffeomorphism.*

For a proof see [55, Theorem 10.39].

For an elementary approach to integration of continuous functions $f: [a, b] \longrightarrow E$, where $E$ is a normed space, see [3, 3.2]. The Bochner integral is also called the *Dunford and Schwartz integral* and the *first Dunford integral*. The pioneering work of S. Bochner was published in 1932, and the same concept was reintroduced by N. Dunford in 1935. The Bochner integral is a particular case of the more general integral studied in the classic [20] by N. Dunford and J. T. Schwartz. Theorem 9.7.4 was published by B. J. Pettis in 1938. For a more in-depth study of the Bochner integral, and also of the Bartle and Pettis integrals, see [16, Chapter II].

## 9.9 Exercises

**Exercise 9.9.1** Prove that the map $(a_n)_{n=1}^\infty \in c_0 \mapsto (a_n^2)_{n=1}^\infty \in c_0$ is continuous.

**Exercise 9.9.2** Show that the following map is bounded, but fails to be continuous:

$$\psi: L_1[0, 1] \longrightarrow L_\infty[0, 1], \ \psi(f)(x) = \begin{cases} 1, & \text{if } f(x) > 0, \\ 0, & \text{otherwise.} \end{cases}$$

**Exercise 9.9.3** Let $(\varphi_n: \mathbb{R} \longrightarrow \mathbb{R})_{n \in \mathbb{N}}$ be a sequence of uniformly bounded, continuous functions. Prove that, if $(\varphi_n)_{n \in \mathbb{N}}$ is equicontinuous, then the map $\Phi: c_0 \longrightarrow \ell_\infty$ defined by $\Phi((a_n)_{n \in \mathbb{N}}) = (\varphi_n(a_n))_{n \in \mathbb{N}}$ is continuous.

**Exercise 9.9.4** Prove that the map $(a_n)_{n=1}^\infty \in c_0 \mapsto (|a_n|^{1/n})_{n=1}^\infty \in \ell_\infty$ is not continuous. Why this does not contradict Exercise 9.9.3?

**Exercise 9.9.5** Show that if $f_n \longrightarrow f$ in $L_p(X)$, $1 \le p < \infty$, then there exist a subsequence $(f_{n_j})_{j=1}^\infty$ and a function $g \in L_p(X)$ such that:

(i) $f_{n_j}(x) \longrightarrow f(x)$ $\mu$-almost everywhere, and
(ii) $|f_{n_j}(x)| \le g(x)$ $\mu$-almost everywhere.

**Exercise 9.9.6** Show that the sequence $(f_n)_{n=1}^\infty$ in $L_2[0, \pi]$ given by $f_n(x) = \sin(nx)$ converges weakly to zero.

**Exercise 9.9.7\*** Let $f: \mathbb{R}^n \longrightarrow \mathbb{R}$ be a bounded function with compact support, that is, the closure of the set $\{x \in \mathbb{R}^n : f(x) \ne 0\}$ is compact, and $0 \ne v \in \mathbb{R}^n$. For each positive integer $m$ consider the function $f_m(x) := f(x + mv)$, $x \in \mathbb{R}^n$. Show that the sequence $(f_m)_{m=1}^\infty$ converges weakly to zero in $L_\infty(\mathbb{R}^n)$.

**Exercise 9.9.8\*** Let $\Omega$ be an open bounded subset of $\mathbb{R}^n$. Knowing that every weakly convergent sequence in $L_\infty(\Omega)$ converges almost everywhere (to the same limit), prove that every weakly convergent sequence in $L_\infty(\Omega)$ converges in $L_p(\Omega)$ for every $p > 1$.

**Exercise 9.9.9\*** Let $1 < q, p < \infty$ and let $\varphi: \mathbb{R} \longrightarrow \mathbb{R}$ be a continuous function satisfying $|\varphi(t)| \le C|t|^{p/q}$ for every $t \in \mathbb{R}$. Consider the substitution operator

## 9.9 Exercises

$$N: L_p[0, 1] \longrightarrow L_q[0, 1], \ N(f)(t) = \varphi \circ f(t).$$

Besides, let $(f_n)_{n=1}^\infty$ be a bounded sequence in $L_p[0, 1]$ and suppose that $f_n \longrightarrow f$ almost everywhere. Show that $(N(f_n))_{n=1}^\infty$ converges weakly to $N(f)$ in $L_q[0, 1]$.

**Exercise 9.9.10** Let $(f_n)_{n=1}^\infty$ be a bounded sequence in $L_p[0, 1]$, $p > 1$, and suppose that $f_n \longrightarrow f$ almost everywhere, where $f$ is measurable. Show that $f_n \longrightarrow f$ in $L_q[0, 1]$ for every $1 \le q < p$.

**Exercise 9.9.11** Prove Lemma 9.2.4.

**Exercise 9.9.12** Let $E$ be a Banach space compactly embedded in the Banach space $F$. Prove that the norm of $E$ is equivalent to the norm induced by $F$ if and only if $E$ is finite dimensional.

**Exercise 9.9.13**

(a) Exhibit a coercive and injective function $f: \mathbb{R} \longrightarrow \mathbb{R}$.
(b) Show that the function in item (a) necessarily has an infinite number of discontinuities. In fact, prove that if $E$ is a Banach space and $f: E \longrightarrow \mathbb{R}$ is coercive and injective, then $f$ has infinite discontinuities.

**Exercise 9.9.14** Prove that the completeness of the space is an essential hypothesis in Banach's fixed point theorem. That is, exhibit a metric space $M$ (incomplete, of course) and a contraction $f: M \longrightarrow M$ that does not have a fixed point.

**Exercise 9.9.15** Let $E$ be a Banach space. Without using Banach's fixed point theorem, prove that every non-null linear contraction from $E$ into $E$ has a unique fixed point.

**Exercise 9.9.16** Prove that the metric space $(M, d)$ from the proof of Theorem 9.4.5 is complete.

**Exercise 9.9.17** Let $E$ and $F$ be normed spaces, let $U \subseteq E$ and $\varphi_n, \varphi: U \longrightarrow F$, $n \in \mathbb{N}$, be given. Prove that if each $\varphi_n$ is compact and $\varphi_n \longrightarrow \varphi$ uniformly, then $\varphi$ is compact.

**Exercise 9.9.18*** Let $K: [0, 1] \times [0, 1] \longrightarrow \mathbb{R}$ be a continuous function and let $f \in C[0, 1]$ be given. Show that for every $\lambda$ with $|\lambda| \le \|K\|_\infty$ there exists a unique continuous function $g: [0, 1] \longrightarrow \mathbb{R}$ such that

$$g(t) = f(t) + \lambda \cdot \int_0^1 K(t, s) u(s) \, ds,$$

for every $t \in [0, 1]$.

**Exercise 9.9.19*** Let $E$ be a Banach space and let $A: [0, 1] \longrightarrow \mathcal{L}(E, E)$ be a continuous map. Show that there exists a unique function $X: [0, 1] \longrightarrow \mathcal{L}(E, E)$ such that $X(0) = \mathrm{id}_E$ and

$$\frac{dX}{dt}(t) = A(t) \circ X(t),$$

for $t \in (0, 1)$.

**Exercise 9.9.20** Let $X$ be a Hausdorff topological space and let $f: X \longrightarrow X$ be a continuous function. Prove that the set of fixed points of $f$ is closed in $X$. Conclude that the set of fixed points of $f$ is dense in $X$ if and only if $f$ is the identity on $X$.

**Exercise 9.9.21** Prove that the function $f$ from Example 9.5.7 is continuous but not completely continuous.

**Exercise 9.9.22\*** Let $H$ be a separable Hilbert space and let $\Phi: B_H \longrightarrow B_H$ be a sequentially weakly continuous (nonlinear) map, defined on the closed unit ball $B_H$ of $H$. Show that $\Phi$ has a fixed point.

**Exercise 9.9.23\*** Let $H$ be a Hilbert space and let $\varphi: B_H \longrightarrow B_H$ be a non-expansive map (or weak contraction), that is, $\|\varphi(x) - \varphi(y)\| \leq \|x - y\|$ for all $x, y \in B_H$, defined on the closed unit ball $B_H$ of $H$. Show that $\varphi$ has a fixed point.

**Exercise 9.9.24** Exhibit an isometry, in particular a non-expansive map, from $S_{c_0} := \{x \in c_0 : \|x\|_\infty = 1\}$ to $S_{c_0}$, having no fixed points.

**Exercise 9.9.25\*** Let $C$ be a non-empty, convex, closed and bounded subset of the Banach space $E$ with $0 \in C$. Prove that if $f: C \longrightarrow C$ is non-expansive, then there exists a sequence $(x_n)_{n=1}^\infty \subseteq E$ such that $\lim_n \|x_n - f(x_n)\| = 0$. Such a sequence is called a sequence of approximate fixed points for $f$.

**Exercise 9.9.26** Let $C$ be a non-empty, convex, closed and bounded subset of the Banach space $E$. Prove that if $f: C \longrightarrow C$ is non-expansive, then $\inf\{\|T(x) - x\| : x \in C\} = 0$.

**Exercise 9.9.27** Show that, given any continuous function $g: [0, 1] \longrightarrow \mathbb{R}$, the Initial Value Problem (9.5) from Example 9.5.3 has a unique $C^2$ class solution.

**Exercise 9.9.28** Show that the operator $\Phi$ from Example 9.5.3 is compact.

**Exercise 9.9.29** Let $E_1, E_2$ and $F$ be Banach spaces and let $A: E_1 \times E_2 \longrightarrow F$ be a continuous bilinear map. Prove that $A$ is Fréchet-differentiable and that

$$DA(x_1, x_2)(t_1, x_2) = A(t_1, x_2) + A(x_1, t_2)$$

for all $x_1, t_1 \in E_1$ and $x_2, t_2 \in E_2$.

**Exercise 9.9.30** Prove that the algebraic operations — addition and scalar multiplication — on a Banach space are Fréchet-differentiable.

**Exercise 9.9.31** Let $E$ and $F$ be Banach spaces, let $A$ be an open, non-empty and convex subset of $E$ and let $f: A \longrightarrow F$ be a Fréchet-differentiable map such that $Df$ is constant in $A$. Prove that $f$ is the restriction to $A$ of a continuous affine map,

## 9.9 Exercises

that is, there exist $T \in \mathcal{L}(E; F)$ and $b \in F$ such that $f(x) = T(x) + b$ for every $x \in A$.

**Exercise 9.9.32** Let $u \in L_1(\Omega)$ be a function satisfying $\mu(\{u = 0\}) > 0$. Prove that the $L_1(\Omega)$ norm is not Fréchet-differentiable at $u$. A similar result holds for the $L_\infty$-norm.

In the following exercises, $\mu$ is a finite measure on a $\sigma$-algebra $\Sigma$ of subsets of the set $X$ and $E$ is a separable Banach space.

**Exercise 9.9.33** Prove Proposition 9.7.3.

**Exercise 9.9.34** Let $f: X \longrightarrow E$ be a Bochner integrable function. Prove that

$$\lim_{\mu(A) \to 0} \int_A f \, d\mu = 0.$$

**Exercise 9.9.35** Prove that the Bochner integral is $\sigma$-additive, that is, if $f: X \longrightarrow E$ is Bochner integrable and $(A_n)_{n=1}^\infty$ is a sequence of measurable pairwise disjoint sets, then

$$\int_{\bigcup_{n=1}^\infty A_n} f \, d\mu = \sum_{n=1}^\infty \int_{A_n} f \, d\mu.$$

**Exercise 9.9.36 (Egorov's Theorem)** Let $f_n: X \longrightarrow E$, $n \in \mathbb{N}$, be a sequence of measurable functions with $f_n \longrightarrow f$ $\mu$-almost everywhere. Show that given $\varepsilon > 0$ there exists a measurable set $X_\varepsilon$ such that $\mu(X - X_\varepsilon) < \varepsilon$ and

$$\sup_{x \in X_\varepsilon} \|f_n(x) - f(x)\| < \varepsilon.$$

**Exercise 9.9.37** Let $f: [a, b] \longrightarrow E$ be a continuous function. Prove that

$$\int_{[a,b]} f \, dm = \lim_{n \to \infty} \sum_{i=0}^{n-1} f(t_i^*)(t_{i+1} - t_i),$$

where $m$ is the Lebesgue measure on $[a, b]$, $a = t_0 < t_1 < \cdots < t_n = b$ and $t_i^* \in [t_i, t_{i+1}]$ for every $i = 0, \ldots, n-1$, and $\lim_{n \to \infty} \max\{|t_{i+1} - t_i| : i = 0, \ldots, n-1\} = 0$.

**Exercise 9.9.38*** It is said that a $\mu$-measurable function $f: X \longrightarrow E$ is *weakly integrable* if there exists a vector $I \in E$ such that

$$\int_X \varphi \circ f \, d\mu = \varphi(I),$$

for every $\varphi \in E'$. In this case we write w-$\int_X f d\mu =: I$. Show that if $f$ is Bochner integrable, then $f$ is weakly integrable and

$$\text{w-}\int_X f d\mu = \int_X f d\mu.$$

**Exercise 9.9.39*** Let $f: X \longrightarrow E$ be a Bochner integrable function. Prove that, for every $A \in \Sigma$ with $\mu(A) > 0$, it is true that

$$\frac{1}{\mu(A)} \int_A f d\mu \in \overline{\text{conv}(f(X))}.$$

See the definition of the convex hull in Exercise 1.8.17.

# Chapter 10
# Elements of Banach Space Theory

Since the early days of Functional Analysis during the first decades of the twentieth century, the theory of Banach spaces has undergone quick and deep development, progressing both horizontally and vertically. The term "horizontal" is used to denote the expansion of the theory's boundaries and the establishment of connections with other areas. Illustrations of this include Chap. 9 on Nonlinear Analysis, the development of Distribution Theory and its interconnections, the application of Sobolev spaces to the theory of Partial Differential Equations, the exploration of Banach Algebras (specifically C*-algebras), and the applications of the Hahn–Banach Theorem outlined at the beginning of Chap. 3.

Conversely, by "vertical" we mean advancements in gaining a deeper understanding of the structure of Banach spaces. This is achieved through the proof of far-reaching theorems and the presentation of examples featuring Banach spaces with special properties, the existence of which was doubted for many years. Several instances of this are provided in the commentary and historical notes sections of the preceding chapters, with the Gowers–Maurey theory mentioned in the Introduction standing out as a particularly illuminating example.

The primary objective of this chapter is to illustrate the vertical development of Banach space theory, providing the reader with insight into key concepts, results, techniques, and illustrative examples within the theory.

## 10.1 Series and the Dvoretzky-Rogers Theorem

We begin by recalling some important notions of convergence of series in normed spaces, which are natural extensions of the respective notions found in Real Analysis:

**Definition 10.1.1** Let $(x_n)_{n=1}^{\infty}$ be a sequence in a normed space $E$. We say that the series $\sum_{n=1}^{\infty} x_n$:

(a) *converges* to $x \in E$ if the sequence of partial sums $\left( \sum_{j=1}^{n} x_j \right)_{n=1}^{\infty}$ converges to $x$. In this case, we write $\sum_{n=1}^{\infty} x_n = x$.

(b) is *absolutely convergent* if $\sum_{n=1}^{\infty} \|x_n\| < \infty$.

(c) is *unconditionally convergent* if the series $\sum_{n=1}^{\infty} x_{\sigma(n)}$ is convergent, regardless of the permutation $\sigma : \mathbb{N} \longrightarrow \mathbb{N}$. In this case, from Proposition 5.3.8 we know that $\sum_{n=1}^{\infty} x_n = \sum_{n=1}^{\infty} x_{\sigma(n)}$ for every permutation $\sigma$.

**Remark 10.1.2 (Cauchy Criterion)** A series $\sum_{n=1}^{\infty} x_n$ in a Banach space is convergent if and only if given $\varepsilon > 0$ there exists $n_0 \in \mathbb{N}$ such that $\left\| \sum_{j=n}^{m} x_j \right\| < \varepsilon$ whenever $m > n \geq n_0$.

**Example 10.1.3** Let $(e_j)_{j=1}^{\infty}$ be the canonical vectors of sequence spaces. What was done in Example 1.6.4 shows that if $x = (a_j)_{j=1}^{\infty}$ belongs to $c_0$ or $\ell_p$, $1 \leq p < \infty$, then $x = \sum_{n=1}^{\infty} a_j e_j$.

From a first course in real analysis, the reader should remember that absolutely convergent series of real numbers are convergent, and perhaps also remembers that the proof of this fact depends on the completeness of $\mathbb{R}$. Banach observed that this property actually characterizes complete spaces:

**Proposition 10.1.4** *The following statements are equivalent for a normed space $E$:*

(a) *Every absolutely convergent series in $E$ is unconditionally convergent.*
(b) *Every absolutely convergent series in $E$ is convergent.*
(c) *$E$ is a Banach space.*

**Proof** It is clear that (a) $\Longrightarrow$ (b) is trivial.
(b) $\Longrightarrow$ (c) Let $(x_n)_{n=1}^{\infty}$ be a Cauchy sequence in $E$. Given $k \in \mathbb{N}$, let $n_0^{(k)} \in \mathbb{N}$ be such that $\|x_n - x_m\| < 2^{-k}$ whenever $n, m \geq n_0^{(k)}$. Therefore, there are $n_1 < n_2 < \cdots$ such that

$$\|x_{n_k} - x_{n_{k+1}}\| < 2^{-k} \text{ for every } k.$$

## 10.1 Series and the Dvoretzky-Rogers Theorem

Thus,

$$\sum_{k=1}^{\infty} \|x_{n_{k+1}} - x_{n_k}\| \le \sum_{k=1}^{\infty} 2^{-k} = 1,$$

and we conclude that the series $\sum_{k=1}^{\infty}(x_{n_{k+1}} - x_{n_k})$ is absolutely convergent and, by assumption, it is convergent. As

$$x_{n_{k+1}} = x_{n_1} + \sum_{j=1}^{k}(x_{n_{j+1}} - x_{n_j}) \text{ for every } k,$$

it follows that the sequence $(x_{n_{k+1}})_{k=1}^{\infty}$ is convergent. Thus, $(x_n)_{n=1}^{\infty}$ is a Cauchy sequence having a convergent subsequence, so it is convergent.

(c) $\Longrightarrow$ (a) Let $(x_n)_{n=1}^{\infty}$ be an absolutely convergent sequence in $E$. Due to a result of Dirichlet, we know that, for series of real numbers, absolute and unconditional convergences coincide, so for any bijection $\sigma \colon \mathbb{N} \longrightarrow \mathbb{N}$ the series $\sum_{n=1}^{\infty} \|x_{\sigma(n)}\|$ converges. Therefore, given $\varepsilon > 0$ there exists $n_0 \in \mathbb{N}$ such that

$$\left\| \sum_{k=1}^{n} x_{\sigma(k)} - \sum_{k=1}^{m} x_{\sigma(k)} \right\| = \left\| \sum_{k=m+1}^{n} x_{\sigma(k)} \right\| \le \sum_{k=m+1}^{n} \|x_{\sigma(k)}\| < \sum_{k=n_0}^{\infty} \|x_{\sigma(k)}\| < \varepsilon$$

whenever $n > m > n_0$. The convergence of $\sum_{n=1}^{\infty} x_{\sigma(n)}$ follows from the Cauchy criterion. ∎

As we used in the previous proof, absolutely convergent series of real numbers are precisely the unconditionally convergent series; this fact is easily extended to finite-dimensional spaces. However, in infinite-dimensional Banach spaces the situation is different:

**Example 10.1.5** Let $(e_n)_{n=1}^{\infty}$ be the canonical unit vectors of sequence spaces and define $x_n = \frac{e_n}{n}$. It is clear that the series $\sum_{n=1}^{\infty} x_n$ is not absolutely convergent in $c_0$ and $\ell_p$, $1 \le p \le \infty$. On the other hand, this series converges unconditionally to $x = \left(1, \frac{1}{2}, \frac{1}{3}, \ldots\right)$ in $c_0$: given a permutation $\sigma$ of $\mathbb{N}$ and $\varepsilon > 0$, choose an integer $N \ge \frac{1}{\varepsilon}$. It is clear that, for each $j = 1, \ldots, N$, there exists $n_j \in \mathbb{N}$ such that $\sigma(n_j) = j$. Call $n_0 = \max\{n_1, \ldots, n_N\}$. Then, for each $n \ge n_0$, the vector $\sum_{j=1}^{n} x_{\sigma(j)}$ has its first $N$ coordinates equal to those of $x$. Therefore,

$$\left\| \sum_{j=1}^{n} x_{\sigma(j)} - x \right\| \leq \frac{1}{N+1} < \frac{1}{N} \leq \varepsilon$$

for every $n \geq n_0$. This proves that $\sum_{n=1}^{\infty} x_{\sigma(n)} = x$.

The example above illustrates that there are unconditionally convergent series which are not absolutely convergent in Banach spaces, of infinite dimension, of course. The reader will have no difficulty in checking that the example above does not work in $\ell_1$. So, are there also unconditionally but not absolutely convergent series in $\ell_1$? And in any infinite-dimensional space? In other words, does the existence of unconditionally but not absolutely convergent series characterize infinite-dimensional spaces? The main purpose of this section is to prove the Dvoretzky-Rogers Theorem, which affirmatively settles this issue. We will need, among other things, the following characterizations of unconditionally convergent series.

**Theorem 10.1.6** *The following statements are equivalent for a sequence* $(x_n)_{n=1}^{\infty}$ *in a Banach space $E$:*

(a) *The series* $\sum_{n=1}^{\infty} x_n$ *is unconditionally convergent.*

(b) *For each $\varepsilon > 0$ there exists $n_\varepsilon \in \mathbb{N}$ such that* $\left\| \sum_{n \in M} x_n \right\| < \varepsilon$ *whenever $M$ is a finite subset of $\mathbb{N}$ with $\min M > n_\varepsilon$.*

(c) *The series* $\sum_{n=1}^{\infty} x_n$ *is subseries-convergent, that is, for any strictly increasing sequence $(k_n)_{n=1}^{\infty}$ of positive integers, the series* $\sum_{n=1}^{\infty} x_{k_n}$ *is convergent.*

(d) *The series* $\sum_{n=1}^{\infty} x_n$ *is sign-convergent, that is,* $\sum_{n=1}^{\infty} \varepsilon_n x_n$ *is convergent regardless of the choice of signs $\varepsilon_n \in \{-1, 1\}$, $n \in \mathbb{N}$.*

**Proof** (a) $\implies$ (b) Assume, by contradiction, that there exists $\delta > 0$ such that for every $m \in \mathbb{N}$ there exists $M \subseteq \mathbb{N}$ finite with $\min M > m$ and $\left\| \sum_{n \in M} x_n \right\| \geq \delta$. In this case there exists a sequence $(M_n)_{n=1}^{\infty}$ of finite subsets of $\mathbb{N}$ such that, for every $n \in \mathbb{N}$,

$$\min M_n > \max M_{n-1} + 1 \text{ and } \left\| \sum_{j \in M_n} x_j \right\| \geq \delta.$$

The symbol $|A|$ denotes the number of elements of the finite set $A$. Let $\sigma : \mathbb{N} \longrightarrow \mathbb{N}$ be a bijection that maps the integers of each interval $[\min M_n, \min M_n + |M_n|)$ to

10.1 Series and the Dvoretzky-Rogers Theorem

the set $M_n$. This is possible since both the intervals and the sets $M_n$ are pairwise disjoint. We prove that the sequence $\left(S_n = \sum_{k=1}^{n} x_{\sigma(k)}\right)_{n=1}^{\infty}$ is not Cauchy in $E$. To do so, observe that for every $m \in \mathbb{N}$ we can choose $n \in \mathbb{N}$ such that $\min M_n > m$ and $\left\|\sum_{j \in M_n} x_j\right\| \geq \delta$. Considering $p = \min M_n - 1$ and $q = \min M_n + |M_n| - 1$, we have $q \geq p + 1 > m$ and

$$\|S_q - S_p\| = \left\|\sum_{k=p+1}^{q} x_{\sigma(k)}\right\| = \left\|\sum_{k \in M_n} x_k\right\| \geq \delta.$$

Since it is not Cauchy, the sequence $(S_n)_{n=1}^{\infty}$ is not convergent, which contradicts the hypothesis.

(b) $\Longrightarrow$ (a) Let $\sigma : \mathbb{N} \longrightarrow \mathbb{N}$ be a bijection. Fix $\varepsilon > 0$ and choose $n_\varepsilon \in \mathbb{N}$ according to (b). Let $m_\varepsilon \in \mathbb{N}$ be such that $\{1, \ldots, n_\varepsilon\} \subseteq \{\sigma(1), \ldots, \sigma(m_\varepsilon)\}$. Thus, if $q, p \in \mathbb{N}$ are such that $q \geq p + 1 > m_\varepsilon$, then $\sigma(p+1), \ldots, \sigma(q) > n_\varepsilon$, and

$$\left\|\sum_{k=p+1}^{q} x_{\sigma(k)}\right\| = \left\|\sum_{n \in \{\sigma(p+1),\ldots,\sigma(q)\}} x_n\right\| < \varepsilon.$$

The convergence of the series $\sum_{n=1}^{\infty} x_{\sigma(n)}$ follows from Cauchy's criterion.

(b) $\Longrightarrow$ (c) Let $(k_n)_{n=1}^{\infty}$ be a strictly increasing sequence of natural numbers. It is clear that $k_n \geq n$ for each $n \in \mathbb{N}$. Given $\varepsilon > 0$, choose $n_\varepsilon \in \mathbb{N}$ according to (b). If $q, p$ are integers with $q \geq p + 1 > n_\varepsilon$ and $M_1 = \{k_{p+1}, \ldots, k_q\}$, then

$$\left\|\sum_{j=p+1}^{q} x_{k_j}\right\| = \left\|\sum_{n \in M_1} x_n\right\| < \varepsilon.$$

The convergence of the series $\sum_{n=1}^{\infty} x_{k_n}$ follows from Cauchy's criterion.

(c) $\Longrightarrow$ (d) Let $(\varepsilon_n)_{n=1}^{\infty}$ be a sequence such that $\varepsilon_n = \pm 1$ for every $n$. Consider the sets

$$S^+ = \{n \in \mathbb{N} : \varepsilon_n = 1\} \quad \text{and} \quad S^- = \{n \in \mathbb{N} : \varepsilon_n = -1\},$$

with the order inherited from $\mathbb{N}$. If $S^+$ or $S^-$ is finite, the convergence of the series $\sum_{n=1}^{\infty} \varepsilon_n x_n$ follows immediately from (c). If $S^+$ and $S^-$ are infinite, consider the sequences

$$S_n^{(1)} = \sum_{\substack{k=1 \\ k \in S^+}}^{n} x_k \quad \text{and} \quad S_n^{(2)} = \sum_{\substack{k=1 \\ k \in S^-}}^{n} x_k, \quad n \in \mathbb{N}.$$

Given $\varepsilon > 0$, as the series $\sum_{n \in S^+} x_n$ and $\sum_{n \in S^-} x_n$ are convergent by hypothesis, there exists $m_\varepsilon \in \mathbb{N}$ such that

$$\left\| S_q^{(1)} - S_p^{(1)} \right\| = \left\| \sum_{\substack{k=p+1 \\ k \in S^+}}^{q} x_k \right\| < \frac{\varepsilon}{2} \quad \text{and} \quad \left\| S_q^{(2)} - S_p^{(2)} \right\| = \left\| \sum_{\substack{k=p+1 \\ k \in S^-}}^{q} x_k \right\| < \frac{\varepsilon}{2}$$

whenever $q > p > m_\varepsilon$. Hence,

$$\left\| \sum_{k=p+1}^{q} \varepsilon_k x_k \right\| = \left\| \sum_{\substack{k=p+1 \\ k \in S^+}}^{q} x_k - \sum_{\substack{k=p+1 \\ k \in S^-}}^{q} x_k \right\| \leq \left\| \sum_{\substack{k=p+1 \\ k \in S^+}}^{q} x_k \right\| + \left\| \sum_{\substack{k=p+1 \\ k \in S^-}}^{q} x_k \right\| < \varepsilon$$

whenever $q > p > m_\varepsilon$. The convergence of the series $\sum_{n=1}^{\infty} \varepsilon_n x_n$ follows from the Cauchy criterion.

(d) $\Longrightarrow$ (b) Suppose, by contradiction, that there exist $\delta > 0$ and a sequence $(M_k)_{k=1}^{\infty}$ of finite subsets of $\mathbb{N}$ such that $\max M_k < \min M_{k+1}$ and $\left\| \sum_{n \in M_k} x_n \right\| \geq \delta$ for every $k$. For each $n \in \mathbb{N}$, define $\varepsilon_n = 1$ if $n \in \bigcup_{k=1}^{\infty} M_k$ and $\varepsilon_n = -1$ otherwise. For each positive integer $m$, choose $k_m \in \mathbb{N}$ such that $m < \min M_{k_m}$. Then

$$\left\| \sum_{j=\min M_{k_m}}^{\max M_{k_m}} (1 + \varepsilon_j) x_j \right\| = \left\| \sum_{n \in M_{k_m}} 2 x_n \right\| \geq 2\delta.$$

This implies, by the Cauchy criterion, that the series $\sum_{n=1}^{\infty} (1+\varepsilon_n) x_n$ is not convergent, therefore at least one of the series $\sum_{n=1}^{\infty} x_n$ and $\sum_{n=1}^{\infty} \varepsilon_n x_n$ does not converge. Since this contradicts the hypothesis, the result follows. ∎

**Definition 10.1.7** Let $A = (a_{ij})_{n \times n}$ be a matrix of scalars. The *trace of A* is the number

## 10.1 Series and the Dvoretzky-Rogers Theorem

$$\text{tr}(A) := \sum_{i=1}^{n} a_{ii}.$$

Similarly, if $V$ is a finite-dimensional vector space, $\beta$ is a basis for $V$, $T: V \longrightarrow V$ is a linear operator, then the *trace of* $T$ is the number

$$\text{tr}(T) := \text{tr}\left([T]_\beta\right),$$

where $[T]_\beta$ is the matrix of $T$ with respect to the basis $\beta$. Exercise 10.6.6 ensures that the trace of $T$ is well defined, in the sense that the number $\text{tr}\left([T]_\beta\right)$ does not depend on the basis $\beta$.

**Example 10.1.8** If $P_m \colon \mathbb{K}^n \longrightarrow \mathbb{K}^n$ is the orthogonal projection from $\mathbb{K}^n$ onto a subspace of $\mathbb{K}^n$ of dimension $m \leq n$, then $\text{tr}(P_m) = m$.

The following three lemmas, interestingly stated in the context of finite-dimensional spaces, are fundamental pieces for the proof of the Dvoretzky-Rogers Theorem. By $I_n$ we denote the identity matrix of order $n$.

**Lemma 10.1.9** *Let $A = (a_{ij})_{n \times n}$ be a matrix of scalars. Then for every scalar $\varepsilon$ there is a scalar $c_n(\varepsilon)$ such that*

$$\det(I_n + \varepsilon A) = 1 + \varepsilon \text{tr}(A) + c_n(\varepsilon) \quad \text{and} \quad \lim_{\varepsilon \to 0} \frac{|c_n(\varepsilon)|}{\varepsilon} = 0.$$

***Proof*** We will proceed by induction on $n$. For $n = 2$, as

$$\det(I_2 + \varepsilon A) = (1 + \varepsilon a_{11})(1 + \varepsilon a_{22}) - \varepsilon^2 a_{21} a_{12}$$
$$= 1 + \varepsilon \text{tr} A + \varepsilon^2 (a_{11} a_{22} - a_{21} a_{12}),$$

it suffices to take $c_2(\varepsilon) = \varepsilon^2 (a_{11} a_{22} - a_{21} a_{12})$.

Assume that the desired equality holds for any square matrix of order $n - 1$. Given a square matrix $A = (a_{ij})_{n \times n}$ of order $n \geq 3$, note that

$$\det(I_n + \varepsilon A) = \begin{vmatrix} 1 + \varepsilon a_{11} & \varepsilon a_{12} & \varepsilon a_{13} & \cdots & \varepsilon a_{1n} \\ \varepsilon a_{21} & 1 + \varepsilon a_{22} & \varepsilon a_{23} & \cdots & \varepsilon a_{2n} \\ \varepsilon a_{31} & \varepsilon a_{32} & 1 + \varepsilon a_{33} & \cdots & \varepsilon a_{3n} \\ \vdots & \vdots & \vdots & \ddots & \vdots \\ \varepsilon a_{n1} & \varepsilon a_{n2} & \varepsilon a_{n3} & \cdots & 1 + \varepsilon a_{nn} \end{vmatrix}$$

$$= (1 + \varepsilon a_{11}) \begin{vmatrix} 1 + \varepsilon a_{22} & \varepsilon a_{23} & \cdots & \varepsilon a_{2n} \\ \varepsilon a_{32} & 1 + \varepsilon a_{33} & \cdots & \varepsilon a_{3n} \\ \vdots & \vdots & \ddots & \vdots \\ \varepsilon a_{n2} & \varepsilon a_{n3} & \cdots & 1 + \varepsilon a_{nn} \end{vmatrix} + c_0(\varepsilon),$$

with $\lim_{\varepsilon \to 0} \frac{|c_0(\varepsilon)|}{\varepsilon} = 0$. From the induction hypothesis it follows that

$$\det(I_n + \varepsilon A) = (1 + \varepsilon a_{11})[1 + \varepsilon(a_{22} + a_{33} + \cdots + a_{nn}) + c_{n-1}(\varepsilon)] + c_0(\varepsilon),$$

with $\lim_{\varepsilon \to 0} \frac{|c_{n-1}(\varepsilon)|}{\varepsilon} = 0$. Considering

$$c_n(\varepsilon) = c_{n-1}(\varepsilon) + \varepsilon^2 a_{11}(a_{22} + \cdots + a_{nn}) + \varepsilon a_{11} c_{n-1}(\varepsilon) + c_0(\varepsilon),$$

we have $\lim_{\varepsilon \to 0} \frac{|c_n(\varepsilon)|}{\varepsilon} = 0$ and

$$\det(I_n + \varepsilon A) = 1 + \varepsilon(a_{11} + a_{22} + a_{33} + \cdots + a_{nn}) + c_n(\varepsilon) = 1 + \varepsilon \mathrm{tr}(A) + c_n(\varepsilon).$$

∎

The technique used in the proof of the next lemma is known as the *perturbation argument*.

**Lemma 10.1.10** *If $E$ is a normed space of dimension $2n$, then there exists a norm 1 isomorphism $u \colon (\mathbb{K}^{2n}, \|\cdot\|_2) \longrightarrow E$ such that*

$$\left|\mathrm{tr}\left(u^{-1} \circ v\right)\right| \leq 2n\|v\|$$

*for every linear operator $v \colon (\mathbb{K}^{2n}, \|\cdot\|_2) \longrightarrow E$.*

**Proof** Given the canonical basis in $\mathbb{K}^{2n}$ and any basis in $E$, for a linear operator $v \colon (\mathbb{K}^{2n}, \|\cdot\|_2) \longrightarrow E$, by $\det(v)$ we mean the determinant of the matrix of $v$ with respect to the fixed bases. As the function

$$|\det| \colon \mathcal{L}\left((\mathbb{K}^{2n}, \|\cdot\|_2), E\right) \longrightarrow [0, \infty)$$

is continuous and the set $\{v \in \mathcal{L}\left((\mathbb{K}^{2n}, \|\cdot\|_2), E\right) : \|v\| = 1\}$ is compact, we can consider $u_0 \in \mathcal{L}((\mathbb{K}^{2n}, \|\cdot\|_2), E)$ such that $\|u_0\| = 1$ and

$$|\det(u_0)| = \max\left\{|\det(v)| : v \in \mathcal{L}\left((\mathbb{K}^{2n}, \|\cdot\|_2), E\right), \|v\| = 1\right\}.$$

Let $\theta_0 \in [0, 2\pi)$ be such that $\det(u_0) = |\det(u_0)|e^{i\theta_0}$ and define

$$u \colon \left(\mathbb{K}^{2n}, \|\cdot\|_2\right) \longrightarrow E, \quad u(x) = e^{-\frac{i\theta_0}{2n}} u_0(x).$$

Then $u \in \mathcal{L}\left((\mathbb{K}^{2n}, \|\cdot\|_2), E\right)$, $\|u\| = \|u_0\| = 1$ and

## 10.1 Series and the Dvoretzky-Rogers Theorem

$$\det(u) = \det\left(e^{-\frac{i\theta_0}{2n}}u_0\right) = \left(e^{-\frac{i\theta_0}{2n}}\right)^{2n} \cdot \det(u_0) = e^{-i\theta_0}|\det(u_0)|e^{i\theta_0} = |\det(u_0)|$$

$$= \max\left\{|\det(v)| : v \in \mathcal{L}\left(\left(\mathbb{K}^{2n}, \|\cdot\|_2\right), E\right), \|v\| = 1\right\} > 0.$$

In particular, $u$ is invertible, therefore it is an isomorphism. Let $0 \neq \varepsilon \in \mathbb{K}$ and let $v \in \mathcal{L}\left(\left(\mathbb{K}^{2n}, \|\cdot\|_2\right), E\right)$ be given. By the choice of $u$ we have

$$\frac{|\det(u+\varepsilon v)|}{\|u+\varepsilon v\|^{2n}} = \left|\det\left(\frac{u+\varepsilon v}{\|u+\varepsilon v\|}\right)\right| \leq \det(u)$$

whenever $u + \varepsilon v \neq 0$, that is,

$$|\det(u+\varepsilon v)| \leq \det(u)\|u+\varepsilon v\|^{2n} \leq \det(u)(\|u\|+|\varepsilon|\cdot\|v\|)^{2n}$$
$$= \det(u)(1+|\varepsilon|\cdot\|v\|)^{2n}. \tag{10.1}$$

It is clear that (10.1) is also true when $u + \varepsilon v = 0$. Since $u$ is invertible,

$$|\det(u+\varepsilon v)| = \left|\det\left(u \circ \left(I_{2n} + \varepsilon(u^{-1} \circ v)\right)\right)\right|$$
$$= \left|\det(u)\det\left(I_{2n} + \varepsilon\left(u^{-1} \circ v\right)\right)\right|$$
$$= \det(u)\left|\det\left(I_{2n} + \varepsilon(u^{-1} \circ v)\right)\right|. \tag{10.2}$$

By Lemma 10.1.9 we know that

$$\det\left(I_{2n} + \varepsilon(u^{-1} \circ v)\right) = 1 + \varepsilon\operatorname{tr}\left(u^{-1} \circ v\right) + c_{2n}(\varepsilon) \tag{10.3}$$

with $\lim_{\varepsilon \to 0} \frac{|c_{2n}(\varepsilon)|}{\varepsilon} = 0$. From (10.2) and (10.1),

$$\det(u)|\det\left(I_{2n} + \varepsilon(u^{-1} \circ v)\right)| = |\det(u+\varepsilon v)| \leq \det(u)(1+|\varepsilon|\cdot\|v\|)^{2n},$$

so from (10.3) it follows that

$$\left|1 + \varepsilon\operatorname{tr}\left(u^{-1} \circ v\right) + c_{2n}(\varepsilon)\right| \leq (1+|\varepsilon|\cdot\|v\|)^{2n} = 1 + 2n|\varepsilon|\cdot\|v\| + O\left(|\varepsilon|^2\right),$$

with $\lim_{\varepsilon \to 0} \frac{O(|\varepsilon|^2)}{\varepsilon} = 0$. Therefore, for every $0 \neq \varepsilon \in \mathbb{K}$,

$$|1 + \varepsilon \operatorname{tr}(u^{-1} \circ v)| - |c_{2n}(\varepsilon)| \leq |1 + \varepsilon \operatorname{tr}(u^{-1} \circ v) + c_{2n}(\varepsilon)|$$
$$\leq 1 + 2n|\varepsilon| \cdot \|v\| + O(|\varepsilon|^2). \tag{10.4}$$

Let $\theta_v \in [0, 2\pi)$ be such that $\operatorname{tr}(u^{-1} \circ v) = |\operatorname{tr}(u^{-1} \circ v)| e^{i\theta_v}$. We now restrict ourselves to the scalars $\varepsilon \neq 0$ belonging to the line connecting the origin to the number $e^{-i\theta_v}$, that is, $\varepsilon = |\varepsilon| e^{-i\theta_v}$. For such scalars,

$$\varepsilon \operatorname{tr}(u^{-1} \circ v) = |\varepsilon| e^{-i\theta_v} |\operatorname{tr}(u^{-1} \circ v)| e^{i\theta_v} = \left|\varepsilon \operatorname{tr}(u^{-1} \circ v)\right|. \tag{10.5}$$

From (10.4) and (10.5) it follows that, for such scalars,

$$|\varepsilon| \cdot \left|\operatorname{tr}(u^{-1} \circ v)\right| \leq 2n |\varepsilon| \cdot \|v\| + |c_{2n}(\varepsilon)| + O(|\varepsilon|^2).$$

First divide by $|\varepsilon|$ and then let $\varepsilon \longrightarrow 0$ (both operations can be done along the scalars we are considering) to obtain the result. ∎

**Lemma 10.1.11** *Let $E$ be a normed space of dimension $2n$. Then there exist $n$ vectors $y_1, \ldots, y_n$ in $E$ such that $\frac{1}{2} \leq \|y_j\| \leq 1$ for $j = 1, \ldots, n$, and*

$$\left\| \sum_{j=1}^n a_j y_j \right\| \leq \left( \sum_{j=1}^n |a_j|^2 \right)^{1/2}$$

*for any scalars $a_1, \ldots, a_n$.*

**Proof** Let $u \colon (\mathbb{K}^{2n}, \|\cdot\|_2) \longrightarrow E$ be the operator whose existence is guaranteed by Lemma 10.1.10. If $P \colon (\mathbb{K}^{2n}, \|\cdot\|_2) \longrightarrow (\mathbb{K}^{2n}, \|\cdot\|_2)$ is the orthogonal projection of $(\mathbb{K}^{2n}, \|\cdot\|_2)$ onto a subspace of $(\mathbb{K}^{2n}, \|\cdot\|_2)$ of dimension $m \leq 2n$, then, by Example 10.1.8,

$$m = \operatorname{tr}(P) = \operatorname{tr}\left(u^{-1} \circ u \circ P\right) \leq 2n \|u \circ P\|.$$

As $\|u\| = 1$, there exists $z_1 \in (\mathbb{K}^{2n}, \|\cdot\|_2)$ such that $\|z_1\| = 1$ and $\|u(z_1)\| = 1$. Let $P_1$ be the orthogonal projection from $(\mathbb{K}^{2n}, \|\cdot\|_2)$ onto $[z_1]^\perp$. Since $[z_1]^\perp$ has dimension $2n - 1$,

$$\|u \circ P_1\| \geq \frac{2n-1}{2n}.$$

There then exists $z_2 \in [z_1]^\perp$ such that $\|z_2\| = 1$ and $\|u(z_2)\| = \|u(P_1(z_2))\| \geq \frac{2n-1}{2n}$. Let $P_2$ be the orthogonal projection from $(\mathbb{K}^{2n}, \|\cdot\|_2)$ onto $[z_1, z_2]^\perp$. It follows that

## 10.1 Series and the Dvoretzky-Rogers Theorem

$$\|u \circ P_2\| \geq \frac{2n-2}{2n}.$$

There then exists $z_3 \in [z_1, z_2]^\perp$ such that $\|z_3\| = 1$ and $\|u(z_3)\| = \|u(P_2(z_3))\| \geq \frac{2n-2}{2n}$. Repeating this procedure we obtain $n$ orthonormal vectors $z_1, \ldots, z_n$ in $(\mathbb{K}^{2n}, \|\cdot\|_2)$ such that $\|u(z_j)\| \geq \frac{2n-j+1}{2n}$ for every $j = 1, \ldots, n$. Define $y_j = u(z_j)$ to obtain

$$\frac{1}{2} \leq \frac{2n-j+1}{2n} \leq \|u(z_j)\| = \|y_j\| = \|u(z_j)\| \leq \|u\| \cdot \|z_j\| = 1,$$

for every $j = 1, \ldots, n$. Furthermore, remembering that $\|u\| = 1$ and that $\{z_1, \ldots, z_n\}$ is an orthonormal set, from the Pythagorean Theorem we have

$$\left\| \sum_{j=1}^{n} a_j y_j \right\| = \left\| u\left( \sum_{j=1}^{n} a_j z_j \right) \right\| \leq \left\| \sum_{j=1}^{n} a_j z_j \right\|$$

$$= \left( \sum_{j=1}^{n} \|a_j z_j\|^2 \right)^{1/2} = \left( \sum_{j=1}^{n} |a_j|^2 \right)^{1/2},$$

for any scalars $a_1, \ldots, a_n$. ∎

Finally, we are ready to show that, in the context of Banach spaces, infinite dimension is characterized by the existence of unconditionally convergent series that are not absolutely convergent.

**Theorem 10.1.12 (Dvoretzky-Rogers Theorem)** *Let $E$ be an infinite-dimensional Banach space. For any sequence $(a_n)_{n=1}^\infty \in \ell_2$ there exists an unconditionally convergent series $\sum_{n=1}^{\infty} x_n$ in $E$ such that $\|x_n\| = |a_n|$ for every $n \in \mathbb{N}$. In particular, if $(a_n)_{n=1}^\infty \in \ell_2 - \ell_1$, then the associated series $\sum_{n=1}^{\infty} x_n$ is unconditionally convergent but not absolutely convergent.*

*Proof* Given $(a_n)_{n=1}^\infty \in \ell_2$, let $(n_k)_{k=1}^\infty$ be a sequence of increasing positive integers such that

$$\sum_{n \geq n_k} |a_n|^2 \leq 2^{-2k} \tag{10.6}$$

for every $k \in \mathbb{N}$. Since $E$ is infinite-dimensional, for each $k \in \mathbb{N}$ we can consider a subspace $E_k$ of $E$ with $\dim E_k = 2(n_{k+1} - n_k)$. By Lemma 10.1.11 there exist $(n_{k+1} - n_k)$ vectors $y_{n_k}, y_{n_k+1}, \ldots, y_{n_{k+1}-1}$ with norms between $1/2$ and $1$ such that

$$\left\| \sum_{n=n_k}^{N} b_n y_n \right\| \le \left( \sum_{n=n_k}^{N} |b_n|^2 \right)^{1/2} \tag{10.7}$$

for every integer $N$ between $n_k$ and $n_{k+1} - 1$ and for any scalars $b_{n_k}, \ldots, b_N$. For $j \ge n_1$, define $x_j = \frac{a_j y_j}{\|y_j\|}$. It is clear that $\|x_n\| = |a_n|$ for every $n \ge n_1$. Note also that for every choice of signs $\varepsilon_n = \pm 1$, from (10.7) and (10.6) it follows, for every integer $N$ between $n_k$ and $n_{k+1} - 1$, that

$$\left\| \sum_{n=n_k}^{N} \varepsilon_n x_n \right\| = \left\| \sum_{n=n_k}^{N} \frac{\varepsilon_n a_n y_n}{\|y_n\|} \right\| \le \left( \sum_{n=n_k}^{N} \frac{|\varepsilon_n|^2 |a_n|^2}{\|y_n\|^2} \right)^{1/2}$$

$$\le 2 \left( \sum_{n=n_k}^{N} |a_n|^2 \right)^{1/2} \le 2^{-k+1}.$$

Also note that if $n_k < n < m < n_{k+1}$, setting $b_{n_k} = 0, \ldots, b_{n-1} = 0$, $b_n = \varepsilon_n, \ldots, b_m = \varepsilon_m$, again from (10.7) and (10.6) it follows that

$$\left\| \sum_{j=n}^{m} \varepsilon_j x_j \right\| = \left\| \sum_{j=n_k}^{m} b_j x_j \right\| = \left\| \sum_{j=n_k}^{m} \frac{b_j a_j y_j}{\|y_j\|} \right\|$$

$$\le \left( \sum_{j=n_k}^{m} \frac{|b_j|^2 |a_j|^2}{\|y_j\|^2} \right)^{1/2} \le 2 \left( \sum_{j=n_k}^{m} |a_j|^2 \right)^{1/2} \le 2^{-k+1}.$$

Given $\varepsilon > 0$, let $k_0 = k_0(\varepsilon)$ be a positive integer such that $\sum_{j=k_0}^{\infty} 2^{-j+1} < \varepsilon$. For $m > n > n_{k_0}$ there exist a non-negative integer $t$ and an integer $k \ge k_0$ such that $n_k \le n \le n_{k+1} - 1$ and $n_{k+t} \le m \le n_{k+t+1} - 1$. If $t = 0$, what has already been proven reveals that

$$\left\| \sum_{j=n}^{m} \varepsilon_j x_j \right\| \le 2^{-k+1} < \sum_{j=k_0}^{\infty} 2^{-j+1} < \varepsilon.$$

On the other hand, if $t > 0$, then

$$\left\| \sum_{j=n}^{m} \varepsilon_j x_j \right\| \le \left\| \sum_{j=n}^{n_{k+1}-1} \varepsilon_j x_j \right\| + \left\| \sum_{j=n_{k+1}}^{n_{k+2}-1} \varepsilon_j x_j \right\| + \cdots + \left\| \sum_{j=n_{k+t}}^{m} \varepsilon_j x_j \right\| \le \sum_{j=k_0}^{\infty} 2^{-j+1} < \varepsilon.$$

By the Cauchy criterion, the series $\sum_{n=n_1}^{\infty} \varepsilon_n x_n$ converges, therefore $\sum_{n=n_1}^{\infty} x_n$ is unconditionally convergent by Theorem 10.1.6. To complete the proof, just consider $x_1, \ldots, x_{n_1-1} \in E$ satisfying $\|x_j\| = |a_j|$ for every $j = 1, \ldots, n_1 - 1$. ∎

## 10.2 Grothendieck's Inequality

In some areas of Banach Space Theory, for example, Operator Ideals, Tensor Norms, Local Theory, Absolutely Summing Operators, it has become usual to say that a certain idea or result is implicitly in the *Résumé*. This is a reference to a work by A. Grothendieck [31], published in 1953, which revolutionized Banach space theory in the following decades. The main result of the Résumé, called by Grothendieck the *Fundamental theorem of the metric theory of tensor products*, is known today as *Grothendieck's Inequality*. In 1968, J. Lindenstrauss and A. Pełczyński [41] rewrote many of Grothendieck's results, including the fundamental theorem, making them more accessible and understandable. From then on, the ideas of Grothendieck's Résumé became universally known, admired, generalized, and applied.

In this section, we will prove Grothendieck's inequality, as well as its interpretation in the language of absolutely summing operators, which states that every continuous linear operator from $\ell_1$ to $\ell_2$ is absolutely summing, a result known as *Grothendieck's Theorem*.

**Lemma 10.2.1** *For a function $f: [0, 1] \longrightarrow \mathbb{R}$, define*

$$\widehat{f}: [0, 1] \longrightarrow \mathbb{R}, \ \widehat{f}(t) = \begin{cases} \min\{f(t), 1\}, & \text{if } f(t) \geq 0 \\ \max\{f(t), -1\}, & \text{if } f(t) < 0. \end{cases}$$

*Then $|\widehat{f}(t)| \leq 1$ and $|f(t) - \widehat{f}(t)| \leq \dfrac{f(t)^2}{4}$ for every $t \in [0, 1]$.*

**Proof** Let $t \in [0, 1]$. It is clear that $|\widehat{f}(t)| \leq 1$. The second inequality is obvious if $f(t) = \widehat{f}(t)$. If $f(t) \neq \widehat{f}(t)$, note that

$$|f(t) - \widehat{f}(t)| = |f(t)| - 1. \tag{10.8}$$

Indeed, if $f(t) \geq 0$, then $\widehat{f}(t) = 1 < f(t)$ and, in this case,

$$|f(t) - \widehat{f}(t)| = |f(t) - 1| = f(t) - 1 = |f(t)| - 1.$$

On the other hand, if $f(t) < 0$, then $\widehat{f}(t) = -1 > f(t)$, therefore

$$|f(t) - \widehat{f}(t)| = |f(t) + 1| = -f(t) - 1 = |f(t)| - 1.$$

Since $a - 1 \leq a^2/4$ for any real number $a$, from (10.8) it follows that

$$|f(t) - \widehat{f}(t)| = |f(t)| - 1 \leq \frac{f(t)^2}{4}.$$

∎

**Theorem 10.2.2 (Grothendieck's Inequality)** *There exists a positive constant $K_G$ such that, for every Hilbert space $H$, every $m \in \mathbb{N}$, every square matrix of scalars $(a_{ij})_{m \times m}$ and any vectors $x_1, \ldots, x_m, y_1, \ldots, y_m \in B_H$, it is true that*

$$\left|\sum_{i,j=1}^m a_{ij}\langle x_i, y_j\rangle\right| \leq K_G \sup\left\{\left|\sum_{i,j=1}^m a_{ij}s_i t_j\right| : |s_i|, |t_j| \leq 1\right\}.$$

**Proof** We prove the real case. For each $m$, it is clear that any set of vectors $x_1, \ldots, x_m, y_1, \ldots, y_m$ is contained in a $2m$-dimensional subspace of $H$. As all Hilbert spaces of the same finite dimension are isometrically isomorphic through isomorphisms that preserve the inner product (Exercises 5.8.25 and 5.8.8), we can choose any specific Hilbert space of dimension $n = 2m$. We will consider the subspace $H_n = [r_1, \ldots, r_n]$ of $L_2[0, 1]$ generated by the Rademacher functions $r_1, \ldots, r_n$ (see Exercise 5.8.34). Call

$$\alpha = \sup\left\{\left|\sum_{i,j=1}^m a_{ij}s_i t_j\right| : |s_i|, |t_j| \leq 1\right\} \text{ and } \widetilde{\alpha} = \sup\left\{\left|\sum_{i,j=1}^m a_{ij}\langle z_i, w_j\rangle\right| : z_i, w_j \in B_H\right\}.$$

Let $x_1, \ldots, x_m, y_1, \ldots, y_m \in B_{H_n}$. From Lemma 10.2.1 we know that $|\widehat{x_i}(t)|, |\widehat{y_j}(t)| \leq 1$ for all $i, j = 1, \ldots, m$, and $t \in [0, 1]$. Therefore

$$\left|\sum_{i,j=1}^m a_{ij}\langle \widehat{x_i}, \widehat{y_j}\rangle\right| = \left|\sum_{i,j=1}^m a_{ij}\int_0^1 \widehat{x_i}(t)\widehat{y_j}(t)\,dt\right| \leq \int_0^1 \left|\sum_{i,j=1}^m a_{ij}\widehat{x_i}(t)\widehat{y_j}(t)\right| dt \leq \alpha. \tag{10.9}$$

For simplicity, we will denote the norm of $H_n \subseteq L_2[0, 1]$ by $\|\cdot\|$. Let $x = \sum_{j=1}^n b_j r_j \in H_n = [r_1, \ldots, r_n] \subseteq L_2[0, 1]$ with $\|x\| \leq 1$. From Lemma 10.2.1 and Exercise 5.8.35, we have

$$\|x - \widehat{x}\|^2 = \int_0^1 |x(t) - \widehat{x}(t)|^2 dt \leq \frac{1}{16}\int_0^1 x(t)^4\,dt$$

$$= \frac{1}{16}\int_0^1 \left(\sum_{i=1}^n b_i r_i(t)\right)\left(\sum_{j=1}^n b_j r_j(t)\right)\left(\sum_{l=1}^n b_l r_l(t)\right)\left(\sum_{k=1}^n b_k r_k(t)\right) dt$$

## 10.2 Grothendieck's Inequality

$$= \frac{1}{16} \left( \sum_{i,j,l,k=1}^{n} b_i b_j b_l b_k \int_0^1 r_i(t) r_j(t) r_l(t) r_k(t) \, dt \right)$$

$$= \frac{1}{16} \left( \sum_{i=j=1}^{n} \sum_{l=k=1}^{n} b_i b_j b_l b_k + \sum_{i=l=1}^{n} \sum_{j=k=1}^{n} b_i b_j b_l b_k + \sum_{i=k=1}^{n} \sum_{j=l=1}^{n} b_i b_j b_l b_k \right.$$

$$\left. -2 \sum_{i=1}^{n} b_i^4 \right)$$

$$= \frac{1}{16} \left( 3 \left( \sum_{i,j=1}^{n} b_i^2 b_j^2 \right) - 2 \sum_{i=1}^{n} b_i^4 \right) \leq \frac{3}{16} \left( \sum_{i,j=1}^{n} b_i^2 b_j^2 \right)$$

$$= \frac{3}{16} \left( \sum_{i=1}^{n} b_i^2 \right)^2 = \frac{3}{16} \|x\|^4 \leq \frac{3}{16}.$$

Hence $\|x - \widehat{x}\| \leq \frac{\sqrt{3}}{4}$ and, in particular, $\left\| \frac{x_i - \widehat{x}_i}{\sqrt{3}/4} \right\|, \left\| \frac{y_j - \widehat{y}_j}{\sqrt{3}/4} \right\| \leq 1$. From this, (10.9) and $\|x_i\|, \|\widehat{y}_j\| \leq 1$, we conclude that

$$\left| \sum_{i,j=1}^{m} a_{ij} \langle x_i, y_j \rangle \right| \leq \left| \sum_{i,j=1}^{m} a_{ij} \langle \widehat{x}_i, \widehat{y}_j \rangle \right| + \left| \sum_{i,j=1}^{m} a_{ij} \langle x_i - \widehat{x}_i, \widehat{y}_j \rangle \right|$$

$$+ \left| \sum_{i,j=1}^{m} a_{ij} \langle x_i, y_j - \widehat{y}_j \rangle \right|$$

$$\leq \alpha + \frac{\sqrt{3}}{4} \left| \sum_{i,j=1}^{m} a_{ij} \left\langle \frac{x_i - \widehat{x}_i}{\sqrt{3}/4}, \widehat{y}_j \right\rangle \right| + \frac{\sqrt{3}}{4} \left| \sum_{i,j=1}^{m} a_{ij} \left\langle x_i, \frac{y_j - \widehat{y}_j}{\sqrt{3}/4} \right\rangle \right|$$

$$\leq \alpha + \frac{\sqrt{3}}{4} \widetilde{\alpha} + \frac{\sqrt{3}}{4} \widetilde{\alpha} = \alpha + \frac{\sqrt{3}}{2} \widetilde{\alpha}.$$

Taking the supremum over $x_1, \ldots, x_m, y_1, \ldots, y_m \in B_{H_n}$, we obtain

$$\widetilde{\alpha} \leq \alpha + \frac{\sqrt{3}}{2} \widetilde{\alpha},$$

that is, $\widetilde{\alpha} \leq \left( \frac{2}{2 - \sqrt{3}} \right) \alpha$, and the proof is done with $K_G = \frac{2}{2 - \sqrt{3}}$.

The simplest way to obtain the complex case from the real case is to make the usual decomposition into real and imaginary parts, and to apply the real case to the resulting real matrices. Proceeding in this way, the constant obtained for the complex case will be greater than the constant obtained for the real case. ∎

**Remark 10.2.3** The smallest of the constants $K_G$ satisfying Theorem 10.2.2 is called the *Grothendieck constant*, which, as seen in the proof above, can depend on the field over which one is working. For more details see Sect. 10.5.

Applications of Grothendieck's inequality are countless, and its interactions with other fields are numerous. The interested reader is referred to the excellent survey [52] by G. Pisier.

To prove an important application of Grothendieck's inequality, we need to introduce absolutely summing operators.

Let $E$ be a Banach space. In Exercises 4.5.8 and 6.8.19 we studied the space $\ell_1(E)$ of sequences of vectors in $E$ that are absolutely summable. Besides, in Exercise 2.7.37 we worked with sequences $(x_j)_{j=1}^\infty$ of vectors in $E$ that satisfy the condition

$$\left(\varphi(x_j)\right)_{j=1}^\infty \in \ell_1 \text{ for every } \varphi \in E'.$$

Such sequences are called *weakly summable* and the space formed by them is denoted by $\ell_1^w(E)$. Given a continuous linear operator $T\colon E \longrightarrow F$, the reader can check in Exercise 10.6.9 that $T$ transforms absolutely (weakly) summable sequences in $E$ into absolutely (weakly) summable sequences in $F$, that is,

$$\left(T(x_j)\right)_{j=1}^\infty \in \ell_1(F) \text{ whenever } (x_j)_{j=1}^\infty \in \ell_1(E) \text{ and}$$

$$\left(T(x_j)\right)_{j=1}^\infty \in \ell_1^w(F) \text{ whenever } (x_j)_{j=1}^\infty \in \ell_1^w(E).$$

It follows that $T$ transforms absolutely summable sequences in $E$ into weakly summable sequences in $F$, because $\ell_1(F) \subseteq \ell_1^w(F)$. However not every continuous linear operator transforms weakly summable sequences into absolutely summable sequences:

**Example 10.2.4** We saw in Exercise 4.5.5 that the sequence $(e_j)_{j=1}^\infty$ is weakly summable in $\ell_\infty$. As $\|e_j\| = 1$ for every $j$, it is clear that $(e_j)_{j=1}^\infty$ is not absolutely summable in $\ell_\infty$. This reveals that the identity operator on $\ell_\infty$, which is linear and continuous, does not transform weakly summable sequences into absolutely summable sequences.

As the inclusion $\ell_1(E) \subseteq \ell_1^w(E)$ is always valid, an operator that transforms weakly summable sequences into absolutely summable sequences improves the convergence of series:

**Definition 10.2.5** An operator $T \in \mathcal{L}(E,F)$ between Banach spaces is *absolutely summing* if $T$ transforms weakly summable sequences in $E$ into absolutely summable sequences in $F$, that is,

## 10.2 Grothendieck's Inequality

$(T(x_j))_{j=1}^\infty \in \ell_1(F)$ whenever $(x_j)_{j=1}^\infty \in \ell_1^w(E)$.

We saw above that the identity operator on $\ell_\infty$ is not absolutely summing. Examples of absolutely summing operators are finite rank operators (Exercise 10.6.11).

We will now apply Grothendieck's inequality to prove one of the central results of the theory of absolutely summing operators, which, in particular, provides examples of absolutely summing operators that are not of finite rank:

**Theorem 10.2.6 (Grothendieck's Theorem)** *Every continuous linear operator $T: \ell_1 \longrightarrow \ell_2$ is absolutely summing.*

**Proof** Let $(x_j)_{j=1}^\infty \in \ell_1^w(\ell_1)$ be such that $\sup_{\varphi \in B_{(\ell_1)'}} \sum_{j=1}^\infty |\varphi(x_j)| \leq 1$. For each $n \in \mathbb{N}$, recall Example 10.1.3 to consider the operator (projection onto the first $n$ coordinates)

$$T_n: \ell_1 \longrightarrow \ell_1, \quad T_n\left(\sum_{j=1}^\infty a_j e_j\right) = \sum_{j=1}^n a_j e_j.$$

It is easy to see that each $T_n$ is linear continuous and $\|T_n\| = 1$. Then, for each $\varphi \in B_{(\ell_1)'}$ we have $\varphi \circ T_n \in B_{(\ell_1)'}$, therefore

$$\sup_{\varphi \in B_{(\ell_1)'}} \sum_{j=1}^\infty |\varphi(T_n(x_j))| \leq \sup_{\varphi \in B_{(\ell_1)'}} \sum_{j=1}^\infty |\varphi(x_j)| \leq 1$$

for every $n$. Writing $x_k = (a_{jk})_{j=1}^\infty$ for every $k$, we have

$$x_k = \sum_{j=1}^\infty a_{jk} e_j \text{ and } T_n(x_k) = \sum_{j=1}^n a_{jk} e_j,$$

for all $n, k$. For any positive integer $N$ consider scalars $s_1, \ldots, s_N$ and $t_1, \ldots, t_N$ in $B_\mathbb{K}$. Calling $s = (s_1, \ldots, s_N)$ and considering the linear functional $\varphi_s \in B_{(\ell_1)'}$ defined by

$$\varphi_s(e_j) = s_j \text{ if } j \leq N \text{ and } \varphi_s(e_j) = 0 \text{ if } j > N,$$

we have

$$\left|\sum_{j=1}^N \sum_{k=1}^N a_{jk} s_j t_k\right| \leq \sum_{k=1}^N \left(|t_k| \cdot \left|\sum_{j=1}^N a_{jk} s_j\right|\right) \leq \sum_{k=1}^N \left|\sum_{j=1}^N a_{jk} \varphi_s(e_j)\right|$$

$$= \sum_{k=1}^N |\varphi_s(T_N(x_k))| \leq \sup_{\varphi \in B_{(\ell_1)'}} \sum_{k=1}^\infty |\varphi(T_N(x_k))| \leq 1. \quad (10.10)$$

Let $m, n \in \mathbb{N}$, with $n \geq m$. For each $1 \leq k \leq m$, applying first the Hahn-Banach Theorem in the form of Corollary 3.1.4 and then the Riesz-Fréchet Theorem (Theorem 5.5.2), there exists $y_{k,n} \in \ell_2$ such that

$$\|y_{k,n}\|_2 = 1 \text{ and } \|T(T_n(x_k))\|_2 = \langle T(T_n(x_k)), y_{k,n}\rangle.$$

If $m < n$ we take $y_{(m+1),n} = \cdots = y_{n,n} = 0$. Given $\varepsilon > 0$, we have

$$\sum_{k=1}^{m}\|T(T_n(x_k))\|_2 = \left|\sum_{k=1}^{m}\langle T(T_n(x_k)), y_{k,n}\rangle\right| = \left|\sum_{k=1}^{m}\sum_{j=1}^{n}a_{jk}\langle T(e_j), y_{k,n}\rangle\right|$$

$$\leq \max\left\{\|T(e_j)\|_2 + \varepsilon : j = 1, \ldots, n\right\} \cdot \left|\sum_{k=1}^{m}\sum_{j=1}^{n}a_{jk}\left\langle \frac{T(e_j)}{\|T(e_j)\|_2 + \varepsilon}, y_{k,n}\right\rangle\right|$$

$$\leq (\|T\| + \varepsilon)\left|\sum_{k=1}^{n}\sum_{j=1}^{n}a_{jk}\left\langle \frac{T(e_j)}{\|T(e_j)\|_2 + \varepsilon}, y_{k,n}\right\rangle\right|$$

$$\leq (\|T\| + \varepsilon) K_G \sup\left\{\left|\sum_{j=1}^{n}\sum_{k=1}^{n}a_{jk}s_jt_k\right| : |s_j|, |t_k| \leq 1\right\}$$

$$\leq K_G(\|T\| + \varepsilon),$$

where the next-to-last inequality is precisely the Grothendieck inequality (Theorem 10.2.2) and the last one is a consequence of (10.10). Letting $\varepsilon \longrightarrow 0^+$ we have

$$\sum_{k=1}^{m}\|T(T_n(x_k))\|_2 \leq K_G\|T\|,$$

for all $n \geq m$. As $T_n(x_k) \xrightarrow{n\to\infty} x_k$, letting first $n \longrightarrow \infty$ and then $m \longrightarrow \infty$ in the inequality above, we conclude, as we wanted, that $(T(x_k))_{k=1}^{\infty} \in \ell_1(\ell_2)$.

For any non-null sequence $(z_j)_{j=1}^{\infty} \in \ell_1^w(\ell_1)$, just apply what we just did to the sequence given by $x_j = \dfrac{z_j}{\sup\limits_{\varphi \in B_{(\ell_1)'}}\sum_{j=1}^{\infty}|\varphi(z_j)|}$, $j \in \mathbb{N}$. ∎

Other applications and consequences of the Grothendieck inequality will be presented in Sect. 10.5.

## 10.3 Schauder Bases and Basic Sequences

In Functional Analysis, algebraic bases (or Hamel bases—see Appendix A) of Banach spaces are of little use because, among other reasons, they are never countable:

**Proposition 10.3.1** *Algebraic bases of infinite-dimensional Banach spaces are never countable.*

*Proof* It is clear that algebraic bases of finite-dimensional spaces are finite. Suppose there is a countable algebraic basis $\mathcal{B} = \{v_j : j \in \mathbb{N}\}$ of an infinite-dimensional Banach space $E$. Hence $E = \bigcup_{n=1}^{\infty} F_n$, where each $F_n$ is the subspace generated by $\{v_1, \ldots, v_n\}$. Since it is finite-dimensional, each $F_n$ is closed, therefore by Baire's Theorem (Theorem 2.3.1) there exists $n_0$ such that $F_{n_0}$ has non-empty interior. This is absurd, since $F_{n_0} \neq E$ and proper subspaces of normed spaces always have empty interior (Exercise 1.8.11). ∎

We now introduce the reader to the notion of Schauder basis.

**Definition 10.3.2** A sequence $(x_n)_{n=1}^{\infty}$ in the Banach space $E$ is called a *Schauder basis* of $E$ if each $x \in E$ has a unique representation in the form

$$x = \sum_{n=1}^{\infty} a_n x_n,$$

where $a_n \in \mathbb{K}$ for each $n \in \mathbb{N}$. Using the uniqueness of the representation we can consider the linear functionals

$$x_n^* \colon E \longrightarrow \mathbb{K}, \quad x_n^*\left(\sum_{j=1}^{\infty} a_j x_j\right) = a_n,$$

$n \in \mathbb{N}$, which are called *coefficient functionals* (or *coordinate functionals* or even *associated biorthogonal functionals*).

The uniqueness of the representation also ensures that the vectors of a Schauder basis are linearly independent. For simplicity, algebraic bases of finite-dimensional spaces are also called Schauder bases.

**Example 10.3.3**

(a) From Example 10.1.3 it easily follows that the sequence $(e_j)_{j=1}^{\infty}$ of canonical unit vectors is a Schauder basis for $c_0$ and for $\ell_p$, $1 \leq p < \infty$.
(b) Combining Theorems 5.3.10(a) and 5.4.3 with Exercise 5.8.26 we conclude that complete orthonormal systems in separable Hilbert spaces are Schauder bases.

(c) An argument analogous to the proof of Lemma 1.6.3 proves that spaces with a Schauder basis are separable. Thus, non-separable spaces, in particular $\ell_\infty$, do not have a Schauder basis.

(d) $C[0, 1]$ has a Schauder basis. We begin by considering the sequence $(s_n)_{n=0}^\infty$ of vectors of $C[0, 1]$ given by $s_n : [0, 1] \longrightarrow \mathbb{R}$,

$$s_0(t) = 1, \quad s_1(t) = t,$$

and, for $n \geq 2$, call $m$ the positive integer with $2^{m-1} < n \leq 2^m$ and define

$$s_n(t) = \begin{cases} 2^m \left(t - \left(\frac{2n-2}{2^m} - 1\right)\right), & \text{if } \frac{2n-2}{2^m} - 1 \leq t < \frac{2n-1}{2^m} - 1 \\ 1 - 2^m \left(t - \left(\frac{2n-1}{2^m} - 1\right)\right), & \text{if } \frac{2n-1}{2^m} - 1 \leq t < \frac{2n}{2^m} - 1 \\ 0, & \text{in the other cases.} \end{cases}$$

The sequence $(s_n)_{n=0}^\infty$ is a Schauder basis of $C[0, 1]$. In fact, given $f \in C[0, 1]$, consider the sequence $(p_n)_{n=0}^\infty$ in $C[0, 1]$ defined by

$$p_0 = f(0) s_0,$$
$$p_1 = p_0 + (f(1) - p_0(1)) s_1,$$
$$p_2 = p_1 + (f(1/2) - p_1(1/2)) s_2,$$
$$p_3 = p_2 + (f(1/4) - p_2(1/4)) s_3,$$
$$p_4 = p_3 + (f(3/4) - p_3(3/4)) s_4,$$
$$p_5 = p_4 + (f(1/8) - p_4(1/8)) s_5,$$
$$p_6 = p_5 + (f(3/8) - p_5(3/8)) s_6,$$
$$p_7 = p_6 + (f(5/8) - p_6(5/8)) s_7,$$
$$p_8 = p_7 + (f(7/8) - p_7(7/8)) s_8$$
$$\vdots$$

Observe that $p_0$ coincides with $f$ at 0; $p_1$ coincides with $f$ at 0 and 1 and its graph is a line segment connecting the points $(0, f(0))$ and $(1, f(1))$; $p_2$ coincides with $f$ at 0, 1 and 1/2, and its graph is composed of two line segments connecting the points $(0, f(0))$, $(1/2, f(1/2))$ and $(1, f(1))$, and so on. That is, for every $n \in \mathbb{N}$, the function $p_n$ coincides with $f$ at the first $n + 1$ points of the set $D = \{0, 1, 1/2, 1/4, 3/4, 1/8, 3/8, 5/8, 7/8, \ldots\} \subseteq [0, 1]$, and its graph is a juxtaposition of line segments whose endpoints have abscissas in the set $D$ (Fig. 10.1).

## 10.3 Schauder Bases and Basic Sequences

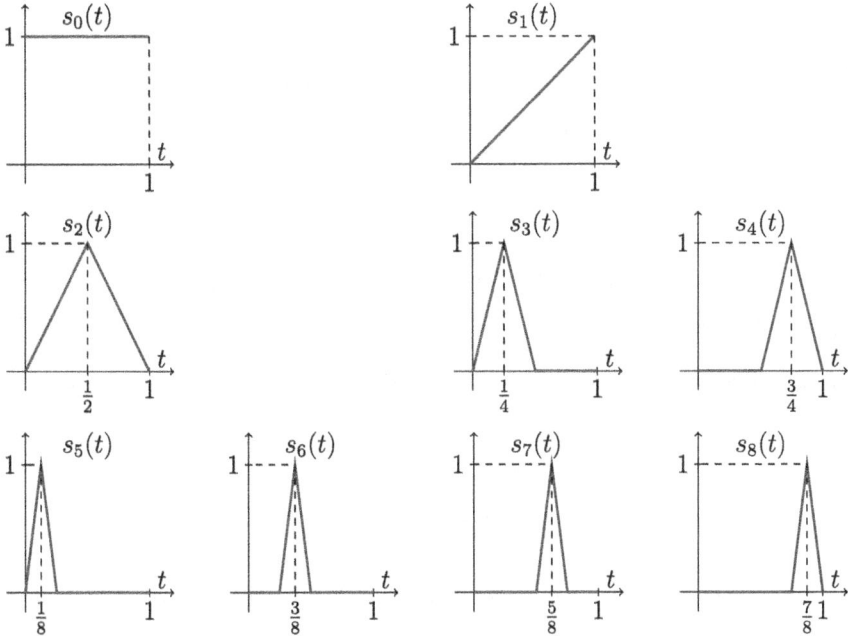

**Fig. 10.1** The first terms of the classical Schauder basis for $C[0, 1]$

For each non-negative integer $m$, call $a_m$ the coefficient of $s_m$ in the expression that defines $p_m$. Thus $p_n = \sum_{m=0}^{n} a_m s_m$ for each positive integer $n$. Let $\varepsilon > 0$. From the uniform continuity of $f$ there exists $\delta > 0$ such that

$$|f(t_1) - f(t_2)| < \varepsilon \text{ whenever } t_1, t_2 \in [0, 1], |t_1 - t_2| < \delta.$$

Consider positive integers $m$ such that $\frac{1}{2^m} < \frac{\delta}{2}$ and $n_0$ such that $f$ and $p_{n_0}$ coincide on the set $\{0, 1, \frac{1}{2}, \frac{1}{4}, \frac{3}{4}, \frac{3}{8}, \ldots, \frac{2^m-1}{2^m}\}$. Given $t \in [0, 1]$, there exists $k \in \{1, \ldots, 2^m - 1\}$ such that $t - \frac{k}{2^m} < \delta$. Then

$$|f(t) - p_n(t)| \leq \left|f(t) - f\left(\frac{k}{2^m}\right)\right| + \left|f\left(\frac{k}{2^m}\right) - p_n(t)\right|$$

$$< \varepsilon + \left|p_n\left(\frac{k}{2^m}\right) - p_n(t)\right| \leq \varepsilon + \left|p_n\left(\frac{k}{2^m}\right) - p_n\left(\frac{k+1}{2^m}\right)\right|$$

$$= \varepsilon + \left|f\left(\frac{k}{2^m}\right) - f\left(\frac{k+1}{2^m}\right)\right| < 2\varepsilon,$$

for each $n > n_0$. We conclude that $\lim_{n\to\infty} \|p_n - f\|_\infty = 0$, therefore $f = \sum_{n=0}^{\infty} a_n s_n$. We leave the verification of the uniqueness of the representation to the reader.

Our next goal is to prove the continuity of the coefficient functionals. It is important to realize that every space with a Schauder basis can be seen as a sequence space.

**Definition 10.3.4** Given a Schauder basis $(x_n)_{n=1}^{\infty}$ of the Banach space $E$, call $\mathcal{L}_E$ the vector space formed by the sequences of scalars $(a_n)_{n=1}^{\infty}$ such that the series $\sum_{n=1}^{\infty} a_n x_n$ is convergent.

**Lemma 10.3.5** *Let $(x_n)_{n=1}^{\infty}$ be a Schauder basis of the Banach space $E$. Then the function*

$$\eta: \mathcal{L}_E \longrightarrow \mathbb{R}, \ \eta\left((a_n)_{n=1}^{\infty}\right) := \sup\left\{\left\|\sum_{i=1}^{n} a_i x_i\right\| : n \in \mathbb{N}\right\},$$

*is a norm on $\mathcal{L}_E$ and $(\mathcal{L}_E, \eta)$ is a Banach space. Furthermore, the operator*

$$T_E: \mathcal{L}_E \longrightarrow E, \ T_E((a_n)_{n=1}^{\infty}) = \sum_{n=1}^{\infty} a_n x_n,$$

*is an isomorphism.*

**Proof** The norm axioms are immediate and they are left to the reader. We prove that the space $(\mathcal{L}_E, \eta)$ is complete. Let $(y_n)_{n=1}^{\infty} = \left((a_n^k)_{k=1}^{\infty}\right)_{n=1}^{\infty}$ be a Cauchy sequence in $(\mathcal{L}_E, \eta)$. For all $k, m, n \in \mathbb{N}$ it is true that

$$\left|a_m^k - a_n^k\right| \cdot \|x_k\| = \left\|\sum_{i=1}^{k}(a_m^i - a_n^i)x_i - \sum_{i=1}^{k-1}(a_m^i - a_n^i)x_i\right\|$$

$$\leq \left\|\sum_{i=1}^{k}(a_m^i - a_n^i)x_i\right\| + \left\|\sum_{i=1}^{k-1}(a_m^i - a_n^i)x_i\right\|$$

$$\leq 2\eta\left((a_m^j - a_n^j)_{j=1}^{\infty}\right) = 2\eta(y_m - y_n).$$

Then, for each $k \in \mathbb{N}$, the sequence $(a_n^k)_{n=1}^{\infty}$ is a Cauchy sequence in $\mathbb{K}$, hence convergent; let $a^k = \lim_{n\to\infty} a_n^k$ and $y = (a^k)_{k=1}^{\infty}$. Given $\varepsilon > 0$, as $(y_n)_{n=1}^{\infty}$ is a Cauchy sequence, there exists $n_\varepsilon \in \mathbb{N}$ such that $\eta(y_m - y_n) \leq \frac{\varepsilon}{4}$ whenever $m, n \geq n_\varepsilon$. Therefore,

## 10.3 Schauder Bases and Basic Sequences

$$\left\|\sum_{i=1}^{k}(a_m^i - a_n^i)x_i\right\| \leq \sup\left\{\left\|\sum_{i=1}^{j}(a_m^i - a_n^i)x_i\right\| : j \in \mathbb{N}\right\}$$

$$= \eta\left((a_m^j - a_n^j)_{j=1}^{\infty}\right) = \eta(y_m - y_n) \leq \frac{\varepsilon}{4}$$

for all $k \in \mathbb{N}$ and $m, n \geq n_\varepsilon$. Letting $m \longrightarrow \infty$ we conclude that

$$\left\|\sum_{i=1}^{k}(a^i - a_n^i)x_i\right\| \leq \frac{\varepsilon}{4} \quad \text{for all } k \in \mathbb{N} \text{ and } n \geq n_\varepsilon. \tag{10.11}$$

By the triangle inequality it follows that

$$\left\|\sum_{i=\ell}^{k}(a^i - a_{n_\varepsilon}^i)x_i\right\| = \left\|\sum_{i=1}^{k}(a^i - a_{n_\varepsilon}^i)x_i - \sum_{i=1}^{\ell-1}(a^i - a_{n_\varepsilon}^i)x_i\right\|$$

$$\leq \left\|\sum_{i=1}^{k}(a^i - a_{n_\varepsilon}^i)x_i\right\| + \left\|\sum_{i=1}^{\ell-1}(a^i - a_{n_\varepsilon}^i)x_i\right\| \leq \frac{\varepsilon}{4} + \frac{\varepsilon}{4} = \frac{\varepsilon}{2}$$

for all $\ell \in \mathbb{N}$ and $k > \ell$. As $y_{n_\varepsilon} = (a_{n_\varepsilon}^k)_{k=1}^{\infty} \in \mathcal{L}_E$, the series $\sum_{i=1}^{\infty} a_{n_\varepsilon}^i x_i$ is convergent, therefore there exists $n_0 \in \mathbb{N}$ such that $\left\|\sum_{i=\ell}^{k} a_{n_\varepsilon}^i x_i\right\| \leq \frac{\varepsilon}{2}$ whenever $k > \ell > n_0$. Thus, for $k > \ell > n_0$, using the triangle inequality again, we have

$$\left\|\sum_{i=\ell}^{k} a^i x_i\right\| \leq \left\|\sum_{i=\ell}^{k}(a^i - a_{n_\varepsilon}^i)x_i\right\| + \left\|\sum_{i=\ell}^{k} a_{n_\varepsilon}^i x_i\right\| \leq \frac{\varepsilon}{2} + \frac{\varepsilon}{2} = \varepsilon.$$

From the Cauchy criterion it follows that the series $\sum_{k=1}^{\infty} a^k x_k$ is convergent, hence $y \in \mathcal{L}_E$. From (10.11) we have

$$\eta(y_n - y) = \eta\left((a_n^k - a^k)_{k=1}^{\infty}\right) = \sup\left\{\left\|\sum_{i=1}^{k}(a_n^i - a^i)x_i\right\| : k \in \mathbb{N}\right\} \leq \frac{\varepsilon}{4}$$

for every $n \geq n_\varepsilon$, that is, $\lim_{n\to\infty} y_n = y$, therefore $(\mathcal{L}_E, \eta)$ is complete.

Since $(x_n)_{n=1}^{\infty}$ is a Schauder basis, the operator $T_E$ is clearly linear and bijective. Moreover, for each $(a_k)_{k=1}^{\infty} \in \mathcal{L}_E$,

$$\left\|T_E\left((a_k)_{k=1}^{\infty}\right)\right\| = \lim_{n\to\infty}\left\|\sum_{k=1}^{n} a_k x_k\right\| \leq \eta((a_k)_{k=1}^{\infty}),$$

which proves the continuity of $T_E$. From the Open Mapping Theorem (Theorem 2.4.2) it follows that $T_E$ is an isomorphism. ∎

**Theorem 10.3.6** *The coefficient functionals associated with a Schauder basis of a Banach space are always continuous.*

*Proof* Let $(x_j)_{j=1}^\infty$ be a Schauder basis of the Banach space $E$ and let $n \in \mathbb{N}$ be given. Given $x \in E$, keep the notation of the statement of Lemma 10.3.5 and consider a sequence $(a_j)_{j=1}^\infty \in \mathcal{L}_E$ such that $x = T_E\left((a_j)_{j=1}^\infty\right) \in E$. Then

$$\left|x_n^*(x)\right| \cdot \|x_n\| = \left\|\sum_{i=1}^n x_i^*(x)x_i - \sum_{i=1}^{n-1} x_i^*(x)x_i\right\|$$

$$\leq \left\|\sum_{i=1}^n a_i x_i\right\| + \left\|\sum_{i=1}^{n-1} a_i x_i\right\| \leq 2\eta\left((a_j)_{j=1}^\infty\right) \leq 2\left\|T_E^{-1}\right\| \cdot \|x\|,$$

from which the continuity of the functional $x_n^*$ results. ∎

**Corollary 10.3.7** *Let $(x_n)_{n=1}^\infty$ be a Schauder basis of a Banach space $E$. Then, for each $n \in \mathbb{N}$, the linear operator*

$$P_n: E \longrightarrow E, \quad P_n\left(\sum_{j=1}^\infty a_j x_j\right) = \sum_{j=1}^n a_j x_j,$$

*is continuous.*

*Proof* Note that $P_n = \sum_{j=1}^n x_j^* \otimes x_j$ and apply Theorem 10.3.6. ∎

It is clear that $P_n^2 = P_n$ for every $n$, so the operators $(P_n)_{n=1}^\infty$ are called *canonical projections* associated to the Schauder basis $(x_n)_{n=1}^\infty$. In particular, $\|P_n\| \geq 1$ for every $n$. The remaining question is whether the canonical projections are uniformly bounded, that is, is it true that $\sup_n \|P_n\| < \infty$? The answer is yes:

**Proposition 10.3.8** *Let $(x_n)_{n=1}^\infty$ be a Schauder basis of a Banach space $E$ and let $(P_n)_{n=1}^\infty$ be their canonical projections. Then $\sup_n \|P_n\| < \infty$.*

The number $\sup_n \|P_n\|$ is called the *basis constant* and denoted by $K_{(x_n)_{n=1}^\infty}$. As $\|P_n\| \geq 1$ for each $n$, it follows that $K_{(x_n)_{n=1}^\infty} \geq 1$.

*Proof* We know that, for each $x \in E$,

$$\lim_{n\to\infty} P_n(x) = \lim_{n\to\infty} \sum_{j=1}^n x_j^*(x)x_j = \sum_{j=1}^\infty x_j^*(x)x_j = x.$$

## 10.3 Schauder Bases and Basic Sequences

Then, being convergent, the sequence $(\|P_n(x)\|)_{n=1}^{\infty}$ is bounded for each $x \in E$. As $E$ is a Banach space and the operators $P_n, n \in \mathbb{N}$, are continuous by Corollary 10.3.7, the result follows from the Banach–Steinhaus Theorem (Theorem 2.3.2). ∎

**Corollary 10.3.9** *Let $(x_n)_{n=1}^{\infty}$ be a Schauder basis of a Banach space $E$ and let $(x_n^*)_{n=1}^{\infty}$ be their coefficient functionals. Then, for every $k \in \mathbb{N}$,*

$$1 \leq \|x_k^*\| \cdot \|x_k\| \leq 2K_{(x_n)_{n=1}^{\infty}}.$$

*Proof* For every $k \in \mathbb{N}$,

$$1 = x_k^*(x_k) \leq \|x_k^*\| \cdot \|x_k\| = \|x_k^* \otimes x_k\| = \left\| \sum_{j=1}^{k} x_j^* \otimes x_j - \sum_{j=1}^{k-1} x_j^* \otimes x_j \right\|$$

$$= \|P_k - P_{k-1}\| \leq \|P_k\| + \|P_{k-1}\| \leq 2K_{(x_n)_{n=1}^{\infty}}.$$

∎

Before moving on, we present another interesting application of Lemma 10.3.5.

**Definition 10.3.10** A Schauder basis $(x_n)_{n=1}^{\infty}$ of a Banach space $E$ is *equivalent* to a Schauder basis $(y_n)_{n=1}^{\infty}$ of a Banach space $F$, and in this case we write $(x_n)_{n=1}^{\infty} \approx (y_n)_{n=1}^{\infty}$, if, for any sequence of scalars $(a_n)_{n=1}^{\infty}$, the series $\sum_{n=1}^{\infty} a_n x_n$ is convergent in $E$ if and only if the series $\sum_{n=1}^{\infty} a_n y_n$ is convergent in $F$.

**Theorem 10.3.11** *Let $E$ and $F$ be Banach spaces with Schauder bases $(x_n)_{n=1}^{\infty}$ and $(y_n)_{n=1}^{\infty}$, respectively. Then $(x_n)_{n=1}^{\infty} \approx (y_n)_{n=1}^{\infty}$ if and only if there exists an isomorphism $T: E \longrightarrow F$ such that $T(x_n) = y_n$ for every $n \in \mathbb{N}$.*

*Proof* Assume that the bases are equivalent. Taking

$$(Z, (z_n)_{n=1}^{\infty}) \in \{(E, (x_n)_{n=1}^{\infty}), (F, (y_n)_{n=1}^{\infty})\},$$

from Lemma 10.3.5 we know that the operator

$$T_Z: (\mathcal{L}_Z, \eta_Z) \longrightarrow Z, \; T_Z((a_n)_{n=1}^{\infty}) = \sum_{n=1}^{\infty} a_n z_n,$$

is an isomorphism. From the equivalence of the bases, it follows that $\mathcal{L}_E$ and $\mathcal{L}_F$ coincide as vector spaces, that is, the identity operator id: $\mathcal{L}_E \longrightarrow \mathcal{L}_F$ is linear and bijective. We leave to the reader the task of showing that id has closed graph. As $\mathcal{L}_E$ and $\mathcal{L}_F$ are Banach spaces by Lemma 10.3.5, we conclude by Theorem 2.5.1 that id is continuous. From the Open Mapping Theorem (Theorem 2.4.2) it follows that id is an isomorphism. Therefore, the operator

$$T := T_F \circ \mathrm{id} \circ T_E^{-1} : E \longrightarrow F$$

is also an isomorphism. For each $n \in \mathbb{N}$,

$$T(x_n) = T_F \left( \mathrm{id} \left( T_E^{-1}(x_n) \right) \right) = T_F \left( \mathrm{id}(e_n) \right) = T_F(e_n) = y_n.$$

Conversely, let $T : E \longrightarrow F$ be an isomorphism with $T(x_n) = y_n$ for each $n$. Given a sequence of scalars $(a_n)_{n=1}^{\infty}$ with $\sum_{n=1}^{\infty} a_n x_n = x \in E$, from the linearity and continuity of $T$ we have

$$T(x) = T \left( \sum_{n=1}^{\infty} a_n x_n \right) = \sum_{n=1}^{\infty} a_n T(x_n) = \sum_{n=1}^{\infty} a_n y_n \in F.$$

It follows that the series $\sum_{n=1}^{\infty} a_n y_n$ converges. The other implication follows in the same way using the inverse isomorphism $T^{-1}$. ∎

Sometimes a sequence $(x_n)_{n=1}^{\infty}$ in $E$ is a Schauder basis for the subspace $\overline{[x_n : n \in \mathbb{N}]}$ but not for the entire space $E$.

**Definition 10.3.12** A sequence $(x_n)_{n=1}^{\infty}$ in the Banach space $E$ is said to be a *basic sequence* if it is a Schauder basis for $\overline{[x_n : n \in \mathbb{N}]}$.

It is important to realize that under the conditions of the definition above, if $x = \sum_{j=1}^{\infty} a_j x_j \in E$, then $a_j = x_j^*(x)$ for every $j$ and, in this case, the coefficient functionals $x_j^*$ are defined on $\overline{[x_n : n \in \mathbb{N}]}$.

The next result is a practical and useful criterion for deciding whether a given sequence is basic or not.

**Theorem 10.3.13 (Banach–Grunblum Criterion)** *A sequence $(x_n)_{n=1}^{\infty}$ of nonzero vectors in a Banach space $E$ is basic if and only if there exists a constant $M \geq 1$ such that for every sequence of scalars $(a_n)_{n=1}^{\infty}$,*

$$\left\| \sum_{i=1}^{m} a_i x_i \right\| \leq M \left\| \sum_{i=1}^{n} a_i x_i \right\| \quad \text{whenever } n \geq m. \tag{10.12}$$

*Proof* Suppose that $(x_n)_{n=1}^{\infty}$ is a basic sequence and consider its canonical projections $(P_n)_{n=1}^{\infty}$. From Proposition 10.3.8 we know that

$$1 \leq K_{(x_n)_{n=1}^{\infty}} = \sup \{ \|P_n\| : n \in \mathbb{N} \} < \infty.$$

## 10.3 Schauder Bases and Basic Sequences

Given a sequence of scalars $(a_n)_{n=1}^\infty$, if $n \geq m$, then

$$\left\| \sum_{i=1}^m a_i x_i \right\| = \left\| P_m \left( \sum_{i=1}^n a_i x_i \right) \right\| \leq \| P_m \| \cdot \left\| \sum_{i=1}^n a_i x_i \right\| \leq K_{(x_n)_{n=1}^\infty} \left\| \sum_{i=1}^n a_i x_i \right\|.$$

Conversely, suppose that (10.12) holds for some $M \geq 1$. Note that the set $\{x_1, x_2, \ldots\}$ is linearly independent. In fact, if $a_1 x_1 + \cdots + a_n x_n = 0$, then

$$\|a_1 x_1\| \leq M \|a_1 x_1 + \cdots + a_n x_n\| = 0,$$

from which it follows that $a_1 = 0$ because $x_1 \neq 0$. Therefore

$$\|a_2 x_2\| = \|a_1 x_1 + a_2 x_2\| \leq M \|a_1 x_1 + \cdots + a_n x_n\| = 0,$$

from which it follows, in the same way, that $a_2 = 0$. Repeating this process, we conclude that $a_1 = \cdots = a_n = 0$.

Consider the subspace $F = [x_n : n \in \mathbb{N}]$ and, for each $n$, the linear functional

$$\varphi_n : F \longrightarrow \mathbb{K} \ , \ \varphi_n \left( \sum_{j=1}^k a_j x_j \right) = a_n,$$

as well as the linear operator

$$T_n : F \longrightarrow \overline{[x_n : n \in \mathbb{N}]} \ , \ T_n \left( \sum_{j=1}^k a_j x_j \right) = \sum_{j=1}^n a_j x_j,$$

defining $a_{k+1} = \cdots = a_n = 0$ if necessary. By (10.12), we have

$$\left\| T_n \left( \sum_{j=1}^k a_j x_j \right) \right\| = \left\| \sum_{j=1}^n a_j x_j \right\| \leq M \left\| \sum_{j=1}^k a_j x_j \right\|,$$

and the continuity of $T_n$ follows from $\|T_n\| \leq M$. For $x = \sum_{j=1}^k a_j x_j \in F$ and $n \in \mathbb{N}$, as $|\varphi_n(x)| = |a_n|$, it follows that

$$|\varphi_n(x)| \cdot \|x_n\| = |a_n| \cdot \|x_n\| = \left\| \sum_{j=1}^n a_j x_j - \sum_{j=1}^{n-1} a_j x_j \right\|$$

$$\leq \left\| \sum_{j=1}^n a_j x_j \right\| + \left\| \sum_{j=1}^{n-1} a_j x_j \right\| \leq 2M \|x\|.$$

This implies the continuity of $\varphi_n$. As $\mathbb{K}$ and $\overline{[x_n : n \in \mathbb{N}]}$ are Banach spaces and $F$ is dense in $\overline{[x_n : n \in \mathbb{N}]}$, there exist unique linear continuous extensions of $\varphi_n$ and $T_n$ to the space $\overline{[x_n : n \in \mathbb{N}]}$, denoted by $\Phi_n$ and $R_n$, respectively. Furthermore, $\|\varphi_n\| = \|\Phi_n\|$ and $\|T_n\| = \|R_n\|$ for every $n \in \mathbb{N}$.

Note also that, as $T_n(z) = \sum_{j=1}^{n} \varphi_j(z) x_j$ for each $z \in F$, then

$$R_n(x) = \sum_{j=1}^{n} \Phi_j(x) x_j \text{ for every } x \in \overline{[x_n : n \in \mathbb{N}]}. \tag{10.13}$$

Given $x \in \overline{[x_n : n \in \mathbb{N}]}$ and $\varepsilon > 0$, we can consider $y = \sum_{j=1}^{m} a_j x_j \in F$ such that $\|x - y\| < \varepsilon$. Thus, for all $n > m$ we have

$$\|x - R_n(x)\| \leq \|x - y\| + \|y - R_n(y)\| + \|R_n(y) - R_n(x)\|$$
$$\leq \|x - y\| + \|y - y\| + \|R_n\| \cdot \|x - y\| < (1 + M)\varepsilon.$$

Combining the inequality above with (10.13) we conclude that

$$x = \lim_{n \to \infty} R_n(x) = \lim_{n \to \infty} \left( \sum_{j=1}^{n} \Phi_j(x) x_j \right) = \sum_{j=1}^{\infty} \Phi_j(x) x_j.$$

It remains to show the uniqueness of the representation above. It is enough to show that if $\sum_{j=1}^{\infty} a_j x_j = 0$ then $a_n = 0$ for each $n \in \mathbb{N}$. Suppose $\sum_{j=1}^{\infty} a_j x_j = 0$ and consider $\varepsilon > 0$. Choosing $n_0 \in \mathbb{N}$ such that

$$\left\| \sum_{j=1}^{n} a_j x_j \right\| < \varepsilon \leq M\varepsilon \text{ whenever } n \geq n_0,$$

from (10.12) it follows that

$$\left\| \sum_{j=1}^{n} a_j x_j \right\| \leq M \left\| \sum_{j=1}^{n_0} a_j x_j \right\| < M\varepsilon \text{ whenever } n \leq n_0.$$

For each $n$,

$$|a_n| \cdot \|x_n\| = \left\| \sum_{j=1}^{n} a_j x_j - \sum_{j=1}^{n-1} a_j x_j \right\| \leq 2M\varepsilon$$

## 10.3 Schauder Bases and Basic Sequences

for every $\varepsilon > 0$. As $x_n \neq 0$, letting $\varepsilon \longrightarrow 0^+$ it follows that $a_n = 0$. ∎

**Corollary 10.3.14** *Every subsequence of a basic sequence in a Banach space is also a basic sequence.*

The Banach–Grunblum criterion is also useful for calculating the basis constant:

**Corollary 10.3.15** *Let $(x_n)_{n=1}^\infty$ be a basic sequence in a Banach space $E$. Then*

$$K_{(x_n)_{n=1}^\infty} = \inf\{M : M \text{ satisfies } (10.12)\},$$

*and the infimum is attained.*

**Proof** In the first part of the proof of Theorem 10.3.13 we saw that $K_{(x_n)_{n=1}^\infty}$ satisfies (10.12), so the infimum is less than or equal to $K_{(x_n)_{n=1}^\infty}$. For the remaining inequality, suppose that $M$ satisfies (10.12). Given $\sum_{j=1}^\infty a_j x_j \in \overline{[x_n : n \in \mathbb{N}]}$, we have

$$\left\| P_n\left(\sum_{j=1}^\infty a_j x_j\right) \right\| = \left\| \sum_{j=1}^n a_j x_j \right\| \leq M \left\| \sum_{j=1}^m a_j x_j \right\| \quad \text{whenever } m \geq n.$$

Therefore

$$\left\| P_n\left(\sum_{j=1}^\infty a_j x_j\right) \right\| \leq M \cdot \lim_{m \to \infty} \left\| \sum_{j=1}^m a_j x_j \right\| = M \left\| \sum_{j=1}^\infty a_j x_j \right\|$$

for every $n \in \mathbb{N}$. It follows that

$$K_{(x_n)_{n=1}^\infty} = \sup\{\|P_n\| : n \in \mathbb{N}\} \leq M.$$

Considering the infimum over the constants $M$ satisfying (10.12), we obtain the inequality that was missing. Once the equality is established, the infimum is attained because, as we have already seen, $K_{(x_n)_{n=1}^\infty}$ satisfies (10.12). ∎

At first glance, the reader may suspect that any countable linearly independent subset of a Banach space is a basic sequence. A second application of the Banach–Grunblum criterion shows that this is not the case in general:

**Example 10.3.16** Consider the sequence $(x_n)_{n=1}^\infty$ in $\ell_p$, $1 < p \leq \infty$, defined by

$$x_1 = e_1, \quad x_j = -\frac{1}{j-1} e_{j-1} + \frac{1}{j} e_j \text{ for } j \geq 2.$$

It is easy to check that the vectors $x_1, x_2, \ldots$ are linearly independent. On the other hand, for every positive integer $n$ it is true that

$$\left\|\sum_{j=1}^{n} x_j\right\|_p = \frac{1}{n}.$$

Thus, there is no constant $M$ that satisfies (10.12), so the sequence $(x_n)_{n=1}^{\infty}$ is not basic in $\ell_p$ by the Banach–Grunblum criterion.

## 10.4 The Bessaga–Pełczyński Selection Principle

On page 238 of the book [6] that inaugurated Functional Analysis, Banach states, without proof and in only two lines, that in every infinite-dimensional Banach space there is an infinite-dimensional subspace with a Schauder basis. Or, equivalently, that every infinite-dimensional Banach space contains an infinite basic sequence. In a famous article [7] from 1958, C. Bessaga and A. Pełczyński obtained this result as an application of their main result, the so-called *Bessaga–Pełczyński selection principle*. The essence of a "selection principle" is the possibility of extracting basic sequences with special properties from certain sequences in Banach spaces. The aim of this section is to prove the Bessaga–Pełczyński selection principle and to apply it to obtain basic sequences in arbitrary Banach spaces.

Some preparatory concepts and results are necessary.

**Definition 10.4.1** A subset $N$ of the dual $E'$ of a Banach space $E$ is said to be *norming for $E$* if

$$\|x\| = \sup\{|\varphi(x)| : \varphi \in N \text{ and } \|\varphi\| \leq 1\} \text{ for every } x \text{ in } E.$$

The Hahn–Banach Theorem in the form of Corollary 3.1.5 says that $E'$ is norming for $E$. It follows from Proposition 4.3.1 that $E = J_E(E)$ is norming for $E'$. It is immediate that every norming set $N$ for $E$ separates points of $E$, that is, for all $x, y \in E$, $x \neq y$, there exists $\varphi \in N$ such that $\varphi(x) \neq \varphi(y)$.

**Lemma 10.4.2** *Let $N$ be a norming set for a Banach space $E$ and let $(x_n)_{n=1}^{\infty}$ be a sequence in $E$ such that $\inf_n \|x_n\| > 0$ and*

$$\lim_{n \to \infty} \varphi(x_n) = 0 \text{ for every } \varphi \in N.$$

*Then, for each $0 < \varepsilon < 1$, each positive integer $k$ and each finite-dimensional subspace $F$ of $E$, there exists an integer $n \geq k$ such that*

$$\|y + ax_n\| \geq (1 - \varepsilon)\|y\| \text{ for all } y \in F \text{ and } a \in \mathbb{K}.$$

**Proof** It is simple to check that it is enough to prove the desired inequality for unit vectors $y \in F$. Suppose, by contradiction, that the result is not true. In this case

## 10.4 The Bessaga–Pełczyński Selection Principle

there are $\varepsilon \in (0, 1)$, a positive integer $k$ and a finite-dimensional subspace $F$ of $E$ with the following property: for every $n \geq k$ there exists a unit vector $y_n \in F$ and a scalar $a_n$ such that

$$\|y_n + a_n x_n\| < 1 - \varepsilon.$$

Since $F$ is finite-dimensional, by compactness there exists a subsequence of $(y_{n_j})_{j=1}^\infty$ that converges to a certain $y_0 \in F$. It is clear that the vector $y_0$ is also unitary. For $n_j \geq k$,

$$|a_{n_j}| \cdot \|x_{n_j}\| = \|a_{n_j} x_{n_j}\| \leq \|y_{n_j} + a_{n_j} x_{n_j}\| + \|y_{n_j}\| < (1 - \varepsilon) + \|y_{n_j}\| = 2 - \varepsilon.$$

It follows that

$$|a_{n_j}| < \frac{2 - \varepsilon}{\|x_{n_j}\|} \leq \frac{2 - \varepsilon}{\inf\{\|x_i\| : i \geq k\}}.$$

Thus, the sequence $(a_{n_j})_{j=1}^\infty$ is bounded. From the assumption and from the convergence $y_{n_j} \longrightarrow y_0$ it follows that

$$\lim_{j \to \infty} \varphi(y_{n_j} + a_{n_j} x_{n_j}) = \varphi(y_0) + \lim_{j \to \infty} \varphi(a_{n_j} x_{n_j})$$

$$= \varphi(y_0) + \lim_{j \to \infty} a_{n_j} \varphi(x_{n_j}) = \varphi(y_0)$$

for every $\varphi \in N$. As $N$ is norming for $E$ and $\|y_0\| = 1$, there exists $\varphi_0 \in N$ such that $\|\varphi_0\| \leq 1$ and $|\varphi_0(y_0)| > 1 - \varepsilon$. On the other hand,

$$1 - \varepsilon > \|y_{n_j} + a_{n_j} x_{n_j}\| = \sup_{\varphi \in B_{E'}} |\varphi(y_{n_j} + a_{n_j} x_{n_j})| \geq |\varphi_0(y_{n_j} + a_{n_j} x_{n_j})|$$

for every $n_j \geq k$, therefore

$$1 - \varepsilon \geq \lim_{j \to \infty} |\varphi_0(y_{n_j} + a_{n_j} x_{n_j})| = |\varphi_0(y_0)| > 1 - \varepsilon,$$

a contradiction that completes the proof. ∎

**Theorem 10.4.3 (Bessaga–Mazur)** *Let $N$ be a norming set for a Banach space $E$ and let $(x_n)_{n=1}^\infty$ be a sequence in $E$ such that $\inf_n \|x_n\| > 0$ and*

$$\lim_{n \to \infty} \varphi(x_n) = 0 \text{ for every } \varphi \in N.$$

*Then $(x_n)_{n=1}^\infty$ contains a basic subsequence $(x_{n_k})_{k=1}^\infty$ with $x_{n_1} = x_1$.*

**Proof** Consider a sequence $(\varepsilon_n)_{n=1}^{\infty}$ of real numbers with $0 < \varepsilon_n < 1$ for every $n \in \mathbb{N}$ satisfying

$$0 < \prod_{n=1}^{\infty}(1-\varepsilon_n) := \lim_{n\to\infty}\prod_{j=1}^{n}(1-\varepsilon_n) < 1.$$

We suggest that the reader convinces herself/himself of the existence of such a sequence $(\varepsilon_n)_{n=1}^{\infty}$. Calling $n_1 = 1$, $F_1 = [x_{n_1}]$, $k = n_1$ and $\varepsilon = \varepsilon_1$, by Lemma 10.4.2 there exists $n_2 > n_1$ such that

$$\|a_1 x_{n_1} + a_2 x_{n_2}\| \geq (1-\varepsilon_1)\|a_1 x_{n_1}\|$$

for any scalars $a_1, a_2$. Inductively, suppose the existence of integers $n_1 < n_2 < \cdots < n_k$ such that

$$\|a_1 x_{n_1} + \cdots + a_k x_{n_k}\| \geq (1-\varepsilon_{k-1})\|a_1 x_{n_1} + \cdots + a_{k-1} x_{n_{k-1}}\|$$

for any scalars $a_1, \ldots, a_k$. Then, calling $F_k = [x_{n_1}, \ldots, x_{n_k}]$, $\varepsilon = \varepsilon_k$, $k = n_k$, by Lemma 10.4.2 there exists $n_{k+1} > n_k$ such that

$$\|a_1 x_{n_1} + \cdots + a_{k+1} x_{n_{k+1}}\| \geq (1-\varepsilon_k)\|a_1 x_{n_1} + \cdots + a_k x_{n_k}\|.$$

We have constructed the subsequence $(x_{n_k})_{k=1}^{\infty}$; it remains to show that it is a basic sequence. For all $k, m \in \mathbb{N}$ and scalars $a_1, \ldots, a_{k+m}$, it is true that

$$\left\|\sum_{i=1}^{k+m} a_i x_{n_i}\right\| \geq (1-\varepsilon_{k+m-1})\left\|\sum_{i=1}^{k+m-1} a_i x_{n_i}\right\|$$

$$\geq (1-\varepsilon_{k+m-1})\cdots(1-\varepsilon_k)\left\|\sum_{i=1}^{k} a_i x_{n_i}\right\|$$

$$\geq \left(\prod_{n=1}^{\infty}(1-\varepsilon_n)\right)\left\|\sum_{i=1}^{k} a_i x_{n_i}\right\|.$$

Considering $M = \left(\prod_{n=1}^{\infty}(1-\varepsilon_n)\right)^{-1}$ in the Banach–Grunblum criterion (Theorem 10.3.13), it follows that the subsequence $(x_{n_k})_{k=1}^{\infty}$ is basic. ∎

**Definition 10.4.4** Let $(x_n)_{n=1}^{\infty}$ be a Schauder basis of a Banach space $E$ and let $(k_n)_{n=0}^{\infty}$ be a strictly increasing sequence of positive integers with $k_0 = 0$. A sequence of non-zero vectors $(y_n)_{n=1}^{\infty}$ in $E$ defined by

## 10.4 The Bessaga–Pełczyński Selection Principle

$$y_n = \sum_{i=k_{n-1}+1}^{k_n} b_i x_i,$$

where each $b_i \in \mathbb{K}$, is called a *block basic sequence of* $(x_n)_{n=1}^\infty$.

The next result justifies the adjective "basic" in the definition above:

**Proposition 10.4.5** *Let* $(x_n)_{n=1}^\infty$ *be a Schauder basis of a Banach space* $E$ *and let* $(y_n)_{n=1}^\infty$ *be a block basic sequence of* $(x_n)_{n=1}^\infty$. *Then* $(y_n)_{n=1}^\infty$ *is a basic sequence in* $E$ *and* $K_{(y_n)_{n=1}^\infty} \leq K_{(x_n)_{n=1}^\infty}$.

**Proof** Let $m$ and $k$ be positive integers and let $a_1, \ldots, a_{m+k}$ be given scalars. From Corollary 10.3.15 it follows that

$$\left\| \sum_{i=1}^m a_i y_i \right\| = \left\| \sum_{i=1}^{k_1} a_1 b_i x_i + \sum_{i=k_1+1}^{k_2} a_2 b_i x_i + \cdots + \sum_{i=k_{m-1}+1}^{k_m} a_m b_i x_i \right\|$$

$$\leq K_{(x_n)_{n=1}^\infty} \left\| \sum_{i=1}^{k_1} a_1 b_i x_i + \cdots + \sum_{i=k_{m-1}+1}^{k_m} a_m b_i x_i + \cdots \right.$$

$$\left. + \sum_{i=k_{m+k-1}+1}^{k_{m+k}} a_{m+k} b_i x_i \right\|$$

$$= K_{(x_n)_{n=1}^\infty} \left\| \sum_{i=1}^{m+k} a_i y_i \right\|.$$

The Banach–Grunblum criterion ensures that the sequence $(y_n)_{n=1}^\infty$ is basic and the definition of the basis constant ensures that $K_{(y_n)_{n=1}^\infty} \leq K_{(x_n)_{n=1}^\infty}$. ∎

One more result is necessary for the proof of the selection principle. We need first the following elementary lemma, whose proof we leave to the reader.

**Lemma 10.4.6** *Let* $(x_n)_{n=1}^\infty$ *be a Schauder basis of a Banach space* $E$, *let* $(y_n)_{n=1}^\infty$ *be a sequence in* $E$ *and let* $T: E \longrightarrow E$ *be an isomorphism with* $T(x_n) = y_n$ *for every* $n$. *Then* $(y_n)_{n=1}^\infty$ *is a Schauder basis of* $E$.

**Theorem 10.4.7 (Bessaga–Pełczyński)** *Let* $E$ *be a Banach space, let* $(x_n)_{n=1}^\infty$ *be a basic sequence in* $E$ *and let* $(x_n^*)_{n=1}^\infty$ *be its coefficient functionals. If* $(y_n)_{n=1}^\infty$ *is a sequence in* $E$ *such that*

$$\sum_{n=1}^\infty \|x_n - y_n\| \cdot \|x_n^*\| =: \lambda < 1, \tag{10.14}$$

then $(y_n)_{n=1}^{\infty}$ is a basic sequence in $E$ equivalent to $(x_n)_{n=1}^{\infty}$. Furthermore, if $(x_n)_{n=1}^{\infty}$ is a Schauder basis of $E$, then $(y_n)_{n=1}^{\infty}$ is also a Schauder basis of $E$.

***Proof*** Given a sequence $(a_n)_{n=1}^{\infty}$ of scalars, from

$$\left\| \sum_{i=1}^{n} a_i(x_i - y_i) \right\| = \left\| \sum_{i=1}^{n} x_i^* \left( \sum_{j=1}^{n} a_j x_j \right) (x_i - y_i) \right\|$$

$$\leq \sum_{i=1}^{n} \left| x_i^* \left( \sum_{j=1}^{n} a_j x_j \right) \right| \cdot \|x_i - y_i\|$$

$$\leq \left\| \sum_{j=1}^{n} a_j x_j \right\| \left( \sum_{i=1}^{n} \|x_i^*\| \cdot \|x_i - y_i\| \right) \leq \lambda \left\| \sum_{i=1}^{n} a_i x_i \right\|,$$

it follows that

$$\left\| \left\| \sum_{i=1}^{n} a_i x_i \right\| - \left\| \sum_{i=1}^{n} a_i y_i \right\| \right\| \leq \left\| \sum_{i=1}^{n} a_i x_i - \sum_{i=1}^{n} a_i y_i \right\| \leq \lambda \left\| \sum_{i=1}^{n} a_i x_i \right\|.$$

Consequently,

$$(1 - \lambda) \left\| \sum_{i=1}^{n} a_i x_i \right\| \leq \left\| \sum_{i=1}^{n} a_i y_i \right\| \leq (1 + \lambda) \left\| \sum_{i=1}^{n} a_i x_i \right\| \tag{10.15}$$

for every $n \in \mathbb{N}$. As $(x_n)_{n=1}^{\infty}$ is a basic sequence, by the Banach–Grunblum criterion (Theorem 10.3.13) there exists $M \geq 1$ such that if $m \geq n$ then

$$\left\| \sum_{i=1}^{n} a_i y_i \right\| \leq (1 + \lambda) \left\| \sum_{i=1}^{n} a_i x_i \right\| \leq (1 + \lambda) M \left\| \sum_{i=1}^{m} a_i x_i \right\| \leq \frac{(1 + \lambda) M}{(1 - \lambda)} \left\| \sum_{i=1}^{m} a_i y_i \right\|.$$

Using once again the Banach–Grunblum criterion, we conclude that $(y_n)_{n=1}^{\infty}$ is a basic sequence of $E$. From (10.15) and the Cauchy criterion (Remark 10.1.2) it immediately follows that $\sum_{n=1}^{\infty} a_n x_n$ converges if and only if $\sum_{n=1}^{\infty} a_n y_n$ converges, that is, $(x_n)_{n=1}^{\infty}$ is equivalent to $(y_n)_{n=1}^{\infty}$.

Now suppose that $(x_n)_{n=1}^{\infty}$ is a Schauder basis of $E$. As $\lambda < \infty$, the following linear operator

$$T: E \longrightarrow E, \ T(x) = \sum_{j=1}^{\infty} x_j^*(x)(x_j - y_j),$$

## 10.4 The Bessaga–Pełczyński Selection Principle

is well defined. Moreover, $\|T\| \leq \lambda < 1$, so the operator $(\mathrm{id}_E - T) \colon E \longrightarrow E$ is an isomorphism by Proposition 7.1.3. As $(\mathrm{id}_E - T)(x_n) = y_n$ for every $n \in \mathbb{N}$, it follows from Lemma 10.4.6 that $(y_n)_{n=1}^{\infty}$ is a Schauder basis of $E$. ∎

Now we are ready to prove the main result of the section:

**Theorem 10.4.8 (Bessaga–Pełczyński Selection Principle)** *Let $(x_n)_{n=1}^{\infty}$ be a Schauder basis of a Banach space $E$ and let $(x_n^*)_{n=1}^{\infty}$ be its coefficient functionals. If $(y_n)_{n=1}^{\infty}$ is a sequence in $E$ such that $\inf_n \|y_n\| > 0$ and*

$$\lim_{n \to \infty} x_i^*(y_n) = 0 \text{ for each } i \in \mathbb{N},$$

*then $(y_n)_{n=1}^{\infty}$ contains a basic subsequence equivalent to a block basic sequence of $(x_n)_{n=1}^{\infty}$.*

**Proof** Let $M$ be the constant of the basis $(x_n)_{n=1}^{\infty}$. Call $\varepsilon = \inf_n \|y_n\|$, $n_1 = 1$ and choose $m_1 \in \mathbb{N}$ such that

$$\left(\frac{4M}{\varepsilon}\right) \left\| \sum_{i=m_1+1}^{\infty} x_i^*(y_{n_1}) x_i \right\| < \frac{1}{2^3}.$$

Now choose an integer $n_2 > n_1$ such that

$$\left(\frac{4M}{\varepsilon}\right) \left\| \sum_{i=1}^{m_1} x_i^*(y_{n_2}) x_i \right\| \leq \frac{1}{2^3}.$$

This is possible because $\lim_{n \to \infty} x_i^*(y_n) = 0$ for every $i \in \mathbb{N}$, in particular for each $i = 1, \ldots, m_1$. We then select an integer $m_2 > m_1$ such that

$$\left(\frac{4M}{\varepsilon}\right) \left\| \sum_{i=m_2+1}^{\infty} x_i^*(y_{n_2}) x_i \right\| < \frac{1}{2^4}.$$

Continuing this process we obtain increasing sequences $(n_k)_{k=1}^{\infty}$ and $(m_k)_{k=1}^{\infty}$ of positive integers such that

$$\left(\frac{4M}{\varepsilon}\right) \left\| \sum_{i=m_k+1}^{\infty} x_i^*(y_{n_k}) x_i \right\| < \frac{1}{2^{k+2}} \text{ and } \left(\frac{4M}{\varepsilon}\right) \left\| \sum_{i=1}^{m_k} x_i^*(y_{n_{k+1}}) x_i \right\| \leq \frac{1}{2^{k+2}}.$$

(10.16)

Defining, for each $k \in \mathbb{N}$,

$$z_k = \sum_{i=m_k+1}^{m_{k+1}} x_i^* \left( y_{n_{k+1}} \right) x_i,$$

it follows that

$$y_{n_{k+1}} = \sum_{i=1}^{m_k} x_i^* \left( y_{n_{k+1}} \right) x_i + z_k + \sum_{i=m_{k+1}+1}^{\infty} x_i^* \left( y_{n_{k+1}} \right) x_i. \qquad (10.17)$$

As $M \geq 1$, from (10.17) and (10.16) it follows that

$$\|z_k\| \geq \|y_{n_{k+1}}\| - \left\| \sum_{i=1}^{m_k} x_i^* \left( y_{n_{k+1}} \right) x_i \right\| - \left\| \sum_{i=m_{k+1}+1}^{\infty} x_i^* \left( y_{n_{k+1}} \right) x_i \right\| \qquad (10.18)$$

$$\geq \varepsilon - \frac{\varepsilon}{4M2^{k+2}} - \frac{\varepsilon}{4M2^{k+3}} \geq \varepsilon - \frac{\varepsilon}{2^{k+4}} - \frac{\varepsilon}{2^{k+5}} \geq \frac{\varepsilon}{2}.$$

It follows that $z_k \neq 0$ for every $k$, therefore $(z_k)_{k=1}^{\infty}$ is a block basic sequence of $(x_n)_{n=1}^{\infty}$. From Proposition 10.4.5 we know that the sequence $(z_k)_{k=1}^{\infty}$ is basic. Calling $(z_k^*)_{k=1}^{\infty}$ the coefficient functionals and $M_1$ the constant of the basic sequence $(z_k)_{k=1}^{\infty}$, from Corollary 10.3.9 and from Proposition 10.4.5 it follows that

$$1 \leq \|z_k^*\| \cdot \|z_k\| \leq 2M_1 \leq 2M,$$

and from (10.18) we have

$$\|z_k^*\| \leq \frac{2M}{\|z_k\|} \leq \frac{4M}{\varepsilon}$$

for every $k \in \mathbb{N}$. Finally, from the inequality above, (10.16) and (10.17), we have

$$\sum_{k=1}^{\infty} \|z_k^*\| \cdot \|z_k - y_{n_{k+1}}\| \leq \sum_{k=1}^{\infty} \left( \frac{4M}{\varepsilon} \right) \left\| \sum_{i=1}^{m_k} x_i^* \left( y_{n_{k+1}} \right) x_i + \sum_{i=m_{k+1}+1}^{\infty} x_i^* \left( y_{n_{k+1}} \right) x_i \right\|$$

$$\leq \sum_{k=1}^{\infty} \left( \frac{4M}{\varepsilon} \right) \left( \frac{\varepsilon}{4M} \left( \frac{1}{2} \right)^{k+2} + \frac{\varepsilon}{4M} \left( \frac{1}{2} \right)^{k+3} \right)$$

$$\leq \sum_{k=1}^{\infty} \left( \left( \frac{1}{2} \right)^{k+2} + \left( \frac{1}{2} \right)^{k+3} \right) < 1.$$

By Theorem 10.4.7 we conclude that $\left( y_{n_{k+1}} \right)_{k=1}^{\infty} \approx (z_k)_{k=1}^{\infty}$. ∎

As a consequence, we have a selection principle with a more concise statement:

## 10.4 The Bessaga–Pełczyński Selection Principle

**Corollary 10.4.9** *Let $E$ be a Banach space and let $(y_n)_{n=1}^\infty$ be a sequence in $E$ with $\inf_n \|y_n\| > 0$ and $y_n \xrightarrow{w} 0$. Then $(y_n)_{n=1}^\infty$ admits a basic subsequence.*

*Proof* Consider the subspace $\overline{[y_n : n \in \mathbb{N}]}$ of $E$, which is separable by Lemma 1.6.3. By the Banach–Mazur Theorem (Theorem 6.5.5) there exists a linear isometry $T \colon \overline{[y_n : n \in \mathbb{N}]} \longrightarrow C[0, 1]$. Note that $\inf_n \|T(y_n)\| = \inf_n \|y_n\| > 0$. Let $(x_n)_{n=1}^\infty$ be a Schauder basis of $C[0, 1]$ — Example 10.3.3(d) — and let $(x_n^*)_{n=1}^\infty$ be the associated coefficient functionals. From $y_n \xrightarrow{w} 0$ and Proposition 6.2.9 we conclude that $T(y_n) \xrightarrow{w} 0$. Hence

$$\lim_{n \to \infty} x_k^*(T y_n) = 0 \text{ for every } k \in \mathbb{N}.$$

By the Bessaga–Pełczyński selection principle there exists a basic subsequence $(T(y_{n_k}))_{k=1}^\infty$ of $(T(y_n))_{n=1}^\infty$. As $T^{-1} \colon T\left(\overline{[y_n : n \in \mathbb{N}]}\right) \longrightarrow \overline{[y_n : n \in \mathbb{N}]}$ is an isometric isomorphism, it follows that $(y_{n_k})_{k=1}^\infty = (T^{-1}(T(y_{n_k})))_{k=1}^\infty$ is a basic sequence. ∎

We finish by applying the selection principle to obtain basic sequences in arbitrary infinite-dimensional Banach spaces:

**Theorem 10.4.10 (Banach–Mazur, Bessaga–Pełczyński, Gelbaum)** *Every infinite-dimensional Banach space contains an infinite-dimensional subspace with a Schauder basis.*

*Proof* As every infinite-dimensional Banach space contains an infinite-dimensional separable subspace (check it!), it is enough to prove the result for separable spaces. As we did in the previous proof, with the help of Theorem 6.5.5 we can assume that $E$ is an infinite-dimensional subspace of $C[0, 1]$. Also as in the previous proof, let $(x_n)_{n=1}^\infty$ be a Schauder basis of $C[0, 1]$ and let $(x_n^*)_{n=1}^\infty$ be their coefficient functionals. For each positive integer $k$, define

$$N_k = \{x \in E : x_1^*(x) = \cdots = x_k^*(x) = 0\} = \bigcap_{j=1}^k \left(\ker(x_j^*) \cap E\right).$$

It is plain that each $N_k$ is a closed subspace of $E$. Moreover, as $N_k$ is the kernel of the linear operator

$$x \in E \mapsto (x_1^*(x), \ldots, x_k^*(x)) \in \mathbb{K}^k,$$

and $\mathbb{K}^k$ has dimension $k$, it follows that $N_k$ is infinite-dimensional for every $k \geq 1$. It is clear that

$$\cdots \subseteq N_3 \subseteq N_2 \subseteq N_1.$$

Suppose that the sequence $(N_k)_{k=1}^\infty$ is eventually stationary, that is, suppose there exists $n_0 \geq 1$ such that $N_j = N_{n_0}$ for each $j > n_0$. Note that, in this case, $N_{n_0} = \{0\}$. Indeed, given $x \in N_{n_0}$, by the definition of $N_{n_0}$ we have

$$x_1^*(x) = \cdots = x_{n_0}^*(x) = 0.$$

Let $j > n_0$. Since $N_j = N_{n_0}$, $x \in N_j$, therefore $x_j^*(x) = 0$. Thus $x_j^*(x) = 0$ for every $j$. From the expression

$$x = \sum_{j=1}^\infty x_j^*(x) x_j,$$

we conclude that $x = 0$. Since $N_{n_0}$ is infinite-dimensional, $N_{n_0} = \{0\}$ does not occur, therefore the sequence $(N_k)_{k=1}^\infty$ is not eventually stationary. This means that there are infinitely many indices $j_1 < j_2 < j_3 < \cdots$ such that the inclusions

$$\cdots \subseteq N_{j_3} \subseteq N_{j_2} \subseteq N_{j_1}$$

are all strict. For each $k \in \mathbb{N}$, considering $y_k \in N_{j_k} - N_{j_{k+1}}$, the sequence $(y_k)_{k=1}^\infty$ is formed by distinct vectors of $E$. It is clear that we can assume $\|y_j\| = 1$ for every $j$. Let $n, k \in \mathbb{N}$ be with $k \geq n$. Then $j_k \geq j_n \geq n$, therefore

$$y_k \in N_{j_k} \subseteq N_{j_n} \subseteq N_n.$$

It follows that $x_n^*(y_k) = 0$ for every $k \geq n$, which reveals that

$$\lim_{k \to \infty} x_n^*(y_k) = 0 \text{ for every } n \in \mathbb{N}.$$

By the Bessaga–Pełczyński selection principle, there exists a basic subsequence $(y_{k_i})_{i=1}^\infty$ of $(y_k)_{k=1}^\infty$. Therefore $\overline{[(y_{k_i})_{i=1}^\infty]}$ is an infinite-dimensional subspace of $E$ with a Schauder basis. ∎

## 10.5 Comments and Historical Notes

It was in 1837 that J. P. G. L. Dirichlet proved that absolute and unconditional convergence coincide for series of real numbers.

Proposition 10.1.4 was proven by Banach in his 1922 thesis and was one of the first indications that completeness is indeed an important property for the study of series convergence.

Banach and his colleagues from Lwów (then in Poland and now in Ukraine) used to discuss mathematics in a place called the Scottish Café. As they wrote on

## 10.5 Comments and Historical Notes

tables and napkins, much was lost at each meeting. Łucja, Banach's wife, provided a book, the so-called *Scottish Book* [43], which was kept in the café, in which participants took notes, mainly of proposed problems and their solutions, as well as prizes offered for their solutions. The prizes were modest, usually a bottle of wine, but one case became particularly well known: in 1936, S. Mazur offered a live goose to whoever solved a problem—Problem 153 of the Scottish Book—which A. Grothendieck showed in 1955 to be closely related to the approximation problem mentioned in Sect. 7.6. After solving the problem in 1972, P. Enflo traveled to Warsaw to receive the goose from Mazur's hands, which they killed and ate the same night in a beautiful feast at W. Zelazko's house (the photo of Enflo receiving the goose from Mazur is famous, see, for example, [58, p. 122]).

The problem of characterizing infinite dimension through the existence of unconditionally convergent non-absolutely convergent series was one of the problems of the Scottish Book and was also mentioned by Banach in [6, p. 240]. The problem proved to be particularly difficult, and even the case of $\ell_1$ had to wait until 1947 to be settled by M. S. Macphail. The general solution by A. Dvoretzky and C. A. Rogers was published in 1950.

The Résumé [31], which contains Grothendieck's inequality, was conceived, written, and published during the period when Grothendieck worked in Brazil, more specifically at the University of São Paulo, from mid-1952 to the end of 1954. It was published in issue 8, year 1953, of the Bulletin of the Mathematical Society of São Paulo. Both the society and the journal were discontinued to give way to the current Brazilian Mathematical Society and its current Bulletin.

The exact value of the Grothendieck constant is still unknown. More precisely, *the values* are not known, as it is known that the constants in the real case and the complex case, denoted by $K_G^{\mathbb{R}}$ and $K_G^{\mathbb{C}}$, respectively, are different. The proof we gave may give the impression that $K_G^{\mathbb{R}} \leq K_G^{\mathbb{C}}$, but, according to the known estimates described below, what happens is the following:

$$1.338 \leq K_G^{\mathbb{C}} \leq 1.4049\ldots < 1.66 \leq K_G^{\mathbb{R}} \leq 1.782\ldots.$$

This does not conflict with what we did, it only shows that the argument we used in the proof is far from obtaining the best constants.

As we have already mentioned, Grothendieck's inequality has many applications. Some of its nice consequences can be found in Banach Space Theory. For example, if $X$ is a complemented subspace of $\ell_1$, then any normalized unconditional Schauder basis of $X$ is equivalent to the canonical basis of $\ell_1$, see [41]. More involved and surprising applications are seen in quantum mechanics, related to Bell's inequality, in graph theory and in computer science, see [52].

The class of absolutely summing operators, which, as it could not be otherwise, has its origins in Grothendieck's Résumé, was formally isolated by A. Pietsch in 1967, and gained relevance already in 1968 with the aforementioned work [41] by Lindenstrauss and Pełczyński. The book [17] is entirely dedicated to the theory of absolutely summing operators.

Schauder bases were introduced by J. Schauder, a member of the Banach circle in Lwów, in 1927. In this same work, Schauder constructed the basis of $C[0, 1]$ shown in Example 10.3.3(d). The functions of this basis had already been studied by G. Faber in 1909, hence the occasional use of the expression *Faber–Schauder basis*. In 1928, Schauder proved that the Haar system (see, for example, [44, Example 4.1.27]) is a Schauder basis for $L_p[0, 1]$, $1 \leq p < \infty$. In his original definition, Schauder required the continuity of the coefficient functionals; a requirement that, according to Theorem 10.3.6 — due to Banach — proved unnecessary.

Theorem 10.3.6 and one of the implications of Theorem 10.3.13 were proven by Banach in his book [6] from 1932. The other implication of Theorem 10.3.13 was proven by M. Grunblum in 1941. This criterion became a usual tool in the theory of Schauder bases since the mention made of it in the classic article by Bessaga and Pełczyński [7] from 1958, which gave rise to the entire theory of Sect. 10.4.

Theorem 10.4.3 was used by Pełczyński [49] to prove that if every subspace with a Schauder basis of a Banach space $E$ is reflexive, then $E$ is reflexive. In addition to what was seen in the text, the Bessaga–Pełczyński selection principle — originally formulated in [7] in the year 1958 — has several remarkable applications, for example: if $E'$ contains a copy of $c_0$, then $E$ contains a copy of $\ell_1$ [44, Lemma 4.4.17]; if $1 \leq q < p < \infty$, then every continuous linear operator from $\ell_p$ to $\ell_q$ is compact (this is Pitt's Theorem — see, for example, [21, Proposition 6.25]). Other applications of the selection principle can be found in [15].

Now we explain the multiple credit attribution in Theorem 10.4.10. As we have already seen, this result appeared without proof in Banach's book [6] from 1932. The first proofs appeared in two works published in 1958, the aforementioned article by Bessaga and Pełczyński and another one by B. Gelbaum, the latter published in the Annals of the Brazilian Academy of Sciences. In 1962, Pełczyński [49] presented a new proof and attributed the idea to Mazur. It is believed that Banach was aware of Mazur's argument.

## 10.6 Exercises

**Exercise 10.6.1** Prove that absolute and unconditional convergence coincide for series in finite-dimensional normed spaces.

**Exercise 10.6.2** Prove the Cauchy criterion stated in Remark 10.1.2.

**Exercise 10.6.3** Let $1 < p < \infty$ and let $(a_n)_{n=1}^{\infty}$ be a sequence that is in $\ell_p$ but not in $\ell_1$. Prove that the series $\sum_{n=1}^{\infty} a_n e_n$ converges unconditionally but not absolutely in $\ell_p$.

**Exercise 10.6.4** Prove Proposition 5.3.8 using Theorem 10.1.6.

**Exercise 10.6.5** Prove that if $A$ and $B$ are two square matrices of order $n$ with scalar entries, then $\operatorname{tr}(AB) = \operatorname{tr}(BA)$.

## 10.6 Exercises

**Exercise 10.6.6** Let $E$ be a finite-dimensional vector space and let $T: E \longrightarrow E$ be a linear operator. Prove that the trace of $T$, $\operatorname{tr}(T)$, does not depend on the basis chosen for $E$.

**Exercise 10.6.7*** Obtain the complex case of Grothendieck's inequality (Theorem 10.2.2) from the real case.

**Exercise 10.6.8** Prove that the space $\ell_1^w(E)$ of weakly summable sequences in the Banach space $E$ is a Banach space with the norm

$$\|(x_j)_{j=1}^\infty\|_w = \sup\left\{\sum_{j=1}^\infty |\varphi(x_j)| : \varphi \in B_{E'}\right\}.$$

Always consider the norm $\|\cdot\|_w$ in the space $\ell_1^w(E)$.

**Exercise 10.6.9** Let $E$ and $F$ be Banach spaces and let $T \in \mathcal{L}(E, F)$ be given. Prove that:

(a) The operator $(x_j)_{j=1}^\infty \in \ell_1(E) \mapsto (T(x_j))_{j=1}^\infty \in \ell_1(F)$ is well defined, linear, continuous, and has the same norm as $T$.
(b) The operator $(x_j)_{j=1}^\infty \in \ell_1^w(E) \mapsto (T(x_j))_{j=1}^\infty \in \ell_1^w(F)$ is well defined, linear, continuous, and has the same norm as $T$.

**Exercise 10.6.10** Prove that the following statements are equivalent for a continuous linear operator $T: E \longrightarrow F$ between Banach spaces:

(a) $T$ is absolutely summing.
(b) There exists a constant $C > 0$ such that

$$\sum_{j=1}^n \|T(x_j)\| \leq C \sup_{\varphi \in B_{E'}} \sum_{j=1}^n |\varphi(x_j)|,$$

for all $n \in \mathbb{N}$ and $x_1, \ldots, x_n \in E$.
(c) The operator $(x_j)_{j=1}^\infty \in \ell_1^w(E) \mapsto (T(x_j))_{j=1}^\infty \in \ell_1(F)$ is well defined, linear, and continuous.

**Exercise 10.6.11** Prove that every continuous linear operator of finite rank between Banach spaces is absolutely summing.

**Exercise 10.6.12** Given Banach spaces $E$ and $F$, let $\Pi(E, F)$ be the subset of $\mathcal{L}(E, F)$ consisting of all absolutely summing operators. Prove that:

(a) $\Pi(E, F)$ is a vector subspace of $\mathcal{L}(E, F)$.
(b) (Ideal property) If $T \in \mathcal{L}(E_0, E)$, $U \in \Pi(E, F)$ and $V \in \mathcal{L}(F, F_0)$, then $V \circ U \circ T \in \Pi(E_0, F_0)$.

**Exercise 10.6.13*** Let $(a_n)_{n=1}^\infty$ be a sequence of scalars such that $\left|\sum_{n \in M} a_n\right| \leq 1$ for every finite set $M \subseteq \mathbb{N}$. Prove that $\sum_{n=1}^\infty |a_n| \leq 4$.

**Exercise 10.6.14** Given a Banach space $E$, define

$$\ell_1^u(E) = \left\{(x_n)_{n=1}^\infty \in \ell_1^w(E) : \lim_{k \to \infty} \|(x_n)_{n=k}^\infty\|_w = 0\right\}.$$

Prove that:

(a) $\ell_1^u(E)$ is a closed subspace of $\ell_1^w(E)$.

(b) A sequence $(x_n)_{n=1}^\infty$ belongs to $\ell_1^u(E)$ if and only if the series $\sum_{n=1}^\infty x_n$ is unconditionally convergent.

**Exercise 10.6.15** Prove that the vectors of a Schauder basis of a Banach space are linearly independent.

**Exercise 10.6.16** Prove the uniqueness of the representation with respect to the Schauder basis of $C[0, 1]$ shown in Example 10.3.3(d).

**Exercise 10.6.17** Prove that, for every Banach space $E$, $\eta$ is a norm on $\mathcal{L}_E$.

**Exercise 10.6.18** Show that every space with a Schauder basis is separable.

**Exercise 10.6.19** Prove that every Banach space with a Schauder basis has the approximation property (see definition in Sect. 7.6).

**Exercise 10.6.20** Show that $(e_1, e_2 - e_1, e_3 - e_2, e_4 - e_3, \ldots)$ is a Schauder basis of $\ell_1$. Also show that the convergence of the series representing a vector of $\ell_1$ with respect to this basis is not always unconditional.

**Exercise 10.6.21**

(a) Let $(x_n)_{n=1}^\infty$ be a Schauder basis of a Banach space $E$ and let $(a_n)_{n=1}^\infty$ be a sequence of non-zero scalars. Prove that $(a_n x_n)_{n=1}^\infty$ is also a Schauder basis of $E$.

(b) Prove that if a Banach space $E$ has a Schauder basis, then $E$ admits a normalized Schauder basis $(x_n)_{n=1}^\infty$, that is, $\|x_n\| = 1$ for every $n$.

**Exercise 10.6.22** Prove that the operator id: $\mathcal{L}_E \longrightarrow \mathcal{L}_F$ from the proof of Theorem 10.3.11 has closed graph.

**Exercise 10.6.23** Let $(x_n)_{n=1}^\infty$ be a Schauder basis of a Banach space $E$ and let $(x_n^*)_{n=1}^\infty$ be its coefficient functionals. Prove that

$$\sup_n \|x_n\| < \infty \iff \inf_n \|x_n^*\| > 0 \text{ and } \inf_n \|x_n\| > 0 \iff \sup_n \|x_n^*\| < \infty.$$

## 10.6 Exercises

**Exercise 10.6.24** Prove that every separable Banach space admits a countable norming set.

**Exercise 10.6.25** Prove Lemma 10.4.6.

**Exercise 10.6.26\*** Let $(x_n)_{n=1}^\infty$ be a Schauder basis of a Hilbert space such that $\|x_n\| = \|x_n^*\| = 1$ for every $n$. Prove that $(x_n)_{n=1}^\infty$ is a complete orthonormal system.

**Exercise 10.6.27\*** Let $1 \leq p < \infty$ and let $(y_n)_{n=1}^\infty$ be a basic sequence with respect to the basis $(e_n)_{n=1}^\infty$ of $\ell_p$. Prove that:

(a) $\overline{[y_n : n \in \mathbb{N}]}$ is isometrically isomorphic to $\ell_p$.
(b) If $0 < \inf_n \|y_n\| \leq \sup_n \|y_n\| < \infty$, then $(y_n)_{n=1}^\infty \approx (e_n)_{n=1}^\infty$.

**Exercise 10.6.28 (Krein–Milman–Rutman Theorem)** Let $E$ be a Banach space with a Schauder basis and let $\mathcal{D}$ be a dense subset of $E$. Prove that there exists a Schauder basis of $E$ consisting solely of elements of $\mathcal{D}$.

**Exercise 10.6.29** Let $1 \leq p \leq \infty$.

(a) For each $t \in (0, 1)$, let $x_t = (t, t^2, t^3, \ldots)$. Show that the set $\{x_t : 0 < t < 1\}$ is linearly independent in $\ell_p$.
(b) Show that $\dim(\ell_p) = c$, where $c$ is the cardinality of the continuum, that is, the cardinality of $\mathbb{R}$.

**Exercise 10.6.30\*** Given a Banach space $E \neq \{0\}$, consider the subspace $c_{00}(E)$ of the Banach space $(\ell_\infty(E), \|\cdot\|_\infty)$ (see Exercise 4.5.8) consisting of all eventually null sequences; that is, a bounded sequence $(x_n)_{n=1}^\infty$ belongs to $c_{00}(E)$ if and only if there is $n_0 \in \mathbb{N}$ such that $x_n = 0$ for every $n \geq n_0$.

(a) Is $c_{00}(E)$ dense in $\ell_\infty(E)$?
(b) Is $c_{00}(E)$ complete with the norm $\|\cdot\|_\infty$?
(c) What is the dimension of $c_{00} = c_{00}(\mathbb{K})$?
(d) Display a Hamel basis of $c_{00}$.
(e) Extending the definition of Schauder basis to normed spaces, is there a Schauder basis for $c_{00}$? If so, display such a basis.
(f) What is the dimension of $c_{00}(E)$?

# Appendix A
# Zorn's Lemma

**Definition A.1**

(a) A *partial order* on the set $P$ is a relation $\leq$ on $P$ that satisfies the following properties:
   - $x \leq x$ for every $x \in P$ (reflexive).
   - If $x, y \in P$, $x \leq y$ and $y \leq x$, then $x = y$ (anti-symmetric).
   - If $x, y, z \in P$, $x \leq y$ and $y \leq z$, then $x \leq z$ (transitive).

   In this case, it is said that $(P, \leq)$ is a *partially ordered set*.
   In what follows, $(P, \leq)$ is a fixed partially ordered set.
(b) An *upper bound* of a subset $Q$ of $P$, if it exists, is an element $p \in P$ such that $q \leq p$ for every $q \in Q$.
(c) A *maximal element of* $P$, if it exists, is an element $m \in P$ such that if $p \in P$ and $m \leq p$, then $m = p$.
(d) A subset $Q$ of $P$ is said to be *totally ordered* if for all $p, q \in Q$, it is true that $p \leq q$ or $q \leq p$.

**Theorem A.2 (Zorn's Lemma)** *Every non-empty partially ordered set $P$, in which every totally ordered subset has an upper bound in $P$, has a maximal element.*

Zorn's Lemma is equivalent to the Axiom of Choice. The proof of this equivalence can be found in any good Set Theory book, for example in [14, Chapter 7]. As we have already said in Sect. 3.5, M. Zorn's proof appeared in 1935. However, the same result had already been proven by K. Kuratowski in 1922. Despite the precedence, credit to Kuratowski is usually omitted.

A simple and well-known application of Zorn's Lemma guarantees that every vector space has a basis. In this book we use a slightly stronger property:

**Definition A.3** Let $V$ be a vector space.

(a) A *generator* of $V$ is a subset $A$ of $V$ such that every element of $V$ can be written as a (finite) linear combination of elements of $A$.

(b) A subset $B$ of $V$ is *linearly independent* if every finite subset of $B$ consists of linearly independent vectors.
(c) A *basis* (or *algebraic basis*, or *Hamel basis*) of $V$ is a linearly independent generator of $V$.

**Proposition A.4** *If $A$ is a generator of a vector space $V \neq \{0\}$, then there is a basis of $V$ contained in $A$.*

**Proof** Consider the set

$$\mathcal{F} = \{(W, B) : W \text{ is a subspace of } V \text{ and } B \subseteq A \text{ is linearly independent} \text{ and generates } W\},$$

equipped with the partial order

$$(W, B), (W_1, B_1) \in \mathcal{F}, \ (W, B) \leq (W_1, B_1) \iff W \subseteq W_1 \text{ and } B \subseteq B_1.$$

Taking $x \in A$ we have $([x], \{x\}) \in \mathcal{F}$, therefore $\mathcal{F}$ is non-empty. Let $\mathcal{G}$ be a totally ordered subset of $\mathcal{F}$, say $\mathcal{G} = (W_i, B_i)_{i \in I}$. Routine arguments show that $\left(\bigcup_{i \in I} W_i, \bigcup_{i \in I} B_i\right)$ belongs to $\mathcal{F}$ and is an upper bound of $\mathcal{G}$. By Zorn's Lemma, $\mathcal{F}$ has a maximal element $(\widetilde{W}, \widetilde{B})$. It is enough to show that $\widetilde{W} = V$. Suppose, by contradiction, that $\widetilde{W} \neq V$. Note that, in this case, $A$ is not contained in $\widetilde{W}$, because if $A \subseteq \widetilde{W}$ we would have

$$V = [A] \subseteq [\widetilde{W}] = \widetilde{W} \subseteq V,$$

which contradicts the assumption that $\widetilde{W} \neq V$. Then we can choose $x_0 \in A - \widetilde{W}$ and define

$$W_N = \widetilde{W} \oplus [x_0], \ B_N = \widetilde{B} \cup \{x_0\}.$$

Since $\widetilde{B}$ generates $\widetilde{W}$, both the elements of $[B_N]$ and the elements of $W_N$ have the form $a_1 x_1 + \cdots + a_n x_n + a_0 x_0$, where $n \in \mathbb{N}$, $a_1, \ldots, a_n, a_0 \in \mathbb{K}$ and $x_1, \ldots, x_n \in \widetilde{W}$. Therefore $[B_N] = W_N$. The linear independence of $B_N$ is clear because $\widetilde{B}$ is linearly independent and $x_0 \notin \widetilde{W}$. Finally, from $\widetilde{B} \subseteq A$ and $x_0 \in A$ we conclude that $B_N \subseteq A$, therefore $(W_N, B_N) \in \mathcal{F}$. From the definitions of $W_N$ and $B_N$ it is clear that $(\widetilde{W}, \widetilde{B}) \leq (W_N, B_N)$. The maximality of $(\widetilde{W}, \widetilde{B})$ implies that $(\widetilde{W}, \widetilde{B}) = (W_N, B_N)$. In particular, $\widetilde{W} = W_N$, hence $x_0 \in \widetilde{W}$. This contradiction ensures that $\widetilde{W} = V$ and completes the proof. ∎

We conclude this appendix with an interesting application to Banach spaces:

**Proposition A.5** *Let $E$ be a separable infinite-dimensional Banach space. Then every dense countable subset of $E$ admits a countable linearly independent subset.*

***Proof*** Let $D$ be a dense countable subset of $E$. It is clear that $D$ is a generator of $[D]$. By Proposition A.4 there exists a basis $\mathcal{B}$ of $[D]$ contained in $D$. Suppose, for a moment, that $\mathcal{B}$ is finite. In this case $[D]$ is finite dimensional, hence $[D]$ is closed in $E$. As $D$ is dense in $E$, we have

$$E = \overline{D} \subseteq \overline{[D]} = [D] \subseteq E,$$

which proves that $E = [D]$, therefore $E$ is finite dimensional. As this contradicts the hypothesis, it follows that the basis $\mathcal{B}$ of $[D]$ is infinite. Being a basis of $[D]$, $\mathcal{B}$ is linearly independent, and being infinite and contained in the countable set $D$, $\mathcal{B}$ is countable. ∎

# Appendix B
# Concepts of General Topology

## Metric Spaces

**Definition B.1** A *metric space* is an ordered pair $(M, d)$ formed by a set $M$ and a function $d \colon M \times M \longrightarrow \mathbb{R}$, called *metric*, satisfying the following conditions for any $x, y, z$ in $M$:

(a) $d(x, y) \geq 0$,
(b) $d(x, x) = 0$,
(c) $d(x, y) = 0$ implies $x = y$,
(d) $d(x, y) = d(y, x)$,
(e) $d(x, z) \leq d(x, y) + d(y, z)$.

If $E$ and $F$ are subsets of $M$, the *distance between $E$ and $F$* is defined by

$$\mathrm{dist}(E, F) = \inf\{d(x, y) : x \in E \text{ and } y \in F\}.$$

**Definition B.2** Let $(x_n)_{n=1}^{\infty}$ be a sequence in the metric space $(M, d)$.

(a) The sequence $(x_n)_{n=1}^{\infty}$ *converges* to $x \in M$ if

$$\lim_{n \to \infty} d(x_n, x) = 0.$$

In this case we write $x = \lim_n x_n = \lim_{n \to \infty} x_n$ or $x_n \longrightarrow x$.

(b) The sequence $(x_n)_{n=1}^{\infty}$ is said to be *convergent* if there exists $x \in M$ such that $x_n \longrightarrow x$. Otherwise, it is said to be *divergent*.

(c) The sequence $(x_n)_{n=1}^{\infty}$ is a *Cauchy sequence* if

$$\lim_{m, n \to \infty} d(x_m, x_n) = 0.$$

It is immediate that every convergent sequence is a Cauchy sequence.

© Sociedade Brasileira de Matemática 2025
G. Botelho et al., *Introduction to Functional Analysis*, Universitext,
https://doi.org/10.1007/978-3-031-81791-5

(d) The metric space $(M, d)$ is a *complete metric space* if every Cauchy sequence in $M$ converges to an element of $M$.

**Definition B.3** Let $(M, d)$ be a metric space.

(a) Given $a \in M$ and $\varepsilon > 0$, the set $B(a; \varepsilon) = \{x \in M : d(x, a) < \varepsilon\}$ is called the *open ball with center a and radius r*.
(b) A subset $A \subseteq M$ is *open* if for each $x \in A$ there exists $\varepsilon > 0$ such that $B(x; \varepsilon) \subseteq A$.
(c) A subset $F \subseteq M$ is *closed* if its complement $F^c := M - F$ is open.

**Definition B.4** Let $(M, d)$ be a metric space and $A \subseteq M$.

(a) The *interior of* $A$ is the set $\text{int}(A) = \bigcup\{B \subseteq M : B \text{ is open and } B \subseteq A\}$.
(b) The *closure of* $A$ is the set $\overline{A} = \bigcap\{F \subseteq M : F \text{ is closed and } A \subseteq F\}$.
(c) It is said that $A$ is *dense* in $M$ if $\overline{A} = M$.

**Proposition B.5** *Let $(M, d)$ be a metric space, $x \in M$ and $A, B \subseteq M$. Then*

(a) *$\text{int}(A)$ is an open set and $\overline{A}$ is a closed set.*
(b) *$A$ is open if and only if $A = \text{int}(A)$.*
(c) *$A$ is closed if and only if $A = \overline{A}$.*
(d) *If $A \subseteq B$, then $\overline{A} \subseteq \overline{B}$.*
(e) *$\overline{A \cup B} = \overline{A} \cup \overline{B}$*
(f) *$x \in \overline{A}$ if and only if there exists a sequence $(x_n)_{n=1}^{\infty}$ in $A$ such that $x_n \longrightarrow x$.*
(g) *$A$ is dense in $M$ if and only if, for all $x \in M$ and $\varepsilon > 0$, we have $A \cap B(x; \varepsilon) \neq \emptyset$.*

When there is no need to specify the metric, we simply write $M$ to denote the metric space $(M, d)$.

**Definition B.6** A metric space $M$ is *compact* if for every collection of open sets $(A_i)_{i \in I}$ such that $M = \bigcup_{i \in I} A_i$, there exist $n \in \mathbb{N}$ and $i_1, \ldots, i_n \in I$ such that

$$M = \bigcup_{k=1}^{n} A_{i_k}.$$

In this book we use two fundamental theorems of metric space theory. The first is Baire's Theorem, stated and proven in Theorem 2.3.1. The second is Ascoli's Theorem, stated below, which characterizes the relatively compact subsets of the complete metric space $C(K)$ of continuous functions $f : K \longrightarrow \mathbb{K}$, where $K$ is a compact metric space, with the metric

$$d(f, g) = \sup\{|f(t) - g(t)| : t \in K\}.$$

**Theorem B.7 (Ascoli's Theorem)** *Let $K$ be a compact metric space and let $A$ be a subset of $C(K)$. Then $\overline{A}$ is compact in $C(K)$ if and only if the following conditions are satisfied:*

(a) *A is equicontinuous, that is, for all $t_0 \in K$ and $\varepsilon > 0$, there exists $\delta > 0$ such that*

$$|f(t) - f(t_0)| < \varepsilon \text{ for all } t \in K \text{ with } d(t, t_0) < \delta \text{ and } f \in A.$$

(b) *The set $\{f(t) : f \in A\}$ is bounded in $\mathbb{K}$ for every $t \in K$.*

## Topological Spaces

**Definition B.8** A *topology* on a set $X$ is a collection $\tau$ of subsets of $X$, called *open sets*, satisfying the following properties:

(a) Any union of elements of $\tau$ is an element of $\tau$.
(b) Any finite intersection of elements of $\tau$ belongs to $\tau$.
(c) $X$ and $\emptyset$ belong to $\tau$.

In this case, we say that $(X, \tau)$ is a *topological space*, which we naturally abbreviate to $X$ when there is no danger of ambiguity or imprecision. A subset $F$ of $X$ is called a *closed set* if its complement is open, that is, if $F^c = X - F \in \tau$.

**Proposition B.9** *In a topological space the following properties hold:*

(a) *Any intersection of closed sets is a closed set.*
(b) *Any finite union of closed sets is a closed set.*
(c) *$X$ and $\emptyset$ are closed sets.*

The open sets — according to Definition B.3(b) — of a metric space $M$ form a topology on $M$, called the *topology on $M$ induced by the metric*. It is in this sense that a metric space will be understood as a topological space. A topological space $X$ is said to be *metrizable* if there exists a metric on $X$ that induces its topology.

The interior $int(A)$, the closure $\overline{A}$, and the denseness of a subset $A$ of a topological space $X$ are defined exactly as in Definition B.4. Items (a)–(e) of Proposition B.5 remain valid exactly as stated there. In addition:

**Proposition B.10** *Let $A$ and $B$ be subsets of a topological space $X$. Then:*

(a) $int(A) = X - \overline{X - A}$ *and* $X - \overline{A} = int(X - A)$.
(b) *If $A \subseteq B$, then $int(A) \subseteq int(B)$.*
(c) $int(A \cap B) = int(A) \cap int(B)$.

## Neighborhoods

**Definition B.11** A *neighborhood* of an element $x$ of the topological space $X$ is a subset $U$ of $X$ that contains an open set $V$ containing $x$, that is, $x \in V \subseteq U$. The collection $\mathcal{U}_x$ of all neighborhoods of $x$ is called the *neighborhood system* of $x$.

**Proposition B.12** *The neighborhood system $\mathcal{U}_x$ of $x$ in a topological space $X$ has the following properties:*

(a) *If $U \in \mathcal{U}_x$, then $x \in U$.*
(b) *If $U, V \in \mathcal{U}_x$, then $U \cap V \in \mathcal{U}_x$.*
(c) *If $U \in \mathcal{U}_x$, then there exists $V \in \mathcal{U}_x$ such that $U \in \mathcal{U}_y$ for each $y \in V$.*
(d) *If $U \in \mathcal{U}_x$ and $U \subseteq V$, then $V \in \mathcal{U}_x$.*
(e) *$A \subseteq X$ is open if and only if $A$ contains a neighborhood of each of its points.*
(f) *Let $A \subseteq X$ and $x \in X$. Then $x \in \overline{A}$ if and only if every neighborhood of $x$ intersects $A$.*

**Proposition B.13** *If to each point $x$ of a set $X$ is associated a non-empty collection $\mathcal{U}_x$ of subsets of $X$ satisfying (a)–(d) of Proposition B.12, then the collection*

$$\tau = \{A \subseteq X : \text{for every } x \in A \text{ there exists } U \in \mathcal{U}_x \text{ such that } x \in U \subseteq A\}$$

*is a topology on $X$ in which $\mathcal{U}_x$ is the neighborhood system of $x$ for every $x \in X$.*

**Definition B.14** Let $X$ be a topological space. A *neighborhood basis* of an element $x$ of $X$ is a subcollection $\mathcal{B}_x$ of $\mathcal{U}_x$ such that each $U \in \mathcal{U}_x$ contains some $V \in \mathcal{B}_x$. In this case $\mathcal{U}_x$ is determined by $\mathcal{B}_x$ in the following way:

$$\mathcal{U}_x = \{U \subseteq X : V \subseteq U \text{ for some } V \in \mathcal{B}_x\}.$$

The elements of $\mathcal{B}_x$ are called *basic neighborhoods* of $x$.

**Theorem B.15** *Let $X$ be a topological space and, for each $x \in X$, let $\mathcal{B}_x$ be a neighborhood basis at $x$. Then:*

(a) *If $V \in \mathcal{B}_x$, then $x \in V$.*
(b) *If $V_1, V_2 \in \mathcal{B}_x$, then there exists $V_3 \in \mathcal{B}_x$ such that $V_3 \subseteq V_1 \cap V_2$.*
(c) *If $V \in \mathcal{B}_x$, then there exists $V_0 \in \mathcal{B}_x$ such that if $y \in V_0$, then there exists $W \in \mathcal{B}_y$ with $W \subseteq V$.*
(d) *$A \subseteq X$ is open if and only if $A$ contains a basic neighborhood of each of its points.*

*Conversely, if to each element $x$ of a set $X$ is associated a collection $\mathcal{B}_x$ of subsets of $X$ satisfying conditions (a)–(c) above, by using (d) to define open sets we obtain a topology on $X$ in which $\mathcal{B}_x$ is a neighborhood basis for each $x \in X$.*

## Topological Basis

**Definition B.16** A *basis* of the topological space $(X, \tau)$ is a subcollection $\mathcal{B}$ of $\tau$ such that every open set can be written as a union of elements of $\mathcal{B}$.

**Proposition B.17** *A collection $\mathcal{B}$ of subsets of the topological space $X$ is a basis for $X$ if and only if for every open set $A \subseteq X$ and every $x \in A$ there exists $B \in \mathcal{B}$ such that $x \in B \subseteq A$.*

**Theorem B.18** *A collection $\mathcal{B}$ of subsets of $X$ is a basis for a topology on $X$ if and only if $X = \bigcup_{B \in \mathcal{B}} B$ and it is true that if $B_1, B_2 \in \mathcal{B}$ and $x \in B_1 \cap B_2$, then there exists $B_3 \in \mathcal{B}$ such that $x \in B_3 \subseteq B_1 \cap B_2$.*

**Theorem B.19** *A collection $\mathcal{B}$ of open sets of the topological space $X$ is a basis for $X$ if and only if $\{B \in \mathcal{B} : x \in B\}$ is a neighborhood basis for each $x \in X$.*

## Subspaces

**Definition B.20** Let $(X, \tau)$ be a topological space and let $A \subseteq X$. The collection $\tau_A := \{B \cap A : B \in \tau\}$ is a topology on $A$, called *relative topology* or *topology on $A$ induced by $\tau$*. With this topology, we say that $A$ is a *subspace of $X$*.

**Theorem B.21** *Let $A$ be a subspace of a topological space $X$. Then:*

(a) *$C \subseteq A$ is open in $A$ if and only if $C = B \cap A$ for some $B$ open in $X$.*
(b) *$F \subseteq A$ is closed in $A$ if and only if $F = K \cap A$ for some $K$ closed in $X$.*
(c) *If $C \subseteq A$, then the closure of $C$ in $A$ coincides with $A \cap \overline{C}$.*
(d) *If $x \in A$, then $V \subseteq X$ is a neighborhood of $x$ in $A$ if and only if $V = U \cap A$ for some neighborhood $U$ of $x$ in $X$.*
(e) *If $x \in A$ and $\mathcal{B}_x$ is a neighborhood basis for $x$ in $X$, then $\{B \cap A : B \in \mathcal{B}_x\}$ is a neighborhood basis for $x$ in $A$.*
(f) *If $\mathcal{B}$ is a basis of $X$, then $\{B \cap A : B \in \mathcal{B}\}$ is a basis of $A$.*

## Continuous Functions

**Definition B.22** A function $f : X \longrightarrow Y$ between topological spaces is *continuous* if $f^{-1}(A) := \{x \in X : f(x) \in A\}$ is open in $X$ for every open $A$ in $Y$.

**Proposition B.23** *The following statements are equivalent for a function $f : X \longrightarrow Y$ between topological spaces:*

(a) *$f$ is continuous.*
(b) *$f^{-1}(F)$ is closed in $X$ for every closed $F$ in $Y$.*
(c) *For every $x \in X$ and every neighborhood $U$ of $f(x)$ in $Y$ there exists a neighborhood $V$ of $x$ in $X$ such that $f(V) \subseteq U$.*
(d) *$f(\overline{A}) \subseteq \overline{f(A)}$ for every $A \subseteq X$.*

*If X and Y are metric spaces, then these statements are also equivalent to:*

(e) $f(x_n) \longrightarrow f(x)$ *for every sequence* $(x_n)_{n=1}^{\infty}$ *in X such that* $x_n \longrightarrow x \in X$.

**Proposition B.24**

(a) *If* $f: X \longrightarrow Y$ *and* $g: Y \longrightarrow Z$ *are continuous functions between topological spaces, then the composite function* $g \circ f: X \longrightarrow Z$ *is continuous.*
(b) *If A is a subspace of the topological space X and the function* $f: X \longrightarrow Y$ *is continuous, then the restriction of f to A,* $f|_A: A \longrightarrow Y$, *is continuous.*

**Definition B.25** A *homeomorphism* between the topological spaces $X$ and $Y$ is a function $f: X \longrightarrow Y$ that is continuous, bijective, and has a continuous inverse.

## Nets

Unlike what happens in metric spaces, convergent sequences do not describe topologies in general. For example, Proposition B.5(f) and item B.23(e) are not true in general topological spaces. In this book, we work with the concept of *nets*, which is a generalization of the concept of sequence very useful in describing topologies in general. Nets were introduced by E. Moore and H. Smith in 1922.

**Definition B.26** A *directed set* is a pair $(\Lambda, \leq)$ where $\leq$ is a *direction* on the set $\Lambda$, that is, it is a relation on $\Lambda$ such that:

(a) $\lambda \leq \lambda$ for every $\lambda \in \Lambda$.
(b) If $\lambda_1, \lambda_2, \lambda_3 \in \Lambda$, $\lambda_1 \leq \lambda_2$ and $\lambda_2 \leq \lambda_3$, then $\lambda_1 \leq \lambda_3$.
(c) For all $\lambda_1, \lambda_2 \in \Lambda$ there exists $\lambda_3 \in \Lambda$ such that $\lambda_1 \leq \lambda_3$ and $\lambda_2 \leq \lambda_3$.

**Definition B.27** A *net* in a set $X$ is a function $P: \Lambda \longrightarrow X$, where $\Lambda$ is a directed set. Usually, $P(\lambda)$ is denoted by $x_\lambda$, and in this case we refer to the net $(x_\lambda)_{\lambda \in \Lambda}$.

**Definition B.28** We say that the net $(x_\lambda)_{\lambda \in \Lambda}$ in the topological space $X$ *converges* to $x \in X$, and in this case we write $x_\lambda \longrightarrow x$, if for each neighborhood $U$ of $x$ there exists $\lambda_0 \in \Lambda$ such that $x_\lambda \in U$ for every $\lambda \geq \lambda_0$.

The following example is fundamental for proving the subsequent propositions and for showing that the convergence of nets characterizes the topology of the space.

**Example B.29** Let $X$ be a topological space, let $x$ be an element of $X$ and let $\mathcal{B}_x$ be a neighborhood basis of $x$. The reverse inclusion relation $U_1 \leq U_2 \iff U_2 \subseteq U_1$ makes $\mathcal{B}_x$ a directed set. In this case, choosing $x_U \in U$ for each $U \in \mathcal{B}_x$, we have a net $(x_U)_{U \in \mathcal{B}_x}$ in $X$ that converges to $x$.

**Definition B.30** A topological space $X$ is a *Hausdorff space* if for all $x, y \in X$, $x \neq y$, there exist neighborhoods $U$ of $x$ and $V$ of $y$ such that $U \cap V = \emptyset$.

The following result contains the results about nets that we use in this book. For the reader's convenience, we include the proofs.

B   Concepts of General Topology                                                299

**Theorem B.31** *Let $X, Y$ be topological spaces, $A \subseteq X$ and $x \in X$.*

(a) *$x \in \overline{A}$ if and only if there exists a net $(x_\lambda)_{\lambda \in \Lambda}$ in $A$ such that $x_\lambda \longrightarrow x$.*
(b) *$A$ is closed if and only if, for every net $(x_\lambda)_{\lambda \in \Lambda}$ in $A$ with $x_\lambda \longrightarrow x$, it follows that $x \in A$.*
(c) *A function $f: X \longrightarrow Y$ is continuous if and only if $f(x_\lambda) \longrightarrow f(x)$ for every net $(x_\lambda)_{\lambda \in \Lambda}$ in $X$ such that $x_\lambda \longrightarrow x \in X$.*
(d) *(Uniqueness of the limit) $X$ is a Hausdorff space if and only if every net in $X$ converges to at most one element of $X$.*

*Proof*

(a) Suppose $x \in \overline{A}$. Then every neighborhood of $x$ intersects $A$. Let $\mathcal{U}_x$ be the set of all neighborhoods of $x$. For each $U \in \mathcal{U}_x$, let $x_U \in U \cap A$. Using the direction given by the reverse inclusion relation (Example B.29), $(x_U)_{U \in \mathcal{U}_x}$ is a net that converges to $x$. By construction, this net is entirely contained in $A$.

   Conversely, suppose there exists a net $(x_\lambda)_{\lambda \in \Lambda}$ in $A$ such that $x_\lambda \longrightarrow x$. Given a neighborhood $U$ of $x$, there exists $\lambda_0 \in \Lambda$ such that $x_\lambda \in U$ for every $\lambda \geq \lambda_0$. In particular, $x_{\lambda_0} \in U \cap A$, proving that $U \cap A \neq \emptyset$ for every neighborhood $U$ of $X$. It follows that $x \in \overline{A}$.

(b) This equivalence follows immediately from (a) and the version of B.5(c) for topological spaces.

(c) Suppose that $f$ is continuous and that $(x_\lambda)_{\lambda \in \Lambda}$ is a net in $X$ such that $x_\lambda \longrightarrow x$. Given a neighborhood $U$ of $f(x)$, by B.23(c) there exists a neighborhood $V$ of $x$ such that $f(V) \subseteq U$. From the convergence $x_\lambda \longrightarrow x$, for this neighborhood $V$ of $x$ there exists $\lambda_0 \in \Lambda$ such that $x_\lambda \in V$ whenever $\lambda \geq \lambda_0$. Then $f(x_\lambda) \in f(V) \subseteq U$ whenever $\lambda \geq \lambda_0$, proving that $f(x_\lambda) \longrightarrow f(x)$. Conversely, suppose that $f(x_\lambda) \longrightarrow f(x)$ for every net $(x_\lambda)_{\lambda \in \Lambda}$ in $X$ such that $x_\lambda \longrightarrow x \in X$. Let $A \subseteq X$ and $x \in \overline{A}$. By (a), there exists a net $(x_\lambda)_{\lambda \in \Lambda}$ in $A$ such that $x_\lambda \longrightarrow x \in X$. Thus $(f(x_\lambda))_{\lambda \in \Lambda}$ is a net in $f(A)$ and, by assumption, $f(x_\lambda) \longrightarrow f(x)$. Again by (a) we conclude that $f(x) \in \overline{f(A)}$. This proves that $f(\overline{A}) \subseteq \overline{f(A)}$ for every $A \subseteq X$. The continuity of $f$ follows from B.23(d).

(d) Suppose that $X$ is a Hausdorff space and that $(x_\lambda)_{\lambda \in \Lambda}$ is a convergent net in $X$. Also suppose that $x_\lambda \longrightarrow x$ and $x_\lambda \longrightarrow y$. If $x \neq y$, we can find neighborhoods $U$ of $x$ and $V$ of $y$ such that $U \cap V = \emptyset$. In this case there exist $\lambda_0, \lambda_1 \in \Lambda$ such that $x_\lambda \in U$ for every $\lambda \geq \lambda_0$ and $x_\lambda \in V$ for every $\lambda \geq \lambda_1$. From condition B.26(c) there exists $\lambda_2 \in \Lambda$ such that $\lambda_2 \geq \lambda_0$ and $\lambda_2 \geq \lambda_1$. It follows that $x_{\lambda_2} \in U \cap V$, a contradiction that proves that $x = y$.

   Conversely, suppose that every net in $X$ converges to at most one element of $X$. Suppose that the diagonal $\Delta := \{(x, x) : x \in X\}$ is not a closed set in $X \times X$ with the product topology, which is the topology that has as a basis the Cartesian products of open sets of $X$. In this case, by (b) there exist $x, y \in X$, $x \neq y$, and a net $(x_\lambda, x_\lambda)_{\lambda \in \Lambda}$ in $\Delta$ that converges to $(x, y)$. As the projections on the first and second coordinates are continuous, by (c) we have $x_\lambda \longrightarrow x$ and $x_\lambda \longrightarrow y$ in $X$. From the assumption it follows that $x = y$, a contradiction that proves that $\Delta$ is closed in $X \times X$. Given $x, y \in X$, $x \neq y$, we have $(x, y) \in \Delta^c$. As $\Delta^c$ is open, there exists a basic neighborhood of $(x, y)$ in $X \times X$, that is, a set of the form $U \times V$ where $U$ is a neighborhood of $x$ and $V$ is a neighborhood of $y$, entirely contained in $\Delta^c$. This means that $(U \times V) \cap \Delta = \emptyset$, or in other words, $U \cap V = \emptyset$. ∎

## Compactness and Tychonoff's Theorem

**Definition B.32** A subset $K$ of the topological space $X$ is *compact* if for every collection $(A_i)_{i \in I}$ of open sets in $X$ such that $K \subseteq \bigcup_{i \in I} A_i$, there exist $n \in \mathbb{N}$ and $i_1, \ldots, i_n \in I$ such that $K \subseteq (A_{i_1} \cup \cdots \cup A_{i_n})$. That is, if every open cover of $K$ admits a finite subcover.

The properties of compact sets that are important in this book are listed next.

**Proposition B.33** *Let $X, Y$ be topological spaces and $K \subseteq X$.*

(a) *If $X$ is compact and $K$ is closed, then $K$ is compact.*
(b) *If $X$ is Hausdorff and $K$ is compact, then $K$ is closed.*
(c) *If $X$ is a metric space, then $K$ is compact if and only if every sequence in $K$ admits a convergent subsequence in $K$.*
(d) *If $f : X \longrightarrow Y$ is continuous and $K$ is compact in $X$, then $f(K)$ is compact in $Y$.*
(e) *$A \subseteq K$ is compact in $K$ if and only if $A$ is compact in $X$.*
(f) *If $A \subseteq K$, $A$ is closed in $X$ and $K$ is compact, then $A$ is compact in $X$.*

**Definition B.34** Let $(X_\alpha)_{\alpha \in \Gamma}$ be a collection of sets.

(a) The *generalized Cartesian product* of the sets $X_\alpha, \alpha \in \Gamma$, is defined as being the following set of functions:

$$\prod_{\alpha \in \Gamma} X_\alpha = \left\{ f : \Gamma \longrightarrow \bigcup_{\alpha \in \Gamma} X_\alpha : f(\alpha) \in X_\alpha \text{ for each } \alpha \in \Gamma \right\}.$$

Denoting $f(\alpha)$ by $x_\alpha$ for each $\alpha \in \Gamma$, we can refer to the element $f \in \prod_{\alpha \in \Gamma} X_\alpha$ as $(x_\alpha)_{\alpha \in \Gamma}$. For a fixed $\beta \in \Gamma$, the function

$$\pi_\beta : \prod_{\alpha \in \Gamma} X_\alpha \longrightarrow X_\beta, \quad \pi_\beta((x_\alpha)_{\alpha \in \Gamma}) = x_\beta,$$

is called the *projection on the $\beta$-th coordinate*.

(b) If each $X_\alpha$ is a topological space, the topology on $\prod_{\alpha \in \Gamma} X_\alpha$ generated by the functions $(\pi_\alpha)_{\alpha \in \Gamma}$ in the sense of Section 6.1 is called the *product topology*.

**Theorem B.35 (Tychonoff's Theorem)** *Let $(X_\alpha)_{\alpha \in \Gamma}$ be a family of topological spaces. The generalized Cartesian product $\prod_{\alpha \in \Gamma} X_\alpha$ is compact in the product topology if and only if $X_\alpha$ is compact for every $\alpha \in \Gamma$.*

# Appendix C
# Measure and Integration

## Measurable Spaces and the Extended Line

**Definition C.1** A *σ-algebra* on the set $X$ is a family $\Sigma$ of subsets of $X$ that satisfies the following properties:

(a) $\emptyset, X \in \Sigma$.
(b) If $A \in \Sigma$, then $A^C := X - A \in \Sigma$.
(c) If $A_n \in \Sigma$ for every $n \in \mathbb{N}$, then $\bigcup_{n=1}^{\infty} A_n \in \Sigma$.

In this case, the pair $(X, \Sigma)$ is called a *measurable space*. Each element of the $\sigma$-algebra is called a *measurable set*.

Given a collection $\mathcal{F}$ of subsets of $X$, the intersection of all the $\sigma$-algebras that contain $\mathcal{F}$ is still a $\sigma$-algebra, called the *$\sigma$-algebra generated by $\mathcal{F}$* and denoted by $\Sigma(\mathcal{F})$. Note that $\Sigma(\mathcal{F})$ is the smallest $\sigma$-algebra in $X$ that contains $\mathcal{F}$.

When $(X, \tau)$ is a topological space, the $\sigma$-algebra $\Sigma(\tau)$ is called the *Borel $\sigma$-algebra of $X$* and is denoted by $\mathcal{B} = \mathcal{B}(X)$. The elements of $\mathcal{B}$ are called *Borel sets* or *Borelians*.

**Definition C.2** The *extended line* is the set

$$\overline{\mathbb{R}} = \mathbb{R} \cup \{\infty\} \cup \{-\infty\},$$

also denoted by $[-\infty, \infty]$, where $\infty$ and $-\infty$ are symbols that have the properties we intuitively expect from them, that is:

$$-\infty < x < \infty \text{ for every } x \in \mathbb{R}.$$

We operate arithmetically with the symbols $\infty$ and $-\infty$ as follows: for $a \in \mathbb{R}$,

- $a + \infty = \infty + a = \infty$, $a - \infty = -\infty + a = -\infty$, $\infty + \infty = \infty$ and $-\infty + (-\infty) = -\infty$.
- $a \cdot \infty = \infty \cdot a = \begin{cases} \infty, & \text{if } a > 0 \\ -\infty, & \text{if } a < 0 \end{cases}$ and $a \cdot (-\infty) = (-\infty) \cdot a = \begin{cases} -\infty, & \text{if } a > 0 \\ \infty, & \text{if } a < 0 \end{cases}$.
- $\infty \cdot \infty = (-\infty) \cdot (-\infty) = \infty$ and $\infty \cdot (-\infty) = (-\infty) \cdot \infty = -\infty$.
- $0 \cdot \infty = 0 \cdot (-\infty) = \infty \cdot 0 = (-\infty) \cdot 0 = 0$.

Note that, as there is no coherent option, the additions $\infty + (-\infty)$ and $(-\infty) + \infty$ are not defined. The multiplication of 0 by the symbols $-\infty$ and $\infty$ defined above may seem artificial, but it is extremely important in Measure Theory.

The following notations are usual:

$$[-\infty, a) = \{x \in \mathbb{R} : x < a\} \cup \{-\infty\} \quad \text{and} \quad (a, \infty] = \{x \in \mathbb{R} : x > a\} \cup \{\infty\}.$$

**Definition C.3** The usual topology of $\mathbb{R}$ induces a topology on $\overline{\mathbb{R}}$ considering as open the subsets $A \subseteq \overline{\mathbb{R}}$ of the form:

(a) $A \subseteq \mathbb{R}$ is open in $\mathbb{R}$, or
(b) $A = [-\infty, a)$ for some $a \in \mathbb{R}$, or
(c) $A = (a, \infty]$ for some $a \in \mathbb{R}$, or
(d) $A$ is a union of sets like those in (a), (b) or (c).

We will consider $\overline{\mathbb{R}}$ as a measurable space with the Borel $\sigma$-algebra $\mathcal{B}(\overline{\mathbb{R}})$ relative to this topology.

## Measurable Functions

**Definition C.4** Let $(X, \Sigma)$ be a measurable space. A function $f : (X, \Sigma) \longrightarrow \overline{\mathbb{R}}$ is *measurable* if $f^{-1}(A) \in \Sigma$ for every Borel set $A \in \mathcal{B}(\overline{\mathbb{R}})$. The set formed by such functions will be denoted by $M(X, \Sigma)$. We will also consider the subset $M^+(X, \Sigma) := \{f \in M(X, \Sigma) : f(x) \geq 0 \text{ for every } x \in X\}$.

In the case where $f(x) \in \mathbb{R}$ for every $x \in X$, $f$ is measurable if $f^{-1}(A) \in \Sigma$ for every Borel set $A$ of $\mathbb{R}$. If $f$ assumes (at least) one of the values $\infty$ and $-\infty$, we have:

**Proposition C.5** *A function $f : (X, \Sigma) \longrightarrow \overline{\mathbb{R}}$ is measurable if and only if the sets $\{x \in X : f(x) = -\infty\}$ and $\{x \in X : f(x) = \infty\}$ belong to $\Sigma$ and the function*

$$f_0 : (X, \Sigma) \longrightarrow \mathbb{R} \ , \quad f_0(x) = \begin{cases} f(x), & \text{if } f(x) \in \mathbb{R} \\ 0, & \text{if } f(x) = -\infty \text{ or } f(x) = \infty \end{cases},$$

*is measurable.*

## C Measure and Integration

**Proposition C.6** *If $f, g: (X, \Sigma) \longrightarrow \overline{\mathbb{R}}$ are measurable functions and $\lambda \in \mathbb{R}$, then the following functions (provided they are well defined) are also measurable: $\lambda f$, $f + g$, $f \cdot g$, $|f|$, $\max\{f, g\}$ and $\min\{f, g\}$.*

**Proposition C.7** *Given a sequence $(f_n)_{n=1}^{\infty}$ in $M(X, \Sigma)$, the following functions defined on $(X, \Sigma)$ are measurable: $f(x) = \inf_{n \in \mathbb{N}} f_n(x)$, $F(x) = \sup_{n \in \mathbb{N}} f_n(x)$, $f^*(x) = \liminf_n f_n(x)$ and $F^*(x) = \limsup_n f_n(x)$. In particular, if $\lim_{n \to \infty} f_n(x) = f(x)$ for every $x \in X$ then $f \in M(X, \Sigma)$.*

**Example C.8** Given a measurable space $(X, \Sigma)$ and $A \subseteq X$, the *characteristic function* of $A$ is defined by

$$\chi_A: X \longrightarrow \mathbb{R}, \quad \chi_A(x) = \begin{cases} 1, & \text{if } x \in A \\ 0, & \text{if } x \notin A \end{cases}.$$

It is clear that $\chi_A$ is measurable if and only if $A \in \Sigma$. A linear combination of measurable characteristic functions is called a *measurable simple function*. Thus, a measurable simple function is a measurable function that assumes only finitely many values. Every measurable simple function $\varphi$ admits a unique representation of the form

$$\varphi = \sum_{i=1}^{n} a_i \chi_{A_i},$$

where $n \in \mathbb{N}$, $a_1, \ldots, a_n$ are non-zero and distinct real numbers, and $A_1, \ldots, A_n$ are non-empty measurable sets that are pairwise disjoint. This is the *canonical representation* of the measurable simple function $\varphi$.

## Measures

**Definition C.9** A *measure* on the measurable space $(X, \Sigma)$ is a function $\mu: \Sigma \longrightarrow [0, \infty]$ that satisfies the following conditions:

(a) $\mu(\emptyset) = 0$.
(b) If $(A_n)_{n=1}^{\infty}$ is a sequence of pairwise disjoint sets in $\Sigma$, then

$$\mu\left(\bigcup_{n=1}^{\infty} A_n\right) = \sum_{n=1}^{\infty} \mu(A_n).$$

The measure $\mu$ is said to be *finite* if $\mu(X) < \infty$, and is said to be *$\sigma$-finite* if there exist sets $(A_n)_{n=1}^{\infty}$ in $\Sigma$ such that $X = \bigcup_{n=1}^{\infty} A_n$ and $\mu(A_n) < \infty$ for every $n$. The triplet $(X, \Sigma, \mu)$ is called a *measure space*.

**Proposition C.10** *Let $(X, \Sigma, \mu)$ be a measure space. Then:*

(a) *If $A, B \in \Sigma$ and $A \subseteq B$, then $\mu(A) \leq \mu(B)$.*
(b) *If $A, B \in \Sigma$, $A \subseteq B$ and $\mu(A) < \infty$, then $\mu(B - A) = \mu(B) - \mu(A)$.*
(c) *If $A_n \in \Sigma$ for every $n \in \mathbb{N}$ and $A_1 \subseteq A_2 \subseteq \cdots$, then $\mu\left(\bigcup_{n=1}^{\infty} A_n\right) = \lim_n \mu(A_n)$.*
(d) *If $A_n \in \Sigma$ for every $n \in \mathbb{N}$, $A_1 \supseteq A_2 \supseteq \cdots$ and $\mu(A_1) < \infty$, then $\mu\left(\bigcap_{n=1}^{\infty} A_n\right) = \lim_n \mu(A_n)$.*
(e) *If $A_n \in \Sigma$ for every $n \in \mathbb{N}$, then $\mu\left(\bigcup_{n=1}^{\infty} A_n\right) \leq \sum_{n=1}^{\infty} \mu(A_n)$.*

**Theorem C.11 (Convergence in Measure)** *The following statements are equivalent for a sequence $(f_n)_{n=1}^{\infty}$ of measurable functions on a measure space $(X, \Sigma, \mu)$:*

(a) *$(f_n)_{n=1}^{\infty}$ converges in measure, that is, there exists a measurable function $f \in M(X, \Sigma)$ such that $\lim_n \mu(\{x \in X : |f_n(x) - f(x)| \geq \varepsilon\}) = 0$ for every $\varepsilon > 0$.*
(b) *$(f_n)_{n=1}^{\infty}$ is Cauchy in measure, that is, for every $\varepsilon > 0$,*

$$\lim_{n,m} \mu(\{x \in X : |f_n(x) - f_m(x)| \geq \varepsilon\}) = 0.$$

**Definition C.12** *Let $(X, \Sigma, \mu)$ be a measure space, $f, g, f_n : X \longrightarrow \overline{\mathbb{R}}$, $n \in \mathbb{N}$. It is said that:*

(a) *$f$ is equal to $g$ $\mu$-almost everywhere if there exists $A \in \Sigma$ such that $\mu(A) = 0$ and $f(x) = g(x)$ for every $x \in A^c$. In this case we write $f = g$ $\mu$-almost everywhere or $f = g$ $\mu$-a.e.*
(b) *$(f_n)_{n=1}^{\infty}$ converges to $f$ $\mu$-almost everywhere if there exists $A \in \Sigma$ such that $\mu(A) = 0$ and $f_n(x) \longrightarrow f(x)$ for every $x \in A^c$. In this case we write $f_n \longrightarrow f$ $\mu$-almost everywhere or $f_n \longrightarrow f$ $\mu$-a.e. or $f = \lim_n f_n$ $\mu$-a.e.*

**Proposition C.13** *If the sequence $(f_n)_{n=1}^{\infty}$ of measurable functions in a measure space $(X, \Sigma, \mu)$ converges in measure to a measurable function $f \in M(X, \Sigma)$, then there exists a subsequence $(f_{n_j})_{j=1}^{\infty}$ that converges $\mu$-almost everywhere to $f$.*

## Integration

**Definition C.14** Let $(X, \Sigma, \mu)$ be a measure space.

(a) The *integral* of a simple function $\varphi \in M^+(X, \Sigma)$, whose canonical representation is $\varphi = \sum_{j=1}^{m} a_j \chi_{A_j}$, with respect to the measure $\mu$ is defined by

$$\int_X \varphi \, d\mu = \sum_{j=1}^{m} a_j \mu(A_j).$$

(b) The *integral* of a function $f \in M^+(X, \Sigma)$ with respect to the measure $\mu$ is defined by

$$\int_X f \, d\mu = \sup \left\{ \int_X \varphi \, d\mu : \varphi \in M^+(X, \Sigma) \text{ is simple and } 0 \leq \varphi \leq f \right\}.$$

(c) For $f \in M^+(X, \Sigma)$ and $A \in \Sigma$, we define

$$\int_A f \, d\mu = \int_X f \chi_A \, d\mu.$$

**Proposition C.15** Let $f, g \in M^+(X, \Sigma)$ and $A, B \in \Sigma$.

(a) If $f \leq g$, then $\int_X f \, d\mu \leq \int_X g \, d\mu$.
(b) If $A \subseteq B$, then $\int_A f \, d\mu \leq \int_B f \, d\mu$.
(c) $\int_X f \, d\mu = 0$ if and only if $f = 0$ $\mu$-a.e.

**Theorem C.16 (Monotone Convergence Theorem)** Let $(f_n)_{n=1}^{\infty}$ be a sequence in $M^+(X, \Sigma)$ such that $0 \leq f_1(x) \leq f_2(x) \leq \cdots$ for every $x \in X$.

(a) If $f_n(x) \longrightarrow f(x)$ for every $x \in X$, then $f \in M^+(X, \Sigma)$ and

$$\int_X f \, d\mu = \lim_{n \to \infty} \int_X f_n \, d\mu.$$

(b) If $f_n \longrightarrow f$ $\mu$-a.e. and $f \in M^+(X, \Sigma)$, then

$$\int_X f \, d\mu = \lim_{n \to \infty} \int_X f_n \, d\mu.$$

**Theorem C.17 (Fatou's Lemma)** If $(f_n)_{n=1}^{\infty}$ is a sequence in $M^+(X, \Sigma)$, then

$$\int_X \liminf_{n \to \infty} f_n \, d\mu \leq \liminf_{n \to \infty} \int_X f_n \, d\mu.$$

**Theorem C.18 (Radon–Nikodým Theorem)** *Let $(X, \Sigma, \mu)$ and $(X, \Sigma, \lambda)$ be $\sigma$-finite measures such that $\lambda(A) = 0$ whenever $A \in \Sigma$ and $\mu(A) = 0$. Then there exists $f \in M^+(X, \Sigma)$ such that*

$$\lambda(A) = \int_A f \, d\mu \text{ for every } A \in \Sigma.$$

**Definition C.19**

(a) Given a function $f : X \longrightarrow \mathbb{R}$, the functions $f^+, f^- : X \longrightarrow [0, \infty)$ are defined by

$$f^+(x) = \max\{f(x), 0\} \quad \text{and} \quad f^-(x) = -\min\{f(x), 0\}.$$

It is easy to see that $f$ is measurable if and only if $f^+$ and $f^-$ are measurable.

(b) Let $(X, \Sigma, \mu)$ be a measure space. A function $f \in M(X, \Sigma)$ is said to be *Lebesgue-integrable* (or *integrable*) if $\int_X f^+ d\mu < \infty$ and $\int_X f^- d\mu < \infty$. In this case we define

$$\int_X f \, d\mu = \int_X f^+ d\mu - \int_X f^- d\mu.$$

**Proposition C.20**

(a) *A measurable function $f : X \longrightarrow \mathbb{R}$ is integrable if and only if $|f|$ is integrable. In this case we have*

$$\left| \int_X f \, d\mu \right| \leq \int_X |f| \, d\mu.$$

(b) *Let $f, g : X \longrightarrow \mathbb{R}$ be integrable and $a \in \mathbb{R}$. Then $af$ and $f + g$ are integrable and*

$$\int_X af \, d\mu = a \int_X f \, d\mu \quad \text{and} \quad \int_X (f + g) \, d\mu = \int_X f \, d\mu + \int_X g \, d\mu.$$

**Definition C.21** Let $(X, \Sigma, \mu)$ be a measure space. A function $f : X \longrightarrow \mathbb{C}$ is *Lebesgue-integrable* (or *integrable*) if the functions $f_1, f_2 : X \longrightarrow \mathbb{R}$ defined by

$$f_1(x) = \text{Re}(f(x)) \quad \text{and} \quad f_2(x) = \text{Im}(f(x)),$$

are integrable. In this case we define

$$\int_X f \, d\mu = \int_X f_1 \, d\mu + i \int_X f_2 \, d\mu.$$

C Measure and Integration

The set of all integrable functions $f: X \longrightarrow \mathbb{K}$ is denoted by $\mathcal{L}_{\mathbb{K}}(X, \Sigma, \mu)$.

**Theorem C.22 (Dominated Convergence Theorem)** *Let $(f_n)_{n=1}^{\infty}$ be a sequence of functions in $\mathcal{L}_{\mathbb{K}}(X, \Sigma, \mu)$ that converges $\mu$-almost everywhere to a function $f: X \longrightarrow \mathbb{K}$. If there exists $g \in \mathcal{L}_{\mathbb{K}}(X, \Sigma, \mu)$ such that $|f_n| \leq |g|$ for every $n$, then $f \in \mathcal{L}_{\mathbb{K}}(X, \Sigma, \mu)$ and*

$$\int_X f d\mu = \lim_{n \to \infty} \int_X f_n d\mu.$$

## General Case of the Riesz Representation Theorem

In Theorem 4.1.2 we state the Riesz Representation Theorem in its general form but our proof only covered the case of $\sigma$-finite measures. Next we present the proof that is missing to complete what was stated, that is, the proof of the case where $p > 1$ and $\mu$ is an arbitrary measure. To save the reader the trouble of going back and forth, we state the theorem again.

**Riesz Representation Theorem** *Let $1 \leq p < \infty$ and let $(X, \Sigma, \mu)$ be a measure space. In the case $p = 1$ we suppose that $\mu$ is a $\sigma$-finite measure. Then the correspondence $g \mapsto \varphi_g$ establishes an isometric isomorphism between $L_{p^*}(X, \Sigma, \mu)$ and $L_p(X, \Sigma, \mu)'$ in which the duality relation is given by*

$$\varphi_g(f) = \int_X fg \, d\mu \text{ for every } f \in L_p(X, \Sigma, \mu).$$

*Proof in the case that $p > 1$ and $(X, \Sigma, \mu)$ is an arbitrary measure space.* The procedure is essentialy a repetition of what was done in the proof of Theorem 4.1.2. Let $\varphi \in L_p(X, \Sigma, \mu)'$. For each $\sigma$-finite set $A$, defining $\Sigma_A$, $\mu_A$ and $\varphi_A$ exactly as it was done in the proof of Theorem 4.1.2, by the already proved case there exists a function $g_A \in L_{p^*}(A, \Sigma_A, \mu_A)$, which is unique up to sets of null measure, such that $\varphi_A(f_A) = \int_A f_A g_A \, d\mu_A$ for every $f_A \in L_p(A, \Sigma_A, \mu_A)$ and $\|g_A\|_{L_{p^*}(A, \Sigma_A, \mu_A)} \leq \|\varphi\|_{L_p(X, \Sigma, \mu)'}$. It is easy to check that if $B$ is a $\sigma$-finite set that contains $A$, then $g_B = g_A$ $\mu$-almost everywhere in $A$, therefore $\|g_A\|_{L_{p^*}(A, \Sigma_A, \mu_A)} \leq \|g_B\|_{L_{p^*}(B, \Sigma_B, \mu_B)}$. Let $K = \sup\{\|g_A\|_{L_{p^*}(A, \Sigma_A, \mu_A)} : A \text{ is } \sigma\text{−finite}\}$ and consider a sequence $(A_n)_{n=1}^{\infty}$ of $\sigma$-finite sets such that $\|g_{A_n}\|_{L_{p^*}(A_n, \Sigma_{A_n}, \mu_{A_n})} \longrightarrow K$. Defining $C = \bigcup_{n=1}^{\infty} A_n$ it follows that $C$ is $\sigma$-finite and $\|g_{A_n}\|_{L_{p^*}(A_n, \Sigma_{A_n}, \mu_{A_n})} \leq \|g_C\|_{L_{p^*}(C, \Sigma_C, \mu_C)}$ for every $n$. It immediately follows that $\|g_C\|_{L_{p^*}(C, \Sigma_C, \mu_C)} = K$. For every $\sigma$-finite set $A$ containing $C$, as $p^* < \infty$, $g_A = g_C$ and $\mu_A = \mu_{A-C}$ in $A - C$ and $g_A = g_{A-C}$ and $\mu_A = \mu_{A-C}$ in $A - C$, we have

$$\int_C |g_C|^{p^*} d\mu_C + \int_{A-C} |g_{A-C}|^{p^*} d\mu_{A-C} = \int_A |g_A|^{p^*} d\mu_A \leq K^{p^*} = \int_C |g_C|^{p^*} d\mu_C,$$

therefore $\int_{A-C} |g_{A-C}|^{p^*} d\mu_{A-C} = 0$. Thus there exists a set $N_A \subseteq (A - C)$ such that $\mu(N_A) = \mu_{A-C}(N_A) = 0$ and $g_A = g_{A-C} = 0$ in $A - (C \cup N_A)$. We still call $g_C$ the obvious extension of $g_C$ to $X$ (that is, $g_C(x) = 0$ for $x \notin C$). To prove that $\varphi = \varphi_{g_C}$, let $f \in L_p(X, \Sigma, \mu)$ be given. Then $A := \{x \in X : f(x) \neq 0\} \cup C$ is a $\sigma$-finite set containing $C$ and we can consider its corresponding set $N_A$. As $f = 0$ in $X - A$, we have $\varphi(f\mathcal{X}_{(X-C)\cap(X-A)}) = 0$. And as $\mu(N_A) = 0$, we have $f\mathcal{X}_{N_A} + f\mathcal{X}_{(X-C)\cap(X-A)} = f\mathcal{X}_{(X-C)\cap(X-A)}$ in $L_p(X, \Sigma, \mu)$, therefore $\varphi(f\mathcal{X}_{N_A} + f\mathcal{X}_{(X-C)\cap(X-A)}) = 0$. From $(X-C) \cap A = N_A \cup [(X-C) \cap (A-N_A)]$, since this last union is disjoint, it follows that

$$\varphi(f) = \varphi(f\mathcal{X}_C + f\mathcal{X}_{X-C}) = \varphi(f\mathcal{X}_C) + \varphi(f\mathcal{X}_{X-C})$$
$$= \varphi(f\mathcal{X}_C) + \varphi(f\mathcal{X}_{(X-C)\cap A} + f\mathcal{X}_{(X-C)\cap(X-A)})$$
$$= \varphi(f\mathcal{X}_C) + \varphi(f\mathcal{X}_{(X-C)\cap A}) + \varphi(f\mathcal{X}_{(X-C)\cap(X-A)})$$
$$= \varphi(f\mathcal{X}_C) + \varphi(f\mathcal{X}_{N_A}) + \varphi(f\mathcal{X}_{(X-C)\cap(A-N_A)}) + \varphi(f\mathcal{X}_{(X-C)\cap(X-A)})$$
$$= \varphi(f\mathcal{X}_C) + \varphi(f\mathcal{X}_{(X-C)\cap(A-N_A)}).$$

But $(X - C) \cap (A - N_A) = A - (C \cup N_A)$, therefore

$$\varphi(f) = \varphi(f\mathcal{X}_C) + \varphi(f\mathcal{X}_{A-(C \cup N_A)}).$$

As $A - (C \cup N_A) \subseteq A$, we have $g_{A-(C \cup N_A)} = g_A = 0$ in $A - (C \cup N_A)$. Finally,

$$\varphi(f) = \varphi_C(f|_C) + \varphi_{A-(C \cup N_A)}(f|_{A-(C \cup N_A)})$$
$$= \int_C f|_C g_C d\mu_C + \int_{A-(C \cup N_A)} f|_{A-(C \cup N_A)} g_{A-(C \cup N_A)} d\mu_{A-(C \cup N_A)}$$
$$= \int_C f|_C g_C d\mu_C = \int_C f g_C d\mu = \int_X f g_C d\mu,$$

proving that $\varphi = \varphi_{g_C}$. ∎

# Appendix D
# Answers/Hints for Selected Exercises

## Exercises from Chapter 1

- Exercise 1.8.3. Choose a basis $\{v_1, \ldots, v_n\}$ for an $n$-dimensional space $E$ and define $\left\|\sum_{j=1}^{n} a_j v_j\right\|_1 = \sum_{j=1}^{n} |a_j|$. Show that $\|\cdot\|_1$ is a norm and that any other norm on $E$ is equivalent to $\|\cdot\|_1$. The compactness of the unit ball may be helpful.
- Exercise 1.8.4. Consider the functions $f_n \in C[0, 1]$, $n \in \mathbb{N}$, given by $f_n(x) = 1 - nx$ if $x \in \left[0, \frac{1}{n}\right]$ and $f_n(x) = 0$ otherwise.
- Exercise 1.8.16. Note that one side is clear: if this property holds, then $C$ is convex. To prove the other implication, note that if $\lambda_3 \neq 1$, we have

$$\sum_{i=1}^{3} \lambda_i x_i = (\lambda_1 + \lambda_2)\left(\frac{\lambda_1}{\lambda_1 + \lambda_2} x_1 + \frac{\lambda_2}{\lambda_1 + \lambda_2} x_2\right) + \lambda_3 x_3.$$

  Proceed by induction.
- Exercise 1.8.23. Use (1.9) to show that $f \in L_p(X, \Sigma, \mu)$ for every $p \geq 1$ and that $\lim_{p \to \infty} \|f\|_p \leq \|f\|_\infty$. For the reverse inequality, given $\varepsilon > 0$, integrate $|f|^p$ over the set $\{x \in X : |f(x)| \geq \|f\|_\infty - \varepsilon\}$, first let $p \longrightarrow \infty$ and then $\varepsilon \longrightarrow 0$.
- Exercise 1.8.32. False: the set $K = \{e_n : n \in \mathbb{N}\}$ is closed in $c_0$, but $[K] = c_{00}$ is not closed in $c_0$.
- Exercise 1.8.33. Use the Riesz Lemma to construct a sequence $(y_n)_{n=1}^{\infty}$ of unit vectors in $E$ such that $\|y_n - x\| \geq 1 - \frac{1}{n}$ for every $x \in M$. Extract a convergent subsequence $(y_{n_k})_{k=1}^{\infty}$ and let $k \longrightarrow \infty$ in the inequality above.
- Exercise 1.8.34. Suppose there exists $g \in E$ with $\|g\| = 1$ such that $\|g - f\| \geq 1$ for every $f \in F$. Given $h \in E - F$, consider $\lambda$ such that $\int_0^1 g(t)dt = \lambda \int_0^1 h(t)dt$. Then $g - \lambda h \in F$ and $1 \leq \|g - (g - \lambda h)\| = |\lambda| \cdot \|h\|$. For each $n$,

consider $h_n(t) = t^{\frac{1}{n}}$. Thus $h_n \in E - F$, $\|h_n\| = 1$ and $\int_0^1 h_n(t)dt \longrightarrow 1$. Conclude that $\left|\int_0^1 g(t)dt\right| \geq 1$ and use the continuity of $g$ at zero to get a contradiction.

- Exercise 1.8.37. Consider the uncountable set

$$\{\chi_A : A \subseteq [a, b] \text{ is Lebesgue-measurable}\}.$$

Observe that the open balls $B(\chi_A; \frac{1}{2})$ are pairwise disjoint.

- Exercise 1.8.39. Call $M_q(g)$ the supremum from the statement. Let $(A_n)_{n=1}^\infty$ be a monotonically increasing sequence of measurable sets such that $X = \bigcup_{n=1}^\infty A_n$, let $(\varphi_n)_{n=1}^\infty$ be a sequence of simple measurable functions such that $\varphi_n(x) \longrightarrow g(x)$ for every $x \in X$ and $|\varphi_n| \leq |g|$ for every $n$, and define $g_n = \varphi_n \chi_{A_n}$. Then $g_n(x) \longrightarrow g(x)$ for every $x \in X$, $|g_n| \leq |g|$ and $g_n \in S$ for every $n$. Define $f_n = \|g_n\|_q^{1-q} \cdot |g_n|^{q-1} \cdot sgn(g)$, where

$$sgn(z) = \begin{cases} \frac{\bar{z}}{|z|}, & \text{if } z \neq 0 \\ 0, & \text{if } z = 0. \end{cases}$$

Then $f_n$ is a simple function, $\|f_n\|_p = 1$ and, from the Fatou Lemma,

$$\|g\|_q \leq \liminf_n \|g_n\|_q = \liminf_n \int_X |f_n g_n| d\mu$$

$$\leq \liminf_n \int_X |f_n g| d\mu = \liminf_n \int_X f_n g \, d\mu \leq M_q(g).$$

In the case $q = \infty$, for each $\varepsilon > 0$ consider the set $A = \{x \in X : |g(x)| \geq M_\infty(g) + \varepsilon\}$. If $\mu(A) > 0$, then choose $B \subseteq A$ with $0 < \mu(B) < \infty$. Defining $f = \frac{1}{\mu(B)} \cdot sgn(g)$, we have $\|f\|_1 = 1$ and $\int_X fg \geq M_\infty(g) + \varepsilon$. This contradiction proves that $\mu(A) = 0$, therefore $\|g\|_\infty \leq M_\infty(g)$.

The reverse inequality follows from Hölder's Inequality.

# Exercises of Chapter 2

- Exercise 2.7.4. Let $z = re^{i\theta} \in \mathbb{C}$ be arbitrary (with $r > 0$). As

$$\sup\{|\varphi(x)| : x \in E \text{ and } \|x\| \leq 1\} = \infty,$$

there exists $x_0 \in E$, $\|x_0\| \leq 1$, such that $|\varphi(x_0)| \geq r$. As $\varphi(x_0) \neq 0$, we have $\left|\varphi\left(\frac{rx_0}{|\varphi(x_0)|}\right)\right| = r$. Hence $\varphi\left(\frac{rx_0}{|\varphi(x_0)|}\right) = re^{i\alpha}$ for some $0 \leq \alpha < 2\pi$. Therefore

D Answers/Hints for Selected Exercises    311

$$\varphi\left(\frac{rx_0 e^{i(\theta-\alpha)}}{|\varphi(x_0)|}\right) = e^{i(\theta-\alpha)} \cdot re^{i\alpha} = re^{i\theta} = z.$$

As $\left\|\frac{rx_0 e^{i(\theta-\alpha)}}{|\varphi(x_0)|}\right\| \leq 1$, it follows that $\{|\varphi(x)| : x \in E \text{ and } \|x\| \leq 1\} = \mathbb{C}$.

- Exercise 2.7.7. (c) Continuous functions do not necessarily map Cauchy sequences into Cauchy sequences, while continuous linear operators are uniformly continuous, therefore map Cauchy sequences into Cauchy sequences.
- Exercise 2.7.13. Given $\varphi \in X'$, there exist real numbers $a, b$ such that

$$\varphi(x, y) = ax + by \text{ for all } x, y \in \mathbb{R}.$$

By definition,

$$\|\varphi\|_{X'} = \max\{ax + by : (x^4 + y^4)^{1/4} = 1\}.$$

Define $g(x, y) = x^4 + y^4$ and $f(x, y) = ax + by$. Applying the Lagrange Multipliers Theorem at the maximum point $(x_m, y_m)$ of the function $f$ under the constraint $g$, we have

$$(a, b) = \nabla f = \lambda \nabla g(x_m, y_m) = \lambda(x_m^3, y_m^3).$$

Then, $\lambda x_m^3 = a$, $\lambda y_m^3 = b$ and

$$\|\varphi\|_{X'} = ax_m + by_m = \lambda x_m^4 + \lambda y_m^4 = \lambda.$$

On the other hand,

$$\lambda^{4/3} x_m^4 + \lambda^{4/3} y_m^4 = |a|^{4/3} + |b|^{4/3}.$$

and, as $x_m^4 + y_m^4 = 1$, it follows that $\|\varphi\|_{X'} = \lambda = (|a|^{4/3} + |b|^{4/3})^{3/4}$.
- Exercise 2.7.14(f). Use item (c).
- Exercise 2.7.15. Use item (f) of Exercise 2.7.14 and the Open Mapping Theorem.
- Exercise 2.7.18. $T : \ell_\infty \longrightarrow \ell_\infty$ given by $T(x) = \left(\frac{x_1}{2}, x_2, x_3, \ldots\right)$.
- Exercise 2.7.19. $(x, y) \in (\mathbb{R}^2, \|\cdot\|_\infty) \mapsto \left(\frac{x-y}{2}, \frac{x+y}{2}\right) \in (\mathbb{R}^2, \|\cdot\|_1)$.
- Exercise 2.7.20. To give an example where it is not closed, consider the operators $u_n \in \mathcal{L}(c_0, c_0)$, $u_n((a_j)_{j=1}^\infty) = \left(a_1, \frac{a_2}{2}, \ldots, \frac{a_n}{n}, 0, 0, \ldots\right)$.
- Exercise 2.7.21. Let $y \notin \ker(T)$. To show that $E = [\ker(T) \cup \{y\}]$, let $z \in E$ be given and consider the following cases:

  (i) $T(z) = 0$. In this case it is clear that $z \in [\ker(T) \cup \{y\}]$.
  (ii) $T(z) \neq 0$. In this case there exists $\lambda \neq 0$ such that $T(z) = \lambda T(y)$, hence $z - \lambda y \in \ker(T)$. Therefore, $z = (z - \lambda y) + \lambda y \in [\ker(T) \cup \{y\}]$.

- **Exercise 2.7.24.** (b) Define $T\colon bs \longrightarrow \ell_\infty$ by $T\left((x_j)_{j=1}^\infty\right) = \left(\sum_{j=1}^n x_j\right)_{n=1}^\infty$ and check that it is linear, bijective and preserves the norm.
- **Exercise 2.7.26.** Proceed by contradiction and use the Banach–Steinhaus Theorem to ensure that $\sup\{\|T\|, \|T_1\|, \|T_2\|, \ldots\} < \infty$.
- **Exercise 2.7.28.** As for all $M \subseteq E$ and $a \in \mathbb{K}$ it holds $\overline{aM} = a\overline{M}$, we have $T(B_E(0; aR)) \supseteq B_F(0; ar)$ for every $a > 0$ positive. Let $0 < \lambda < 1$ and define $\lambda_1 = \lambda$ and $\lambda_j = \frac{(1-\lambda)}{2^{j-1}}$ for $j \geq 2$. Thus all the $\lambda_j$ are less than 1 and $\sum_{j=1}^\infty \lambda_j = 1$.
Let $y \in B_F(0; \lambda r)$. There exists $x_1 \in B_E(0; \lambda R)$ such that $\|y - Tx_1\| < \lambda_2 r$, that is, $y - Tx_1 \in B_F(0; \lambda_2 r)$. There also exists $x_2 \in B_E(0; \lambda_2 R)$ such that $\|(y - Tx_1) - Tx_2\| < \lambda_3 r$. For each $n$, consider $x_n \in B_E(0; \lambda_n R)$ such that $\|y - Tx_1 - \cdots - Tx_n\| < \lambda_{n+1} r$. Then,

$$\sum_{n=1}^\infty \|x_n\| < \lambda R + \sum_{n=2}^\infty \frac{(1-\lambda) R}{2^{n-1}} = R.$$

Now use the convergence of the numerical series $\sum_{n=1}^\infty \|x_n\|$ and the fact that $E$ is a Banach space to conclude that the series $\sum_{n=1}^\infty x_n$ converges to a certain $x \in E$ (see the definition of convergent series in Definition 5.3.7). Then $\|x\| \leq \sum_{n=1}^\infty \|x_n\| < R$, therefore $x \in B_E(0; R)$. Let $n \longrightarrow \infty$ in the inequality $\|y - Tx_1 - \cdots - Tx_n\| < \lambda_{n+1} r$ to obtain $y = T(x)$, hence $y \in T(B_E(0; R))$. It follows that $T(B_E(0; R)) \supseteq B_F(0; \lambda r)$ for every $0 < \lambda < 1$. Therefore, $T(B_E(0; R)) \supseteq \bigcup_{0 < \lambda < 1} B_F(0; \lambda r) = B_F(0; r)$.
- **Exercise 2.7.33.** Apply the Closed Graph Theorem in the following way: let $x_n \longrightarrow x \in E$ be such that $T(x_n) \longrightarrow \varphi \in E'$. Let $y \in E$. Letting $n \longrightarrow \infty$ in $(T(x_n) - T(y))(x_n - y) = T(x_n - y)(x_n - y) \geq 0$, it follows that $(\varphi - T(y))(x - y) \geq 0$. This holds for every $y \in E$; therefore, for all $a \in \mathbb{R}$ and $z \in E$, letting $y = x - az$ it follows that

$$a(\varphi(z) - T(x)(z)) + a^2 T(z)(z) = (\varphi - T(x - az))(az) \geq 0.$$

Given $z \in E$, we have $a(\varphi(z) - T(x)(z)) + a^2 T(z)(z) \geq 0$ for every $a \in \mathbb{R}$. Thus $0 \geq \Delta = (\varphi(z) - T(x)(z))^2$, which gives $\varphi(z) = T(x)(z)$. As this holds for every $z \in E$, it follows that $\varphi = T(x)$.
- **Exercise 2.7.37.** Define $u\colon E' \longrightarrow \ell_1$ by $u(\varphi) = (\varphi(x_j))_{j=1}^\infty$. From the assumption, $u$ is well defined, and in this case it is linear. We will show that $u$ has a closed graph. Suppose that $\varphi_n \longrightarrow \varphi$ in $E'$ and that $u(\varphi_n) \longrightarrow y = (y_j)_{j=1}^\infty \in \ell_1$. As $u(\varphi_n) = (\varphi_n(x_j))_{j=1}^\infty$, it follows that $\varphi_n(x_j) \longrightarrow y_j$ for every

# D  Answers/Hints for Selected Exercises

j. On the other hand, as $\varphi_n \longrightarrow \varphi$, we have $\varphi_n(x_j) \longrightarrow \varphi(x_j)$ for every $j$. It follows that $\varphi(x_j) = y_j$ for every $j$. Then we have

$$y = (y_j)_{j=1}^{\infty} = (\varphi(x_j))_{j=1}^{\infty} = u(\varphi).$$

Therefore $u$ has a closed graph and it follows that $u$ is continuous. The continuity of $u$ immediately provides the desired result.

## Exercises of Chapter 3

- Exercise 3.6.4. $E = \ell_1$, $G = \{(0, a_1, a_2, \ldots) : a_j \in \mathbb{K} \text{ for } j \geq 1\}$ and $\varphi \colon G \longrightarrow \mathbb{K}$ given by $\varphi(x) = x_2$.
- Exercise 3.6.7. Prove that $F$ is isometrically isomorphic to a closed subspace of $\mathcal{L}(E, F)$ by means of the operator $y \in F \mapsto \varphi \otimes y \in \mathcal{L}(E, F)$, where $\varphi \in E'$ is a norm 1 functional whose existence is guaranteed by Exercise 3.6.1.
- Exercise 3.6.11. Let $(x_n)_{n=1}^{\infty}$ be dense a sequence in $E$. By Corollary 3.1.4, for each $n$ there exists $\varphi_n \in E'$ such that $\|\varphi_n\| = 1$ and $\varphi_n(x_n) = \|x_n\|$. Given $x \in E$, by Corollary 3.1.5, $\|x\| = \sup_{\varphi \in B_{E'}} |\varphi(x)| \geq \sup_n |\varphi_n(x)|$. On the other hand, the denseness ensures that $x = x_j$ for some $j$ or there exists a subsequence $x_{n_k} \longrightarrow x$. In the first case, $\|x\| = \|x_j\| = \varphi_j(x_j) = |\varphi_j(x_j)| \leq \sup_n |\varphi_n(x_j)| = \sup_n |\varphi_n(x)|$. In the second case, $x - x_{n_k} \longrightarrow 0$, therefore $\varphi_{n_k}(x - x_{n_k}) \longrightarrow 0$, and $\varphi_{n_k}(x_{n_k}) = \|x_{n_k}\| \longrightarrow \|x\|$. Thus, $\varphi_{n_k}(x) = \varphi_{n_k}(x - x_{n_k}) + \varphi_{n_k}(x_{n_k}) \longrightarrow \|x\|$. It follows that $|\varphi_{n_k}(x)| \longrightarrow \|x\|$, which combined with the inequality $\|x\| \geq \sup_n |\varphi_n(x)|$ provides the equality.
- Exercise 3.6.12. Associate each $\varphi \in F'$ with a Hahn–Banach extension of $\varphi$ to $E$.
- Exercise 3.6.13. (b) Define $T \colon E'/M^\perp \longrightarrow M'$ by $T([\varphi])(x) = \varphi(x)$ for all $\varphi \in E'$ and $x \in M$. Prove that $T$ is well defined, is linear, isometry and surjective. Use Corollary 3.1.3.
  (c) Define $T \colon (E/M)' \longrightarrow M^\perp$ by $T(\varphi)(x) = \varphi([x])$ for all $\varphi \in (E/M)'$ and $x \in E$. Check that $T$ is well defined, is linear, isometry and surjective. For the isometry use Exercise 2.7.14(c).
- Exercise 3.6.14. Use Proposition 3.3.1.
- Exercise 3.6.15. Combine Exercises 1.8.38 and 3.6.13(b).
- Exercise 3.6.16. For the convexity prove that the set coincides with $\{\varphi \in E' : \|\varphi\| \leq \|x_0\| \text{ and } \varphi(x_0) = \|x_0\|^2\}$.
- Exercise 3.6.17. Adapt the proof of Theorem 3.2.7.
- Exercise 3.6.18. Use the Isomorphism Theorem (Exercise 2.7.15).
- Exercise 3.6.22. (b) Define $T_n((a_j)_{j=1}^{\infty}) = \frac{a_1 + \cdots + a_n}{n}$,

$$G = \{x \in \ell_\infty : \text{there exists } \lim_n T_n(x)\},$$

$p(x) = \limsup_n T_n(x)$ and apply the Hahn-Banach Theorem.

- Exercise 3.6.23. Let $K > 0$ be such that $\|y\| \leq K$ for every $y \in C$. Given $x \in E$, choose $a > 0$ such that $\frac{x}{a} \in C$. If $a < \frac{\|x\|}{L}$, then $\|\frac{x}{a}\| > L$ – absurd. Therefore $\|x\| \leq L p_C(x)$.
- Exercise 3.6.24. Call $C = B(0; 1)$. If $a > 0$ is such that $\frac{x}{a} \in C$ then $\|x\| \leq a$; therefore $\|\cdot\| \leq p_C$. For every $\varepsilon > 0$, $\frac{p_C(x)}{\|x\|+\varepsilon} = p_C(\frac{x}{\|x\|+\varepsilon}) < 1$, therefore $p_C(x) < \|x\| + \varepsilon$. Letting $\varepsilon \longrightarrow 0$ it follows that $p_C \leq \|\cdot\|$.
- Exercise 3.6.25. The set $G := B(0; \delta) + B$ is open, convex, and $G \cap A = \emptyset$. By Theorem 3.4.8 there exists $\varphi \in E'$ such that $\varphi(x) < \varphi(y)$ for all $x \in A$ and $y \in G$. Dividing by $\|\varphi\|$ if necessary, we can assume $\|\varphi\| = 1$. Let $\alpha = \inf \varphi(A)$ and $\beta = \sup \varphi(B)$. Thus $\alpha - \beta \leq \varphi(x) - \varphi(y) \leq \|x - y\|$ for all $x \in A$ and $y \in B$, therefore $\alpha - \beta \leq \delta$. If $z \in B(0; \delta)$, then $\varphi(y + z) \leq \alpha$ for every $y \in B$, therefore $\beta \leq \alpha - \varphi(z)$. As $\sup \varphi(B(0; \delta)) = \delta$ it follows that $\delta \leq \alpha - \beta$.
- Exercise 3.6.26. Use the part of the proof of Theorem 3.1.2 that teaches how to extend a functional that is real-linear to a functional that is complex-linear.
- Exercise 3.6.27. Let $F$ be a closed and convex subset of the real normed space $E$. It is clear that $F$ is contained in the intersection of all closed half-spaces that contain it. For the other inclusion, let $x \notin F$. Since $F^c$ is open, there exists $\delta > 0$ such that the open ball $B_E(x; \delta)$ is contained in $F^c$, that is, $F \cap B_E(x; \delta) = \emptyset$. By Theorem 3.4.8 there exist $\varphi \in E'$ and $a \in \mathbb{R}$ such that $\varphi(y) < a \leq \varphi(z)$ for all $y \in B_E(x; \delta)$ and $z \in F$. In particular, $\varphi(x) < a$, therefore $x$ does not belong to the closed half-space $\varphi^{-1}([a, \infty))$ that contains $F$.
- Exercise 3.6.30. Prove that, for each $\varphi \in E'$, $\varphi(x_i) = 0$ for every $i \in I$ if and only if $\varphi(y) = 0$ for every $y \in \overline{[x_i : i \in I]}$. Then apply Corollary 3.4.11.
- Exercise 3.6.31. Let $K := \sum_{i=1}^n c_i K_i$ and prove that $K$ is closed and convex. As $\{x\}$ is closed and compact and $K \cap \{x\} = \emptyset$, by the Second Geometric Form of the Hahn-Banach Theorem, there exist $c \in \mathbb{R}$ and $\Phi \in E'$ such that $\Phi(y) < c < \Phi(x)$ for every $y \in K$. It is clear that $0 \in K$, so $c > 0$. Check that $\varphi := \frac{\Phi}{c}$ satisfies the desired conditions.

# Exercises of Chapter 4

- Exercise 4.5.2. Given $\varphi = (b_j)_{j=1}^\infty \in \ell_{p^*} = (\ell_p)'$, choose $x = (a_j)_{j=1}^\infty$ where $a_j = 0$ if $b_j = 0$ and $a_j = \frac{\|\varphi\|_{p^*}^{1-p^*} \cdot |b_j|^{p^*}}{b_j}$ otherwise.
- Exercise 4.5.3. For each $n$, define $\varphi_n : \ell_p \longrightarrow \mathbb{K}$ by $\varphi_n\left((b_j)_{j=1}^\infty\right) = \sum_{j=1}^n a_j b_j$, and prove that $\varphi_n \in (\ell_p)'$ and that $\|\varphi_n\| = \max_{1 \leq j \leq n} |a_j|$ if $p = 1$ and $\|\varphi_n\|^{p^*} =$

# D  Answers/Hints for Selected Exercises 315

$\sum_{j=1}^{n} |a_j|^{p^*}$ if $p > 1$. Use the convergence hypothesis to show that the family $(\varphi_n)_{n=1}^{\infty}$ is pointwise bounded and then use the Banach–Steinhaus Theorem to ensure that it is uniformly bounded.

- Exercise 4.5.4. $(a_j)_{j=1}^{\infty} \in c_0 \mapsto (a_1 + a_{j+1})_{j=1}^{\infty} \in c$.
- Exercise 4.5.5. Restrict $\varphi$ to $c_0$ and apply Proposition 4.2.3.
- Exercise 4.5.6. Combine Exercise 3.6.15 with the fact that $\ell_\infty = (\ell_1)'$ is not separable.
- Exercise 4.5.7. For the first part, the same argument that works for $\ell_p$ works in this case. In fact, $\ell_p$ is the particular case where $E_n = \mathbb{K}$ for every $n$. For the second part, use the operator

$$\varphi \in \left(\left(\sum E_n\right)_p\right)' \mapsto (\varphi_n)_{n=1}^{\infty},$$

where $\varphi_n \in E_n'$ for every $n$ and $\varphi((x_n)_{n=1}^{\infty}) = \sum_{n=1}^{\infty} \varphi_n(x_n)$.

- Exercise 4.5.8. Make $E_n = E$ for every $n$ in Exercise 4.5.7.
- Exercise 4.5.9. Consider the subspace $\mathcal{P}_n \subseteq L_1[0, 1]$ of polynomials of degree less than or equal to $n$. The functional $\varphi(p) := p(0) + p'(0) - 3p''(1)$ is linear and continuous, because $\mathcal{P}_n$ is finite-dimensional. The Hahn–Banach Theorem guarantees the existence of an extension of $\varphi$ to $L[0, 1]$, preserving its norm. As linear functionals on $L_1[0, 1]$ are represented by functions of $L_\infty[0, 1]$, the first part of the exercise is done. For the second part, just use the first part with the functions $p_n(t) = \frac{t^n}{n+1} \in L[0, 1]$, to obtain

$$|\varphi(p_n)| = \frac{3n}{n+1} \leq \frac{1}{n+1} \int_0^1 |g_n(t)| dt.$$

- Exercise 4.5.10. First convince yourself that $L \cup S \subseteq L_1[0, 1]$. Check that: (i) $L$ is convex and, by the Ascoli Theorem, compact in $L_1[0, 1]$; (ii) $S$ is convex and closed in $L_1[0, 1]$; (iii) $L \cap S = \emptyset$. For (iii) use that $0 \notin L$ and that

$$\frac{|g(t) - g(0)|}{|t|} = \sum_{i=2}^{100} \lambda_i t^{1/i - 1} \longrightarrow \infty,$$

unless all the $\lambda_i$ are null. The result now follows from the second geometric form of the Hahn–Banach Theorem.

- Exercise 4.5.12. Use the Banach–Steinhaus Theorem and Proposition 4.3.1.
- Exercise 4.5.14. $T \in \mathcal{L}(\widehat{E}, F) \mapsto T|_E \in \mathcal{L}(E, F)$ and Exercise 2.7.5.
- Exercise 4.5.15. (b) By item (a) it is enough to show that $(M^\perp)^\perp \subseteq J_E(M)$. Given $f \in (M^\perp)^\perp$, we have $f \in E''$ and $f(\varphi) = 0$ for every $\varphi \in E'$ such that $\varphi|_M = 0$. As $E$ is reflexive, there exists $x \in E$ such that $J_E(x) = f$. Then,

$\varphi(x) = J_E(x)(\varphi) = f(\varphi) = 0$ for every $\varphi \in E'$ such that $\varphi|_M = 0$. It follows from Corollary 3.4.10 that $x \in M$, therefore $f = J_E(x) \in J_E(M)$.
- Exercise 4.5.21. Use Exercise 4.5.20.
- Exercise 4.5.22. Use Exercise 4.5.19.
- Exercise 4.5.23. It is easy to see that $T'$ is a projection if $T$ is a projection. For the other implication, use the previous implication and Exercise 4.5.20.
- Exercise 4.5.24. (a) $^\perp B = \bigcap \{\ker(\varphi) : \varphi \in B\}$.
- Exercise 4.5.25.

    (a) $(J_E)' \circ J_{E'}(\varphi)(x) = (J_E)' (J_{E'}(\varphi))(x) = J_{E'}(\varphi)(J_E(x)) = J_E(x)(\varphi) = \varphi(x)$.
    
    (b) Use item (a).
    
    (c) The sought projection is the operator $J_{E'} \circ (J_E)'$: $(J_{E'} \circ (J_E)')^2 = J_{E'} \circ (J_E)' \circ J_{E'} \circ (J_E)' = J_{E'} \circ \mathrm{id}_{E'} \circ (J_E)' = J_{E'} \circ (J_E)'$. And from (b) it follows that $J_{E'} \circ (J_E)'(E''') = J_{E'}((J_E)'(E''')) = J_{E'}(E') = E'$.

- Exercise 4.5.26.

    (a) Use Exercise 4.5.14.
    
    (b) Use Exercise 4.5.25.

- Exercise 4.5.27. Let $E$ be reflexive and let $T : E \longrightarrow F$ be an isomorphism. Given $\Phi \in F''$, use Proposition 4.3.11 twice to ensure that $(T'')^{-1}(\Phi) \in E''$. Therefore, there exists $x \in E$ such that $J_E(x) = (T'')^{-1}(\Phi)$, that is, $T''(J_E(x)) = \Phi$. It is enough to check that $J_F(T(x)) = T''(J_E(x))$.
- Exercise 4.5.28. Use Corollary 3.1.5.
- Exercise 4.5.29. Consider the functional $\varphi \in E'$ given by $\varphi(f) = \int_0^1 f(t)\,dt$ and apply Exercise 4.5.28.
- Exercise 4.5.30.

    (a) Consider the norm $\|\cdot\|_1$ on $E \times F$, the norm $\|\cdot\|_\infty$ on $E' \times F'$ and prove that $T_{E,F}(\varphi, \psi)(x, y) = \varphi(x) + \varphi(y)$ is an isomorphism.
    
    (b) $J_{E \times F} = [(T_{E,F})']^{-1} \circ T_{E',F'} \circ (J_E \times J_F)$.

- Exercise 4.5.31. (c) Exercise 4.5.15 and $J_M = (T_M)' \circ [T^{M^\perp}]^{-1} \circ J_E|_M$.
- Exercise 4.5.32. (b) Fix $x \in E$ of norm 1 and consider the operator $y \in F \mapsto T(y) \in (\mathcal{L}(E, F'))'$ given by $T(y)(u) = u(x)(y)$. It is easy to see that $\|T(y)\| \le \|y\|$. To prove the equality use Corollary 3.1.4 to consider functionals $\varphi \in E'$ and $\psi \in F'$ such that $\varphi(x) = \|\psi\| = 1$ and $\psi(y) = \|y\|$.
- Exercise 4.5.33. Use Exercises 4.5.27, 4.5.31(c) and 4.5.32.

## Exercises from Chapter 5

- Exercise 5.8.2. (b) Consider $y = ix$.
- Exercise 5.8.3. Use the Parallelogram Law.

D  Answers/Hints for Selected Exercises    317

- Exercise 5.8.8. Expand $\langle T(x + \lambda y), T(x + \lambda y)\rangle$ in two ways and then consider $\lambda = 1$ in the real case and $\lambda = i$ in the complex case.
- Exercise 5.8.9.

  (a) Given $x, y \in E$, expand $\langle T(ax+by), ax+by\rangle$ and apply for the cases $a = 1$, $b = i$, and $a = i$, $b = 1$.
  (b) Consider the rotation by an angle of 90° in $\mathbb{R}^2$.

- Exercise 5.8.10. In the real case define $\langle x, y\rangle_\mathbb{R} = \frac{1}{4}\left(\|x+y\|^2 - \|x-y\|^2\right)$ and in the complex case $\langle x, y\rangle_\mathbb{C} = \langle x, y\rangle_\mathbb{R} + i\langle x, iy\rangle_\mathbb{R}$.
- Exercise 5.8.11. Parallelogram Law.
- Exercise 5.8.14.

  (a) Just rework the proof of Theorem 5.2.2. Note that all the steps remain valid with these new assumptions.
  (b) Choose $x = 0$ in item (a).

- Exercise 5.8.15. Closed Graph Theorem.
- Exercise 5.8.19. Let $S_n = \sum_{i=1}^{n}\langle x, y_i\rangle y_i$ and $R_n = \sum_{i=1}^{n}\langle x, z_i\rangle z_i$. It is known that $S_n$ converges to a certain $s \in E$ and that $R_n$ converges to a certain $t \in E$. It must be shown that $s = t$. Given $\varepsilon > 0$, there exist natural numbers $m_0$ and $n_0$ such that

$$\sum_{i=m+1}^{\infty} |\langle x, y_i\rangle|^2 = \|S_m - s\|^2 \leq \varepsilon^2 \text{ for } m \geq m_0,$$

$$\sum_{i=n+1}^{\infty} |\langle x, z_i\rangle|^2 = \|R_n - t\|^2 \leq \varepsilon^2 \text{ for } n \geq n_0.$$

Fixing $m_1 \geq m_0$ and taking $n_1 \geq n_0$ such that $\{y_1, \ldots, y_{m_1}\} \subseteq \{z_1, \ldots, z_{n_1}\}$, we have $R_{n_1} - S_{m_1} = \sum_{j \in J}\langle x, y_j\rangle y_j$, where $J \subseteq \mathbb{N} - \{1, \ldots, m_1\}$. Hence

$$\|R_{n_1} - S_{m_1}\|^2 = \sum_{j\in J}|\langle x, y_j\rangle|^2 \leq \sum_{i=m_1+1}^{\infty}|\langle x, y_i\rangle|^2 \leq \varepsilon^2.$$

Therefore, $\|t - s\| \leq \|t - R_{n_1}\| + \|R_{n_1} - S_{m_1}\| + \|S_{m_1} - s\| \leq 3\varepsilon$.
- Exercise 5.8.20. As $[x_1] = [y_1]$, it follows that $y_1 = a_1 x_1$. As $\|x_1\| = \|y_1\| = 1$, it follows that $|a_1| = 1$. Now, we proceed by induction. Suppose that

$$y_1 = a_1 x_1 \text{ with } |a_1| = 1, \ldots, y_n = a_n x_n \text{ with } |a_n| = 1.$$

We will show that $y_{n+1} = a_{n+1}x_{n+1}$ with $|a_{n+1}| = 1$. As $[x_1, \ldots, x_{n+1}] = [y_1, \ldots, y_{n+1}]$, we have

$$y_{n+1} = \lambda_1 x_1 + \cdots + \lambda_n x_n + \lambda_{n+1} x_{n+1}. \tag{D.1}$$

For any $j = 1, \ldots, n$, using the induction hypothesis,

$$0 = \langle y_{n+1}, y_j \rangle = \langle \lambda_1 x_1 + \cdots + \lambda_n x_n + \lambda_{n+1} x_{n+1}, a_j x_j \rangle = \lambda_j \overline{a_j}.$$

As $|a_j| = 1$, it follows that $\lambda_j = 0$. Therefore $\lambda_1 = \cdots = \lambda_n = 0$. From (D.1) we conclude that $y_{n+1} = \lambda_{n+1} x_{n+1}$. As $\|y_{n+1}\| = \|x_{n+1}\| = 1$, it follows that $|\lambda_{n+1}| = 1$.

- Exercise 5.8.27. Induction on $n$.
- Exercise 5.8.28. Consider $x, y \in M^\perp$ and study the vector $v = \varphi(x)y - \varphi(y)x$.
- Exercise 5.8.30. Consider the inclusion of $\{(a, 0, 0, \ldots) : a \in \mathbb{K}\}$ in $\ell_2$.
- Exercise 5.8.31. Use the Riesz–Fréchet Theorem twice, once for existence and once for uniqueness.
- Exercise 5.8.32. As $G$ is a Hilbert space, by the Riesz–Fréchet Theorem there exists $g_\varphi \in G$ such that $\varphi(x) = \langle x, g_\varphi \rangle$. Let $\widetilde{\varphi} \colon \longrightarrow \mathbb{K}$ be the unique linear and continuous extension of $\varphi$ that preserves the norm. By the Riesz–Fréchet Theorem, there exists $x_0 \in H$ such that $\widetilde{\varphi}(x) = \langle x, x_0 \rangle$. Write $x_0 = x_0^G + x_0^{G^\perp}$. For $x \in G$,

$$\varphi(x) = \widetilde{\varphi}(x) = \langle x, x_0^G + x_0^{G^\perp} \rangle = \langle x, x_0^G \rangle.$$

Therefore, $g_\varphi = x_0^G$, so $\mathrm{card}(\mathcal{F}) = \mathrm{card}(G^\perp)$. Calculating the cardinality of $\mathcal{G}$ is easier. Let $B$ be a (Hamel) basis of $G^\perp$. As $H = G \oplus G^\perp$, to define a linear extension of $\varphi$ it is enough to define $\widetilde{\varphi}(b)$ for $b \in B$. It follows that $\mathrm{card}(\mathcal{G}) = \mathrm{card}(F^B)$. The calculation of the cardinality of $\mathcal{H}$ is even easier. As linearity is not even required, $\mathrm{card}(\mathcal{H}) = \mathrm{card}(F^{H-G})$.
- Exercise 5.8.33. (a), (b) and (c): analogous to the case of $\ell_p$. For (d) use Theorem 5.3.10. For (e) use item (d).
- Exercise 5.8.34. (a) Draw the graphs of $r_1, r_2, r_3$.
  (b) The function $r_1 \cdot r_2$ is orthogonal to $r_n$ for every $n$.
- Exercise 5.8.36. Check that the system $(h_{mn})_{m,n=1}^\infty$ is orthonormal. For $f \in L_2(\mu_1 \otimes \mu_2)$, use the condition of Theorem 5.3.10(a) twice to justify the steps

$$\left( \sum_{m,n=1}^\infty \langle f, h_{mn} \rangle h_{mn} \right)(x_0, y_0)$$

$$= \sum_{m=1}^\infty \sum_{n=1}^\infty \left( \int_{X_1 \times X_2} f(x, y) h_{mn}(x, y) d\mu_1(x) d\mu_2(y) \right) h_{mn}(x_0, y_0)$$

$$= \sum_{m=1}^\infty \sum_{n=1}^\infty \left( \int_{X_2} \left( \int_{X_1} f(x, y) f_n(x) g_m(y) d\mu_1(x) \right) d\mu_2(y) \right) f_n(x_0) g_m(y_0)$$

$$= \sum_{m=1}^{\infty} \left( \int_{X_2} \left( \sum_{n=1}^{\infty} \int_{X_1} f(x,y) f_n(x) g_m(y) d\mu_1(x) \right) d\mu_2(y) \right) f_n(x_0) g_m(y_0)$$

$$= \sum_{m=1}^{\infty} \left( \int_{X_2} \left( \left( \sum_{n=1}^{\infty} \langle f(\cdot, y) g_m(y), f_n \rangle f_n \right) (x_0) \right) d\mu_2(y) \right) g_m(y_0)$$

$$= \sum_{m=1}^{\infty} \left( \int_{X_2} f(x_0, y) g_m(y) d\mu_2(y) \right) g_m(y_0)$$

$$= \left( \sum_{m=1}^{\infty} \langle f(x_0, \cdot), g_m \rangle g_m \right) (y_0) = f(x_0, y_0),$$

and once more to conclude the result.

## Exercises of Chapter 6

- Exercise 6.8.7(b). Consider the sequences $(e_1, e_2, e_3, \ldots)$ and $(e_2, e_2, e_4, e_4, \ldots)$ where $(e_n)_{n=1}^{\infty}$ is orthonormal.
- Exercise 6.8.9. Use Lemma 6.3.5.
- Exercise 6.8.10. Consider the closure of the convex hull of the set $\{x_n : n \in \mathbb{N}\}$ and use Mazur's Theorem (Theorem 6.2.11).
- Exercise 6.8.13. Note that the canonical basis of $\ell_1$ converges to zero in the weak-star topology.
- Exercise 6.8.14. One can assume

$$W = \{h \in E'' : |h(\varphi_i) - f(\varphi_i)| < \varepsilon, i = 1, \ldots, n\},$$

where $\varepsilon > 0$ and $\varphi_1, \ldots, \varphi_n \in E'$. Define

$$W_0 := \{h \in E'' : |h(\varphi_i)| < \varepsilon, i = 1, \ldots, n\}$$

and check that $W = f + W_0$.
- Exercise 6.8.15.
    (a) Use the Banach-Steinhaus Theorem.
    (b) Show that the linear functional $f : E' \longrightarrow \mathbb{K}$ given by $f(\varphi) = \lim_{n \to \infty} \varphi(x_n)$ is continuous and use the reflexivity of $E$.
- Exercise 6.8.17. Use Exercise 3.6.13(c) and Corollary 4.4.6.
- Exercise 6.8.18. Assume that $E$ is not reflexive. One can assume that $\|\varphi_0\| = 1$. In this case, it is clear that $\{x \in E : \varphi_0(x) = 1\} \cap \{x \in E : \|x\| \leq 1\} = \emptyset$. Since

$1 = \|\varphi_0\| = \sup\{\varphi_0(x) : \|x\| \le 1\}$, there exists a sequence $(x_n)_{n=1}^\infty$ such that $1 > \varphi_0(x_n) \ge 1 - \frac{1}{n}$ for every $n$. If there existed a subsequence $(y_n)_{n=1}^\infty$ of $(x_n)_{n=1}^\infty$ such that $y_n \xrightarrow{w} x_0$, it would be true that $\varphi_0(x_0) = 1$ and $\|x_0\| \le \liminf_n \|x_n\| = 1$.

- Exercise 6.8.19. (c) For (i) $\implies$ (ii) use Exercise 4.5.8, and for (iii) $\implies$ (i) use item (a) and Corollary 6.4.6.
- Exercise 6.8.21. (c) Use Theorems 6.5.4 and 1.5.4.
- Exercise 6.8.23.

    (a) If $E = \overline{[K]}$, then $E = \overline{[\operatorname{conv}(K \cup -K)]}$.
    (b) $E = [B_E]$.
    (c) Consider a sequence $(x_n)_{n=1}^\infty$ dense in the unit sphere $S_E$ and the compact set (therefore weakly compact) $K = \left(\frac{1}{n} x_n\right)_{n=1}^\infty \cup \{0\}$.

- Exercise 6.8.24. Let $(I, \le)$ be a directed set and let $(x_\lambda)_{\lambda \in I}$ be a Cauchy net in the Banach space $E$. For $\lambda, \alpha \in I$, the notation $\lambda < \alpha$ means that $\lambda \le \alpha$ and $\lambda \ne \alpha$. Choose $\lambda_0, \lambda_1 \in I$, $\lambda_0 \ne \lambda_1$. Since $I$ is directed, there exists $\lambda_2 \in I$ such that $\lambda_0 \le \lambda_2$ and $\lambda_1 \le \lambda_2$. Then $\lambda_0 < \lambda_2$ or $\lambda_1 < \lambda_2$. If $\lambda_0 < \lambda_2$, call $\alpha_1 = \lambda_0$ and $\alpha_2 = \lambda_2$; otherwise call $\alpha_1 = \lambda_1$ and $\alpha_2 = \lambda_2$. In both cases, we have $\alpha_1 < \alpha_2$. Since $\alpha_1 \ne \alpha_2$, the same procedure produces $\alpha_3 \in I$ such that $\alpha_1 < \alpha_2 < \alpha_3$. Repeat the process to construct a sequence $(\alpha_j)_{j=1}^\infty$ in $I$ such that $\alpha_1 < \alpha_2 < \cdots$. Check that the sequence $(x_{\alpha_j})_{j=1}^\infty$ is a Cauchy sequence in $E$, therefore $x_{\alpha_j} \xrightarrow{j} x \in E$. Conclude that $x_\alpha \xrightarrow{\alpha} x$.
- Exercise 6.8.26. For one implication use Theorem 6.6.6 and for the reverse implication use Exercise 6.8.25.
- Exercise 6.8.27. Use Exercise 6.8.26.
- Exercise 6.8.28. Combine Theorem 6.6.6 with Exercise 4.5.27.
- Exercise 6.8.29. (c) $\implies$(b) Let $x_n, y_n \in E$, $n \in \mathbb{N}$, with $(x_n)_{n=1}^\infty$ bounded and $\lim_{n \to \infty} \left(2\|x_n\|^2 + 2\|y_n\|^2 - \|x_n + y_n\|^2\right) = 0$. Using that

$$2\|x_n\|^2 + 2\|y_n\|^2 - \|x_n + y_n\|^2 \ge 2\|x_n\|^2 + 2\|y_n\|^2 - (\|x_n\| + \|y_n\|)^2$$
$$= (\|x_n\| - \|y_n\|)^2 \ge 0,$$

it follows that $\lim_{n \to \infty} (\|x_n\| - \|y_n\|) = 0$, therefore $(y_n)_{n=1}^\infty$ is bounded. Passing to a subsequence, one can assume $\lim_{n \to \infty} \|x_n\| = \lim_{n \to \infty} \|y_n\| = a$. For $a = 0$ the desired limit follows immediately. If $a > 0$, we can assume, without loss of generality, $x_n, y_n$ non-zero, and in this case we have $\lim_{n \to \infty} \|x_n + y_n\| = 2a$. Therefore

$$\lim_{n \to \infty} \left\| \frac{x_n}{\|x_n\|} + \frac{y_n}{\|y_n\|} \right\| = 2,$$

and from (c) it follows that

$$\lim_{n \to \infty} \left\| \frac{x_n}{\|x_n\|} - \frac{y_n}{\|y_n\|} \right\| = 0.$$

- Exercise 6.8.30. (a) $\Longrightarrow$ (b) The case where $x = 0$ or $y = 0$ is trivial. Suppose $x, y \neq 0$. From

$$0 = 2\|x\|^2 + 2\|y\|^2 - \|x+y\|^2 \geq 2\|x\|^2 + 2\|y\|^2 - (\|x\| + \|y\|)^2$$
$$= (\|x\| - \|y\|)^2 \geq 0,$$

it follows that $\|x\| = \|y\|$. As $2\|x\|^2 + 2\|y\|^2 - \|x+y\|^2 = 0$, we have

$$\left\| \frac{x}{\|x\|} + \frac{y}{\|y\|} \right\|^2 = 4.$$

Calling $x_0 = \frac{x}{\|x\|}$ and $y_0 = \frac{y}{\|y\|}$, we get $2 = \|x_0 + y_0\|$ and by item (a) we conclude that $x_0 = y_0$, hence $x = y$.

(a) $\Longrightarrow$ (c) Suppose $\|x+y\| = \|x\| + \|y\|$ for $x, y \neq 0$. Suppose, without loss of generality, $\|y\| \geq \|x\| > 0$. Then

$$2 \geq \left\| \frac{x}{\|x\|} + \frac{y}{\|y\|} \right\| \geq \left\| \frac{x}{\|x\|} + \frac{y}{\|x\|} \right\| - \left\| \frac{y}{\|x\|} - \frac{y}{\|y\|} \right\|$$

$$= \frac{1}{\|x\|} (\|x+y\|) - \left\| \left( \frac{1}{\|x\|} - \frac{1}{\|y\|} \right) y \right\|$$

$$= \frac{1}{\|x\|} (\|x\| + \|y\|) - \|y\| \left( \frac{1}{\|x\|} - \frac{1}{\|y\|} \right) = 2.$$

It follows that $\frac{x}{\|x\|} = \frac{y}{\|y\|}$ and the result follows.
- Exercise 6.8.33. (b) Use Exercise 6.8.30.
(c) Consider, for every $n \in \mathbb{N}$, $f_n$ the constant function equal to 1 and $g_n$ the function that joins the points $(0, 1)$, $(\frac{1}{n}, 1)$ and $(1, 1)$ by line segments, and apply Exercise 6.8.29.

## Chapter Exercises 7

- Exercise 7.7.2. Be inspired by the end of the proof of Proposition 7.5.2.
- Exercise 7.7.10. Use the following results: Theorem 6.5.4, Proposition 6.2.9, Theorem 6.2.12 and Proposition 7.2.3.
- Exercise 7.7.11. Use Exercise 7.7.10 and Theorem 7.2.7.

- Exercise 7.7.12. $(I-T)^n = \sum_{j=0}^{n}(-1)^j\binom{n}{j}T^j = I + \sum_{j=1}^{n}(-1)^j\binom{n}{j}T^j$.
- Exercise 7.7.13. If $f$ is continuous then $f(\overline{A}) \subseteq \overline{f(A)}$.
- Exercise 7.7.14. $T: \ell_2 \longrightarrow \ell_2$, $T((a_n)_{n=1}^{\infty}) = (\lambda_n a_n)_{n=1}^{\infty}$.
- Exercise 7.7.15. Consider the identity operator on $\ell_1$.
- Exercise 7.7.17. Given $\varepsilon > 0$, the collection $\left(B\left(y, \frac{\varepsilon}{2}\right)\right)_{y \in T(B_E)}$ is an open cover of the compact $\overline{T(B_E)}$. There then exist $y_1, \ldots, y_n \in T(B_E)$ such that $T(B_E) \subseteq \bigcup_{j=1}^{n} B\left(y_j, \frac{\varepsilon}{2}\right)$. Let $M = [y_1, \ldots, y_n]$, let $P$ be the orthogonal projection from $H$ onto $M$ (Theorem 5.2.5) and $P_0 = P \circ T$. Then $P_0$ is linear, continuous, has finite rank and $\|T - P_0\| < \varepsilon$ because, for every $x \in B_E$,

$$\|T(x) - P_0(x)\| = \|T(x) - P(T(x))\| = \inf_{y \in M} \|T(x) - y\| \leq \inf_{j=1,\ldots,n} \|T(x) - y_j\| < \varepsilon.$$

- Exercise 7.7.18. Prove that $T = \sum_{n=1}^{\infty} \varphi_n \otimes b_n$ in $\mathcal{L}(E, F)$ and apply Propositions 7.2.2(b) and 7.2.5.
- Exercise 7.7.19.

  (a) From $\|T^{-1} \circ (T - S)\| \leq \|T^{-1}\| \cdot \|T - S\|$ and Proposition 7.1.3 it follows that $I - T^{-1} \circ (T - S) = T^{-1} \circ S$ is an isomorphism, therefore $S$ is an isomorphism. Moreover, $(I - T^{-1} \circ (T - S))^{-1} = \sum_{n=0}^{\infty} (T^{-1} \circ (T - S))^n$, hence

  $$S^{-1} = (T - (T-S))^{-1} = \left(T \circ \left(I - T^{-1} \circ (T-S)\right)\right)^{-1}$$
  $$= \sum_{n=0}^{\infty} \left(T^{-1} \circ (T-S)\right)^n \circ T^{-1},$$

  therefore

  $$\left\|S^{-1} - T^{-1}\right\| \leq \sum_{n=1}^{\infty} \left\|\left(T^{-1} \circ (T-S)\right)^n \circ T^{-1}\right\| \leq \left\|T^{-1}\right\|$$
  $$\cdot \sum_{n=1}^{\infty} \left(\|T-S\| \left\|T^{-1}\right\|\right)^n.$$

D  Answers/Hints for Selected Exercises

(b) Use item (a).

- Exercise 7.7.20. Suppose that $S \circ T$ is an isomorphism. Work with $S \circ T$ to prove that $T$ is injective, and with $T \circ S$ to prove that $S$ is injective. If $S$ were not surjective, we would have $(S \circ T)(E) = S(T(E)) \subseteq S(E) \neq X$, which contradicts the fact that $S \circ T$ is an isomorphism. The case of $T$ is analogous.
- Exercise 7.7.21. Use Exercise 7.7.20.
- Exercise 7.7.22. Given $\lambda \in \mathbb{C}$, let $\lambda_1, \ldots, \lambda_n \in \mathbb{C}$ be the roots of the polynomial $t^n - \lambda$. Obtain the factorization $(T^n - \lambda I) = (T - \lambda_1 I) \circ \cdots \circ (T - \lambda_n I)$ and use Exercise 7.7.21.
- Exercise 7.7.30. For $n = 2$ use Exercise 7.7.24(c). As $T$ is self-adjoint, from Exercise 7.7.27 it follows that $T^k$ is self-adjoint for every positive integer $k$. It follows that $\left\|T^{2^k}\right\| = \|T\|^{2^k}$ for every $k$. If $1 \leq n < 2^k$, then

$$\|T\|^{2^k} = \left\|T^{2^k}\right\| = \left\|T^n \circ T^{2^k-n}\right\| \leq \|T^n\| \cdot \|T\|^{2^k-n} \leq \|T\|^n \cdot \|T\|^{2^k-n} = \|T\|^{2^k}.$$

- Exercise 7.7.33. $T_1 = \frac{1}{2}(T + T^*)$, $T_2 = -\frac{1}{2}i(T - T^*)$.
- Exercise 7.7.34. Revise the proof of Theorem 7.5.6.
- Exercise 7.7.37. $T \circ T^* = T^* \circ T \iff T(T^*(x)) = T^*(T(x))$ for every $x \iff \langle T(T^*(x)), y \rangle = \langle T^*(T(x)), y \rangle$ for all $x, y \iff \langle T^*(x), T^*(y) \rangle = \langle T^*(T(x)), y \rangle$ for all $x, y \iff \langle T^*(x), T^*(y) \rangle = \overline{\langle y, T^*(T(x)) \rangle}$ for all $x, y \iff \langle T^*(x), T^*(y) \rangle = \overline{\langle T(y), T(x) \rangle}$ for all $x, y \iff \langle T^*(x), T^*(y) \rangle = \langle T(x), T(y) \rangle$ for all $x, y$.

## Exercises from Chapter 8

- Exercise 8.6.5. Check that $T(0) = 0$ and that $T(nx) = nT(x)$ for every positive integer $n$. If $p, q$ are integers with $q \neq 0$, from

$$pT(x) = T(px) = T\left(q\left(\frac{p}{q}x\right)\right) = qT\left(\frac{p}{q}x\right),$$

it follows that $T\left(\frac{p}{q}x\right) = \frac{p}{q}T(x)$. Given $r \in \mathbb{R}$, choose a sequence $(q_n)_{n=1}^\infty$ of rationals such that $q_n \longrightarrow r$. Thus $q_n x \longrightarrow rx$, therefore $T(q_n x) \longrightarrow T(rx)$. On the other hand, $T(q_n x) = q_n T(x) \longrightarrow rT(x)$. It follows that $T(rx) = rT(x)$ because $F$ is Hausdorff.
- Exercise 8.6.6(c). Let $(f_n)_{n=1}^\infty$ be a Cauchy sequence in $L_p(X, \Sigma, \mu)$. For every $\varepsilon > 0$,

$$\mu(\{x \in X : |f_n(x) - f_m(x)| > \varepsilon\}) = \int_{\{x \in X : |f_n(x) - f_m(x)| > \varepsilon\}} 1 \, d\mu$$

$$\leq \varepsilon^{-p} \int_{\{x \in X : |f_n(x) - f_m(x)| > \varepsilon\}} |f_n - f_m|^p \, d\mu$$

$$\leq \varepsilon^{-p} d(f_n, f_m),$$

which proves that the sequence $(f_n)_{n=1}^\infty$ is Cauchy in measure. By Theorem C.11 there exists a measurable function $f$ to which $(f_n)_{n=1}^\infty$ converges in measure. From Proposition C.13 there exists a subsequence $(f_{n_j})_{j=1}^\infty$ that converges to $f$ $\mu$-almost everywhere. Given $\varepsilon > 0$, let $N_0$ be a positive integer such that $d(f_m, f_n) < \varepsilon$ whenever $m, n \geq N_0$. If $n \geq N_0$, then

$$\int_X |f_n - f|^p \, d\mu = \int_X \lim_j |f_{n_j}(x) - f(x)|^p \, d\mu$$

$$\leq \liminf_j \int_X \int_X |f_{n_j}(x) - f(x)|^p \, d\mu = \liminf_j d(f_{n_j}, f) \leq \varepsilon.$$

Therefore $f = (f - f_{N_0}) + f_{N_0} \in L_p(X, \Sigma, \mu)$ and the result follows.

- Exercise 8.6.7. Apply Exercise 8.6.6 to the $\sigma$-algebra of parts of $\mathbb{N}$ with the counting measure.
- Exercise 8.6.10. Given $x, y \in \overline{A}$, consider nets of the form $(x - x_\lambda)_\lambda$ and $(y - y_\lambda)_\lambda$, with $x_\lambda, y_\lambda \in A$, both converging to zero and indexed by the neighborhood system of zero.
- Exercise 8.6.22. Use Theorem B.15.
- Exercise 8.6.24. Take a normed space $E$ equipped with the weak topology $\sigma(E, E')$, $0 \neq x_0 \in E$ and the sequence $(x_n)_{n=1}^\infty = ((-1)^n x_0)_{n=1}^\infty$.
- Exercise 8.6.27. Use the sequences $((-1)^n)_{n=1}^\infty$ and $((-1)^{n+1})_{n=1}^\infty$ to prove that addition is not continuous.
- Exercise 8.6.30. Corollary 8.4.2 and Theorem 8.3.3.
- Exercise 8.6.31.

  (a) Suppose $f_n \xrightarrow{\tau_2} f$. It is clear that $f_n(x) \longrightarrow f(x)$ for every $x \in [0, 1]$. For each $n$, define $g_n \in C[0, 1]$ by

  $$g_n(x) = \frac{|f_n(x) - f(x)|}{1 + |f_n(x) - f(x)|}.$$

  Then $g_n \longrightarrow 0$ and $|g_n(x)| \leq 1$ for all $n \in \mathbb{N}$ and $x \in [0, 1]$. By the Dominated Convergence Theorem it follows that $d(f_n, f) = \int_0^1 g_n(x) dx \longrightarrow 0$.

  (b) Suppose, by contradiction, that it is continuous. In this case, given $0 < \varepsilon < \frac{1}{2}$ there exist $\delta > 0$, $n \in \mathbb{N}$ and $x_1, \ldots, x_n \in [0, 1]$ such that

  $$A := \{f \in C[0, 1] : |f(x_i)| \leq \delta \text{ for all } i = 1, \ldots, n\}$$

D  Answers/Hints for Selected Exercises                                              325

is contained in the open ball of center 0 and radius $\varepsilon$ of $(C[0, 1], \tau_1)$. For each positive integer $k$ define $g_k(x) = k(x - x_1) \cdots (x - x_n)$. As $g_k \in A$ for every $k$, it follows that

$$\int_0^1 \frac{|g_k(x)|}{1 + |g_k(x)|} dx = d(g_k, 0) < \varepsilon < \frac{1}{2} \tag{D.2}$$

for every $k$. Also define, for each $k$, $h_k(x) = \frac{|g_k(x)|}{1+|g_k(x)|}$. As $h_k \longrightarrow 1$ $m$-almost everywhere, where $m$ is the Lebesgue measure on $[0, 1]$, and $|h_k(x)| \le 1$ for all $k \in \mathbb{N}$ and $x \in [0, 1]$, the Dominated Convergence Theorem ensures that $\int_0^1 h_k(x) dx \longrightarrow \int_0^1 1 = 1$. This contradicts (D.2).

- Exercise 8.6.32.

  (d) Theorem C.11.
  (e) Prove that the metric is translation invariant, that is, $d(f + h, g + h) = d(f, g)$, and proceed as was done in Example 8.1.2(e).
  (f) $f = \sum_{k=0}^{\infty} k\chi_{[k, k+1)}$.
  (g) Suppose $a_n \longrightarrow a$ in $\mathbb{K}$ and $f_n \longrightarrow f$ in $L_0(X, \Sigma, \mu)$. Choose $(b_n)_{n=1}^{\infty}$ a decreasing sequence of positive real numbers converging to zero such that $|a_n - a| \le b_n$ for every $n$ and also a constant $C$ such that $|a_n| \le C$ for every $n$. By $[f > \delta]$ we denote the set $\{x \in X : f(x) > \delta\}$. For all $\varepsilon > 0$ and $n \in \mathbb{N}$,

  $$[|a_n f_n - af| > \varepsilon] \subseteq [|a_n f_n - a_n f| > \varepsilon/2] \cup [|a_n f - af| > \varepsilon/2].$$

  Let $n \longrightarrow \infty$ in the inequality

  $$\mu([|a_n f_n - af| > \varepsilon]) \le \mu([|a_n f_n - a_n f| > \varepsilon/2]) + \mu([|a_n f - af| > \varepsilon/2])$$
  $$\le \mu([C|f_n - f| > \varepsilon/2]) + \mu([b_n|f| > \varepsilon/2])$$
  $$= \mu([|f_n - f| > \varepsilon/2C]) + \mu([|f| > \varepsilon/2b_n]).$$

  (h) Given a non-empty convex neighborhood $V$ of the origin, let $\varepsilon$ be such that $B(0; \varepsilon) \subseteq V$. Let $f \in L_0[0, 1]$. Choose $n$ natural such that $1/n < \varepsilon/2$. For each $j$ define $f_j(t) = nf(t)\chi_{[\frac{j-1}{n}, \frac{j}{n})}$. From $m([|f_j| > \varepsilon/2]) \le 1/n < \varepsilon/2$ for each $j$, it follows each $f_j \in V$. As $V$ is convex and $f = \sum_{j=1}^{n} n^{-1} f_j$, we have $f \in V$, therefore $V = L_0[0, 1]$.

## Exercises of Chapter 9

- Exercise 9.9.5. Adapt the arguments used in the proof of Theorem 1.2.3.
- Exercise 9.9.12. For one implication, let $(x_n)_{n=1}^\infty$ be a sequence in $B_E$. As the inclusion in $F$ is compact, there exists a subsequence $(x_{n_j})_{j=1}^\infty$ convergent in the norm of $F$, hence Cauchy in the norm of $F$. As the norms are equivalent by assumption, it is also Cauchy in the norm of $E$, hence convergent in $E$. This proves that the ball $B_E$ is compact, therefore $E$ is finite-dimensional by Theorem 1.5.4. For the other implication use Exercise 1.8.3.
- Exercise 9.9.13(b). Suppose that $f$ has only a finite number of discontinuities. In this case there exists $r > 0$ such that $f$ is continuous in $E - B(0; r)$. Then the function

$$g : \mathbb{R} \longrightarrow \mathbb{R}, \ g(t) = f(tx_0),$$

where $x_0 \in E$ is a fixed unit vector, is continuous in $(-\infty, -r]$ and in $[r, \infty)$. Choose $y > \max\{g(r), g(-r)\}$. As $\lim_{t \to +\infty} g(t) = \infty$, there exists $t_1 > r$ such that $g(t_1) > y$; and as $\lim_{t \to -\infty} g(t) = \infty$, there exists $t_2 < -r$ such that $g(t_2) > y$. From the continuity of $g$ in $[r, \infty)$, as $g(r) < y < g(t_1)$, there exists $c_1 \geq r > 0$ such that $g(c_1) = y$. And from the continuity of $g$ in $(-\infty, -r]$, as $g(-r) < y < g(t_2)$, there exists $c_2 \leq -r < 0$ such that $g(c_2) = y$. As $c_2 < 0 < c_1$, it is clear that $c_2 x_0 \neq c_1 x_0$. The equality $f(c_1 x_0) = g(c_1) = y = g(c_2) = f(c_2 x_0)$ contradicts the injectivity of $f$.
- Exercise 9.9.14. $f : (0, 1] \longrightarrow (0, 1]$, $f(x) = \frac{x}{2}$.
- Exercise 9.9.15. Use Proposition 7.1.3.
- Exercise 9.9.17. See the proof of Proposition 7.2.5.
- Exercise 9.9.20. Prove and apply the following more general result: if $X$ is a topological space, $Y$ is a Hausdorff topological space and $f, g : X \longrightarrow Y$ are continuous functions, then the set of points where $f$ and $g$ coincide is closed in $X$.
- Exercise 9.9.21. The continuity of the norm ensures that the function

$$g : B \longrightarrow \mathbb{R}, \ g(a) = \left(1 - \|a\|_2^2\right)^{1/2},$$

is continuous. Given $a \in B$ and $\varepsilon > 0$, consider $0 < \delta < \frac{\varepsilon}{2}$ such that $|g(a) - g(x)| < \frac{\varepsilon}{2}$ whenever $x \in B$ and $\|x - a\|_2 < \delta$. It follows that $\|f(x) - f(a)\| < \varepsilon$ whenever $x \in B$ and $\|x - a\|_2 < \delta$, proving that $f$ is continuous. And $f$ is not completely continuous because $e_n \xrightarrow{w} 0$ in $B$ but $\|f(e_n) - f(0)\|_2 = \|e_{n+1} - e_1\|_2 = \sqrt{2}$ for every $n$.
- Exercise 9.9.23. Check and use that, for every $t < 1$, the map $t\varphi$ has a fixed point.
- Exercise 9.9.24. $T(x_1, x_2, \ldots) = (1, x_1, x_2, \ldots)$.

D  Answers/Hints for Selected Exercises

- Exercise 9.9.25. Given $0 < \varepsilon < 1$, call $C_\varepsilon = \{(1-\varepsilon)x : x \in C\}$ and note that $C_\varepsilon$ is closed. Given $x \in C_\varepsilon$, $x = \varepsilon \cdot 0 + (1-\varepsilon)y$ for some $y \in C$. As $0, y \in C$, $0 < \varepsilon < 1$ and $C$ is convex, it follows that $x \in C$. We have proven that $C_\varepsilon \subseteq C$, hence we can consider the function

$$f_\varepsilon : C_\varepsilon \longrightarrow C_\varepsilon, \quad f_\varepsilon(x) = (1-\varepsilon)f(x).$$

It is immediate that $f_\varepsilon$ is a contraction, therefore it has a fixed point, say $x_\varepsilon \in C_\varepsilon \subseteq C$, by the Banach fixed point theorem. Then

$$\|f(x_\varepsilon) - x_\varepsilon\| = \|f(x_\varepsilon) - f_\varepsilon(x_\varepsilon)\| = \|\varepsilon \cdot f(x_\varepsilon)\| = \varepsilon \cdot \|f(x_\varepsilon)\|.$$

The result follows by taking $\varepsilon = \frac{1}{n}, n \in \mathbb{N}$.
- Exercise 9.9.26. Choose $x_0 \in C$ and apply Exercise 9.9.25 to the function $g : x_0 - C \longrightarrow x_0 - C$, $g(x_0 - x) = x_0 - f(x)$.
- Exercise 9.9.27. Make $y = f'$ and use the Cauchy–Picard theorem.
- Exercise 9.9.28. Use the Ascoli theorem (Theorem B.7).
- Exercise 9.9.29. Use the norm $\|\cdot\|_\infty$ on $E_1 \times E_2$ and Exercise 2.7.25.
- Exercise 9.9.30. Addition is linear and scalar multiplication is bilinear, and both are continuous.
- Exercise 9.9.31. Use Corollary 9.6.9.
- Exercise 9.9.34. Use that $\lim_{\mu(A)\to 0} \int_A \|f\|\,d\mu = 0$ and apply Corollary 9.7.8.
- Exercise 9.9.35. First prove that the Bochner integral is finitely additive. Use the $\sigma$-additivity in the scalar case and Corollary 9.7.8 to ensure that the series $\sum_{n=1}^{\infty} \int_{A_n} f\,d\mu$ is absolutely convergent, hence convergent. From finite additivity it follows that, for every $m$,

$$\left\| \int_{\bigcup_{n=1}^{\infty} A_n} f\,d\mu - \sum_{n=1}^{m} \int_{A_n} f\,d\mu \right\| = \left\| \int_{\bigcup_{n=m+1}^{\infty} A_n} f\,d\mu \right\|.$$

As $\lim_m \mu\left(\bigcup_{n=m+1}^{\infty} A_n\right) = 0$, Exercise 9.9.34 ensures that $\lim_m \left\| \int_{\bigcup_{n=m+1}^{\infty} A_n} f\,d\mu \right\| = 0$.
- Exercise 9.9.36. Apply the scalar case of Egorov's Theorem to the sequence $(\|f_n - f\|)_{n=1}^{\infty}$.
- Exercise 9.9.37. Adapt the proof of the characterization of the Riemann integral as the limit of Riemann sums (see, for example, [57, Theorem 6.7(c)]).
- Exercise 9.9.39. Suppose there exists $A \in \Sigma$, with $0 < \mu(A) < \infty$, not satisfying the thesis. By the Second Geometric Form of the Hahn-Banach Theorem there exist $\varphi \in E'$ and $\alpha \in \mathbb{R}$ such that

$$\varphi\left(\frac{1}{\mu(A)}\int_A f\,d\mu\right) < \alpha \le \varphi(f(x))$$

for every $x \in E$. From Proposition 9.7.9 it follows that

$$\frac{1}{\mu(A)}\int_A \varphi \circ f\,d\mu < \alpha \le \varphi(f(x))$$

for every $x \in E$. Integrating the second inequality above over $A$ we obtain the contradiction

$$\int_A \varphi \circ f\,d\mu < \alpha\mu(A) = \int_A \alpha\,d\mu \le \int_A \varphi \circ f\,d\mu.$$

## Exercises of Chapter 10

- Exercise 10.6.5. Let $A = (a_{ij})_{n\times n}$ and $B = (b_{ij})_{n\times n}$ be two matrices of order $n \times n$. We have $AB = (c_{ij})_{n\times n}$ with $c_{ij} = \sum_{l=1}^{n} a_{il}b_{lj}$. Consequently, $\operatorname{tr}(AB) = \sum_{k=1}^{n} c_{kk} = \sum_{k=1}^{n}\sum_{l=1}^{n} a_{kl}b_{lk}$. Similarly $BA = (d_{ij})_{n\times n}$ with $d_{ij} = \sum_{l=1}^{n} b_{il}a_{jl}$. Then $\operatorname{tr}(BA) = \sum_{l=1}^{n} d_{ll} = \sum_{l=1}^{n}\sum_{k=1}^{n} b_{lk}a_{kl}$ and the result follows.
- Exercise 10.6.6. Let $\alpha$ and $\alpha'$ be bases for $E$. Consider $A$ and $A'$ as the matrices of $T$ with respect to $\alpha$ and $\alpha'$, respectively. There exists a matrix $B$ such that $A' = B^{-1}AB$. Hence

$$\operatorname{tr}(A') = \operatorname{tr}\left(B^{-1}AB\right) = \operatorname{tr}\left(ABB^{-1}\right) = \operatorname{tr}(A).$$

- Exercise 10.6.13. For each $k \in \mathbb{N}$ define

$$M_{\operatorname{Re}}^+ = \{1 \le n \le k : \operatorname{Re}(a_n) > 0\},\ M_{\operatorname{Re}}^- = \{1 \le n \le k : \operatorname{Re}(a_n) < 0\},$$
$$M_{\operatorname{Im}}^+ = \{1 \le n \le k : \operatorname{Im}(a_n) > 0\},\ M_{\operatorname{Im}}^- = \{1 \le n \le k : \operatorname{Im}(a_n) < 0\},$$

to obtain

$$\sum_{n=1}^{k} |a_n| \le \sum_{n=1}^{k} |\operatorname{Re}(a_n)| + \sum_{n=1}^{k} |\operatorname{Im}(a_n)| = \sum_{n \in M_{\operatorname{Re}}^+} \operatorname{Re}(a_n) +$$

$$+ \sum_{n \in M_{\text{Re}}^-} (-\operatorname{Re}(a_n)) + \sum_{n \in M_{\text{Im}}^+} \operatorname{Im}(a_n) + \sum_{n \in M_{\text{Im}}^-} (-\operatorname{Im}(a_n)).$$

Since $\left| \sum_{n \in M} a_n \right| \leq 1$ for any finite $M$, we have

$$\left| \sum_{n \in M_{\text{Re}}^+} \operatorname{Re}(a_n) \right| = \left| \operatorname{Re} \left( \sum_{n \in M_{\text{Re}}^+} a_n \right) \right| \leq \left| \sum_{n \in M_{\text{Re}}^+} a_n \right| \leq 1.$$

In a similar way, the numbers $\left| \sum_{n \in M_{\text{Re}}^-} (-\operatorname{Re}(a_n)) \right|$, $\left| \sum_{n \in M_{\text{Im}}^+} \operatorname{Im}(a_n) \right|$ and $\left| \sum_{n \in M_{\text{Im}}^-} (-\operatorname{Im}(a_n)) \right|$ are all less than or equal to 1. The result follows because $k$ is arbitrary.

- Exercise 10.6.14(b). Use Exercise 10.6.13.
- Exercise 10.6.16. Let $f = \sum_{n=0}^{\infty} a_n s_n \in C[0, 1]$. If $(b_n)_{n=1}^{\infty}$ is a sequence of scalars such that $f = \sum_{n=0}^{\infty} b_n s_n$, then $\sum_{n=0}^{\infty} (b_n - a_n) s_n(t) = 0$ for every $t \in [0, 1]$. Note that using this for $t = 0, 1, \frac{1}{2}, \frac{1}{4}, \frac{3}{4}, \frac{1}{8}, \frac{3}{8}, \frac{5}{8}, \frac{7}{8}, \ldots$ it follows that $a_n = b_n$ for every $n$. In fact, for $t = 0$ we have $\sum_{n=0}^{\infty} (b_n - a_n) s_n(0) = 0$, therefore $b_0 - a_0 = 0$. It follows that $\sum_{n=1}^{\infty} (b_n - a_n) s_n(t) = 0$, and applying at $t = 1$ we get $\sum_{n=1}^{\infty} (b_n - a_n) s_n(1) = 0$, therefore $b_1 - a_1 = 0$. Just continue the process.
- Exercise 10.6.17.

$$\eta(a) = 0 \implies 0 = \sup \left\{ \left\| \sum_{i=1}^{n} a_i x_i \right\| : n \in \mathbb{N} \right\} \geq \|a_1 x_1\| \implies a_1 = 0.$$

It follows that

$$0 = \sup \left\{ \left\| \sum_{i=1}^{n} a_i x_i \right\| : n \in \mathbb{N} \right\} \geq \|a_1 x_1 + a_2 x_2\| = \|a_2 x_2\| \implies a_2 = 0.$$

Repeating the procedure we conclude that $a_i = 0$ for every $i \in \mathbb{N}$.

- Exercise 10.6.19. Let $(P_n)_{n=1}^{\infty}$ be the canonical projections associated with a Schauder basis of $E$. As $P_n(x) \longrightarrow x$ for every $x \in E$, from Exercise 2.7.26 we know that $\sup_{x \in K} \|P_n(x) - x\| \longrightarrow 0$ for any compact $K \subseteq E$. Now use Exercise 7.7.16.
- Exercise 10.6.20. Choose

$$x_1 = \frac{1}{2}, \; x_2 = \frac{1}{2} - \frac{1}{3}, \; \ldots, \; x_n = \frac{1}{n} - \frac{1}{n+1}, \; \ldots$$

Then $x = (x_j)_{j=1}^{\infty} \in \ell_1$ and its representation in this given Schauder basis is $x = 1e_1 + \frac{1}{2}(e_2 - e_1) + \frac{1}{3}(e_3 - e_2) + \cdots$. If this convergence were unconditional, then, in particular, the series $1e_1 + \frac{1}{3}(e_3 - e_2) + \frac{1}{5}(e_5 - e_4) + \cdots$ would be convergent in $\ell_1$, but this does not occur.

- Exercise 10.6.22. Consider a sequence of the form $(\alpha_n, \alpha_n)_{n=1}^{\infty} \in \mathcal{L}_E \times \mathcal{L}_F$ converging to $(\beta, \gamma) \in \mathcal{L}_E \times \mathcal{L}_F$, where $\beta = (b_n)_{n=1}^{\infty}$ and $\gamma = (c_n)_{n=1}^{\infty}$, that is,

$$\lim_{n \to \infty} \eta_E(\alpha_n - \beta) = 0 = \lim_{n \to \infty} \eta_F(\alpha_n - \gamma).$$

Given $\varepsilon > 0$, there exists $n_\varepsilon \in \mathbb{N}$ such that $\eta_E(\alpha_n - \beta) \leq \varepsilon$ and $\eta_F(\alpha_n - \gamma) \leq \varepsilon$ for every $n \geq n_\varepsilon$. Writing $\alpha_j = \left(\alpha_j^n\right)_{n=1}^{\infty}$ for each $j$, we have

$$\sup\left\{\left\|\sum_{i=1}^{m} (\alpha_i^n - b_i) x_i\right\| : m \in \mathbb{N}\right\} \leq \varepsilon \text{ and}$$

$$\sup\left\{\left\|\sum_{i=1}^{m} (\alpha_i^n - c_i) y_i\right\| : m \in \mathbb{N}\right\} \leq \varepsilon$$

for every $n \geq n_\varepsilon$. Hence, for every $m$, we have

$$\left|\alpha_m^{n_\varepsilon} - b_m\right| \cdot \|x_m\| = \left\|\sum_{i=1}^{m} (\alpha_i^{n_\varepsilon} - b_i) x_i - \sum_{i=1}^{m-1} (\alpha_i^{n_\varepsilon} - b_i) x_i\right\| \leq 2\varepsilon,$$

$$\left|\alpha_m^{n_\varepsilon} - c_m\right| \cdot \|y_m\| = \left\|\sum_{i=1}^{m} (\alpha_i^{n_\varepsilon} - c_i) y_i - \sum_{i=1}^{m-1} (\alpha_i^{n_\varepsilon} - c_i) y_i\right\| \leq 2\varepsilon.$$

For all $m \in \mathbb{N}$ and $\varepsilon > 0$ it follows that

$$|b_m - c_m| = \left|b_m - \alpha_m^{n_\varepsilon} + \alpha_m^{n_\varepsilon} - c_m\right|$$

$$\leq \left|\alpha_m^{n_\varepsilon} - b_m\right| + \left|\alpha_m^{n_\varepsilon} - c_m\right| \leq \frac{2\varepsilon}{\|x_m\|} + \frac{2\varepsilon}{\|y_m\|}.$$

Letting $\varepsilon \longrightarrow 0$ we have $b_m = c_m$ for every $m \in \mathbb{N}$, therefore $\gamma = \beta$.
- Exercise 10.6.23. Corollary 10.3.9.
- Exercise 10.6.24. By Exercise 1.8.36 we can consider a sequence $(x_n)_{n=1}^{\infty}$ dense in the unit sphere $S_E = \{x \in E : \|x\| = 1\}$. By Corollary 3.1.4, for each $n$ there exists $\varphi_n \in E'$ such that $\|\varphi_n\| = 1 = \|x_n\| = \varphi(x_n)$. Let $x \in E$ with

$\|x\| = 1$ be given. It is clear that $\sup_n |\varphi_n(x)| \leq \|x\|$. Given $\varepsilon > 0$, let $n_0$ be such that $\|x - x_{n_0}\| < \varepsilon$. Thus $|\|x\| - \|x_{n_0}\|| \leq \|x - x_{n_0}\| < \varepsilon$. Therefore, $\|x\| - \varepsilon \leq \|x_{n_0}\| = \varphi(x_{n_0})$. This proves that $\sup_n |\varphi_n(x)| = \|x\|$ if $\|x\| = 1$. The transition to any element of $E$ is immediate.

- Exercise 10.6.26. Suppose there exist $k, j$ such that $\langle x_j, x_k \rangle \neq 0$. In this case there exists $y \in E$ such that $\|y\| = 1$, $y \in [x_k, x_j]$ and $y \perp x_k$. Note that $|\langle x_j, y \rangle| > 0$. Writing $x_j = \langle x_j, y \rangle y + \langle x_j, x_k \rangle x_k$, we have $1 = |\langle x_j, y \rangle|^2 + |\langle x_k, x_j \rangle|^2$, therefore $|\langle x_j, y \rangle| < 1$. From $1 = |x_j^*(x_j)| = |\langle x_j, y \rangle| \cdot |x_j^*(y)|$ it follows that $\|x_j^*\| \geq |x_j^*(y)| > 1$.

- Exercise 10.6.28. Let $(x_n)_{n=1}^\infty$ be a Schauder basis of $E$ and let $(x_n^*)_{n=1}^\infty$ be its coefficient functionals. For each $k \in \mathbb{N}$, choose $y_k \in \mathcal{D}$ such that $\|x_k - y_k\| \leq \dfrac{1}{2^{k+1}\|x_k^*\|}$. Therefore,

$$\sum_{k=1}^\infty \|x_k - y_k\| \cdot \|x_k^*\| \leq \sum_{k=1}^\infty \frac{1}{2^{k+1}} < 1,$$

and Theorem 10.4.7 guarantees that $(y_n)_{n=1}^\infty$ is a Schauder basis of $E$.

- Exercise 10.6.29.

(a) If $x_{t_1}, \ldots, x_{t_n}$ are distinct vectors and $a_1 x_{t_1} + \cdots + a_n x_{t_n} = 0$, then

$$a_1(t_1, t_1^2, \ldots) + a_2(t_2, t_2^2, \ldots) + \cdots + a_n(t_n, t_n^2, \ldots) = 0.$$

Therefore

$$\begin{cases} t_1 a_1 + t_2 a_2 + \cdots t_n a_n = 0 \\ \quad \vdots \\ t_1^n a_1 + t_2^n a_2 + \cdots t_n^n a_n = 0 \\ \quad \vdots \end{cases}$$

For the infinite system to have a solution, in particular the first $n$ equalities must be satisfied. We will see that there is no non-trivial solution for the system formed by the first $n$ equalities (note that the variables are $a_1, \ldots, a_n$). There would only be a non-trivial solution if

$$\det \begin{bmatrix} t_1 & t_1^2 & \cdots & t_1^n \\ t_2 & t_2^2 & \cdots & t_2^n \\ \vdots & \vdots & \vdots & \vdots \\ t_n & t_n^2 & \cdots & t_n^n \end{bmatrix} = 0.$$

But

$$\det \begin{bmatrix} t_1 & t_1^2 & \cdots & t_1^n \\ t_2 & t_2^2 & \cdots & t_2^n \\ \vdots & \vdots & \vdots & \vdots \\ t_n & t_n^2 & & t_n^n \end{bmatrix} = t_1 \cdots t_n \det \begin{bmatrix} 1 & t_1 & \cdots & t_1^{n-1} \\ 1 & t_2 & \cdots & t_2^{n-1} \\ \vdots & \vdots & \vdots & \vdots \\ 1 & t_n & & t_n^{n-1} \end{bmatrix}$$

is non-zero, as it is a Vandermonde determinant with distinct $t_1, \ldots, t_n$.

(b) Remember that the dimension of a vector space is the cardinality of any of its Hamel basis (all Hamel bases of the same vector space have the same cardinality). From item (a) it follows that $\dim(\ell_p) \geq \mathrm{card}(\mathbb{R})$ and use that $\mathrm{card}(\mathbb{R} \times \mathbb{R} \times \cdots \times \cdots) = \mathrm{card}(\mathbb{R})$.

- Exercise 10.6.30(f). For each $x \in E$ and $j$ natural, define $e_j(x) = (0, \ldots, 0, x, 0, \ldots)$, where $x$ appears in the $j$-th coordinate. If $\mathcal{A}$ is a Hamel basis of $E$, show that $\mathcal{B} = \{e_j(x) : j \in \mathbb{N} \text{ and } x \in \mathcal{A}\}$ is a Hamel basis of $c_{00}(E)$. Conclude that $\mathrm{card}(\mathcal{B}) = \mathrm{card}(\mathbb{N}) \cdot \mathrm{card}(\mathcal{A})$.

# Bibliography

1. M.D. Acosta, D. García, M. Maestre, A multilinear Lindenstrauss theorem. J. Funct. Anal. **235**, 122–136 (2006)
2. F. Albiac, N. Kalton, *Topics in Banach Space Theory*, Graduate Texts in Mathematics (Springer, 2005)
3. G.R. Allan, *Introduction to Banach Spaces and Algebras*, Oxford Graduate Texts in Mathematics, vol. 20 (Oxford University Press, 2011)
4. S. Argyros, R. Haydon, A hereditarily indecomposable $\mathcal{L}_\infty$-space that solves the scalar-plus-compact problem. Acta Math. **206**, 1–54 (2011)
5. S. Banach, On operations in abstract sets and their application to integral equations. Fundam. Math. **3**, 133–181 (1922)
6. S. Banach, *Theory of Linear Operations*, Monografie Matematyczne 1 (Warsaw, 1932)
7. C. Bessaga, A. Pełczyński, On basis and unconditional convergence of series in Banach spaces. Studia Math. **17**, 151–164 (1958)
8. N. Bourbaki, *Topological Vector Spaces*, Chapters 1–5, Elements of Mathematics (Springer, 1987)
9. H. Brézis, *Functional Analysis, Sobolev Spaces and Partial Differential Equations* (Springer, 2011)
10. D. Carando, D. García, M. Maestre, Homomorphisms and composition operators on algebras of analytic functions of bounded type. Adv. Math. **197**, 607–629 (2005)
11. N.L. Carothers, *A Short Course on Banach Space Theory*, London Mathematical Society Student Texts 64 (Cambridge University Press, 2005)
12. K. Ciesielski, On Stefan Banach and some of his results. Banach J. Math. Anal. **1**, 1–10 (2007)
13. J.B. Conway, *A Course in Functional Analysis*, 2nd edn. (Springer, 1990) (Corrected fourth printing, 1997)
14. C.A. Di Prisco, *Una Introduccíion a la Teoría de Conjuntos y los Fundamentos de las Matemáticas*, Coleção CLE-UNICAMP, vol. 20 (Campinas, 1997)
15. J. Diestel, *Sequences and Series in Banach Spaces* (Springer, 1984)
16. J. Diestel, J.J. Uhl, *Vector Measures*, American Mathematical Society Mathematical Surveys, vol. 15 (American Mathematical Society, 1977)
17. J. Diestel, H. Jarchow, A. Tonge, *Absolutely Summing Operators*, Cambridge Studies in Advanced Mathematics 43 (Cambridge University Press, 1995)
18. R. Duda, The discovery of Banach spaces, in *European Mathematics in the Last Centuries*, ed. by W. Wieslaw (Wroclaw, 2005), pp. 37–46
19. R.M. Dudley, *Real Analysis and Probability*, Cambridge Studies in Advanced Mathematics, vol. 74 (Cambridge University Press, 2002)

20. N. Dunford, J.T. Schwartz, *Linear Operators*, Part I (Interscience, New York, 1958)
21. M. Fabian, P. Habala, P. Hajek, V. Montesinos Santalucia, J. Pelant, V. Zizler, *Functional Analysis and Infinite-Dimensional Geometry*, CMS Books in Mathematics (Springer, 2001)
22. M. Feinberg, R. Lavine, Thermodynamics based on the Hahn–Banach theorem: The Clausius inequality. Arch. Rat. Mech. and Anal. **82**, 203–293 (1983)
23. H. Fetter, B. Gamboa, *The James Forest*, London Mathematical Society Lecture Notes Series, vol. 236 (1997)
24. D. G. Figueiredo, *Lectures on the Ekeland Variational Principle with Applications and Detours* (Springer, 1989)
25. G.B. Folland, *Real Analysis, Modern Techniques and Their Applications* (John Wiley & Sons, 1984)
26. D.J.H. Garling, *Inequalities: A Journey Into Linear Analysis* (Cambridge University Press, 2007)
27. D. Gilbarg, N.S. Trundinger, *Elliptic Partial Differential Equations of Second Order* (Springer, 2001)
28. K. Goebel, W.A. Kirk, *Topics in Metric Fixed Point Theory*, Cambridge Studies in Advanced Mathematics, vol. 28 (Cambridge University Press, 1990)
29. W.T. Gowers, Decompositions, approximate structure, transference, and the Hahn–Banach theorem. Bull. Lond. Math. Soc. **42**, 573–606 (2010)
30. A. Granas, J. Dugundji, *Fixed Point Theory*, Springer Monographs in Mathematics (Springer, New York, 2010)
31. A. Grothendieck, Résumé de la théorie métrique des produits tensoriels topologiques. Bol. Soc. Mat. São Paulo **8**, 1–79 (1953)
32. J. Hennefeld, A nontopological proof of the uniform boundedness theorem. Amer. Math. Mon. **87**, 217 (1980)
33. R.C. James, A non-reflexive Banach space isometric with its second conjugate space. Proc. Nat. Acad. Sci. U. S. A. **37**, 174–177 (1951)
34. R.C. James, Weak compactness and reflexivity. Israel J. Math. **2**, 101–119 (1964)
35. H. Jarchow, *Locally Convex Spaces* (Teubner, Stuttgart, 1981)
36. W.B. Johnson, J. Lindenstrauss (ed.), *Handbook of the Geometry of Banach Spaces*, vol. 1 (North-Holland, 2001)
37. W.B. Johnson, J. Lindenstrauss (ed.), *Handbook of the Geometry of Banach Spaces*, vol. 2 (North-Holland, 2003)
38. N. Kalton, Quasi-Banach spaces, in *Handbook of the Geometry of Banach Spaces*, vol. 2 (North-Holland, 2003), pp. 1099–1130
39. E. Kreyszig, *Introductory Functional Analysis with Applications*, Wiley Classics Library Edition (1989)
40. D. Li, H. Queffélec, *Introduction à l'Étude des Espaces de Banach, Analyse et Probabilités*, Cours Spécialisés 12 (Société Matématique de France, Paris, 2004)
41. J. Lindenstrauss, A. Pełczyński, Absolutely summing operators in $\mathcal{L}_p$ spaces and their applications. Stud. Math. **29**, 275–326 (1968)
42. J. Lindenstrauss, L. Tzafriri, On the complemented subspace problem. Israel J. Math. **9**, 263–269 (1971)
43. R. Maudlin, *The Scottish Book* (Birkhauser, Boston, 1981)
44. R. Megginson, *An Introduction to Banach Space Theory*, Graduate Texts in Mathematics, vol. 183 (Springer, 1998)
45. F.J. Murray, On complementary manifolds and projections in spaces $L_p$ and $\ell_p$. Trans. Amer. Math. Soc. **41**, 138–152 (1937)
46. L. Nachbin, A theorem of Hahn-Banach type for linear transformations. Trans. Amer. Math. Soc. **68**, 28–46 (1950)
47. L. Narici, E. Beckenstein, The Hahn-Banach theorem: the life and times. Topol. Appl. **77**, 193–211 (1997)
48. L. Narici, E. Beckenstein, *Topological Vector Spaces*, 2nd edn. (CRC Press, 2011)

49. A. Pełczyński, A note on the paper of I. Singer "Basic sequences and reflexivity of Banach spaces". Studia Math. **21**, 371–374 (1962)
50. R.S. Phillips, On linear transformations. Trans. Amer. Math. Soc. **48**, 516–541 (1940)
51. A. Pietsch, *History of Banach Spaces and Linear Operators* (Birkhäuser, 2007)
52. G. Pisier, Grothendieck's Theorem, past and present. Bull. Amer. Math. Soc. **49**, 237–323 (2012)
53. S. Ponnusamy, *Foundations of Functional Analysis* (Alpha Science International, 2002)
54. H.L. Royden, P.M. Fitzpatrick, *Real Analysis*, 4th edn. (Prentice-Hall, 2010)
55. W. Rudin, *Functional Analysis* (Tata McGraw-Hill Publishing Company, 1976)
56. W. Rudin, *Real and Complex Analysis* (McGraw Hill Book Company, 1986)
57. W. Rudin, *Principles of Mathematical Analysis*, 3rd edn. (McGraw Hill Book Company, 1989)
58. K. Saxe, *Beginning Functional Analysis* (Springer, 2002)
59. H.H. Schaefer, *Espacios Vectoriales Topológicos* (Editorial Teide, Barcelona, 1974)
60. J. Shapiro, *A Fixed-point Farrago* (Springer, 2016)
61. Y. Sonntag, *Topologie et Analyse Fonctionelle* (Ellipses, 1998)
62. E. Teixeira, Strong solutions for differential equations in abstract spaces. J. Differential Equations **214**, 65–91 (2005)
63. V.A. Trenoguin, B.M. Pisarievski, T.S. Soboleva, *Problemas y Ejercícios de Análisis Funcional* (Editora Mir Moscú, 1987)
64. D. van Dulst, *Reflexive and Superreflexive Banach Spaces*, Mathematical Centre Tracts, vol. 102 (Mathematisch Centrum, Amsterdam, 1978)
65. J. von Neumann, The mathematician, in *The Works of the Mind*, ed. by R. B. Heywood (Chicago Univ. Press, 1947), pp. 180–196
66. D. Werner, *Functional Analysis* (Springer, 1997)
67. S. Willard, *General Topology* (Dover Publications, 2004)

# Index

**A**
Adjoint, 76
Annihilator, 66
Application
 bilinear, 34
  coercive, 107
  continuous, 34
  non-degenerate, 107
  separately continuous, 34
  symmetric, 107
 completely continuous, 212
 demicontinuous, 212
 quotient, 42
 sequentially weakly continuous, 212
Approximation property, 174

**B**
Banach-Grumblum criterion, 270
Base
 equivalent, 269
 de Schauder, 263
Basis
 algebraic or Hamel, 290
 of neighborhoods, 296
 of topological space, 296
Bessaga–Pełczyński selection principle, 279
Best approximation, 95

**C**
Canonical embedding, 74
Chain rule, 231
Complemented subspace, 52
Complete orthonormal system, 94

Constant
 of Grothendieck, 260
 of the basis, 268
Contraction, 217
Convergence in measure, 304
Convex hull, 23
Copy
 isometric, 74
 isomorphic, 74

**D**
Diffeomorphism, 239

**E**
Eigenspace, 151
Eigenvalue, 151
Eigenvector, 151
Equivalent norms, 21

**F**
Fourier coefficient, 111
Fredholm Alternative, 162
Function
 Bochner integrable, 235
 bounded, 2
  almost everywhere, 10
 characteristic, 303
 coercive, 215
 convex, 215
 Fréchet-differentiable, 225
 integrable, 306
 Lipschitz, 28

Function (*cont.*)
  measurable, 302
    simple, 303
  of Rademacher, 115
  uniformly
    continuous, 28
    convex, 215
Functional
  continuous linear, 27
  of Minkowski, 58

## G
Gram–Schmidt orthogonalization process, 101
Graph, 38

## H
Hilbert cube, 24
Homeomorphism, 298
Homothety, 183
Hyperplane, 57
  affine, 57

## I
Inequality
  Bessel's, 96
  of Cauchy–Schwarz, 87
  of Clarkson
    first, 141
    second, 142
  of Grothendieck, 258
  of Hölder
    generalized, 23
    for integrals, 7
    reciprocal, 25
    for sequences, 12
  mean value, 231
  of Minkowski
    for integrals, 7
    reversed, 19
    for sequences, 13
  of Young, 19
Inner product, 86
Integral, 305
Isomorphic spaces, 27
  isometrically, 27
Isomorphism, 27
  isometric, 27

## K
Kronecker delta, 94

## L
Lemma
  Fatou's, 305
  of Helly, 131
  of Zorn, 289
Linear isometry, 27

## M
Mapping
  open, 35
Maximal element, 289
Measure, 303
  $\sigma$-finite, 304
  finite, 304
Metric, 293

## N
Neighborhood, 295
Net, 298
  Cauchy, 138
  convergent, 298
Norm, 2
  induced by the inner product, 86
  Lipschitz, 210

## O
Operator
  diagonal, 31
  linear
    $\varphi \otimes y$, 30
    absolutely summable, 260
    adjoint of, 76
    that attains the norm, 40
    bounded, 29
    compact, 154
    continuous, 27
    of finite rank, 43
    image of, 42
    integral, 155
    kernel of a, 42
    normal, 178
    positive, 177
    self-adjoint, 165
    symmetric, 39
    unitary, 178
  multiplication, 31
  nonlinear
    compact, 220
    substitution or Nemytskii, 210
Orthogonal complement, 92

# Index

## P
Parallelogram law, 89
Parseval's Identity, 98
Partial order, 289
$p$-norm, 204
Polarization formula
  complex case, 89
  real case, 89
Principle
  of uniform boundedness, 40
Projection, 52
  canonical, 268
  orthogonal, 92

## R
Regular value, 152
Riesz's lemma, 14

## S
*Scottish book*, 283
Seminorm, 192
Sequence
  basic, 270
  block basic, 277
  weakly Cauchy, 147
  weakly summable, 260
Series
  absolutely convergent, 246
  convergent, 97
  unconditionally convergent, 97
Set
  absolutely convex, 206
  absorbing, 184
  balanced, 184
  Borel or Borelian, 301
  compact, 294
  convex, 22
  directed, 298
  measurable, 301
  norming, 274
  orthonormal, 94
  partially ordered, 289
  resolvent, 152
  totally ordered, 289
$\sigma$-algebra, 301
Space
  Banach, 2
  bidual, 74
  $c_o$, 5
  $c_{00}$, 6
  compactly immersed, 213
  completion of, 75
  of continuous linear operators, 27
  dual, 27
  with finite deficiency, 43
  of functions
    continuous, 3
    continuously differentiable, 3
  Hausdorff, 298
  Hilbert, 89
  with inner product, 86
  of James, 79
  $L_\infty$, 10, 13
  of Lipschitz continuous functions, 20
  locally convex, 188
  $L_p$, 6, 12
  measurable, 301
  of measure, 304
  metric, 293
  metrizable, 295
  normed, 2
  product, 22
  quotient, 22
  reflexive, 75
  Schur, 144
  separable, 16
  strictly convex, 149
  topological, 295
  topological vector, 180
  uniformly convex, 137
  weakly
    compactly generated, 148
    sequentially complete, 147
Spectral decomposition of compact and self-adjoint operators, 172
Spectrum, 152

## T
Theorem
  of Ascoli, 294
  of Baire, 32
  of Banach–Alaoglu–Bourbaki, 129
  of Banach–Mazur, 136
  of Banach–Mazur, Bessaga– Pełczyński, Gelbaum, 281
  Banach–Steinhaus, 33
  of Bessaga–Mazur, 275
  of Bessaga–Pełczyński, 277
  of Bishop–Phelps, 40
  of Cauchy–Picard, 219
  Closed Graph, 38
  of Dixmier, 82
  of dominated convergence, 307
  Dvoretzky-Rogers, 255
  Eberlein–Smulian, 145

Theorem (*cont.*)
    fixed point
        of Banach, 217
        Brouwer, 223
        of Schauder, 223
    Goldstine's, 132
        complex case, 203
    of Grothendieck, 261
    of Hahn–Banach
        complex case of geometric forms, 68
        first geometric form, 61
        for locally convex spaces, 198
        geometric form for locally convex spaces, 201
        real and complex cases, 49
        real case, 47
        second geometric form, 62
    of the inverse function, 239
    of isomorphism, 43
    Josefson-Nissenzweig, 145
    of Kakutani, 133
    Krein–Milman–Rutman, 287
    of Lax–Milgram, 108
    of Mazur, 124
    Milman–Pettis, 139
    of monotone convergence, 305
    Hellinger–Toeplitz, 113
    of Open Mapping, 36
    orthogonal projection, 90
    of Peano, 224
    Pettis' Measurability, 233
    of Phillips, 54
    Pythagorean, 90
    Radon–Nikodým, 306
    of Riesz–Fischer, 103
    of Riesz-Fréchet, 105
    of Riesz representation, 70
    of Schauder, 157
    of Schauder regularity, 20
    of Schur, 124
    Sobczyk's, 67
    spectral for compact and self-adjoint operators, 172
    of Tychonoff, 300
    of Weierstrass approximation, 18
Topology, 295
    of convergence in measure, 208
    determined or generated by a family of seminorms, 193
    generated by a family of functions, 118
    of pointwise convergence, 195
    product, 300
    weak, 119
    weak-star, 126
Trace, 250
Translation, 183

**U**
Upper bound, 289

**V**
Vectors
    orthogonal, 90

The manufacturer's authorised representative in the EU is Springer Nature Customer Service Centre GmbH, Europaplatz 3, 69115 Heidelberg, Germany. If you have any concerns regarding our products, please contact ProductSafety@springernature.com

Printed and bound by CPI Group (UK) Ltd, Croydon, CR0 4YY

25/03/2026

02078192-0016